# Role of Plant Growth Promoting Microorganisms in Sustainable Agriculture and Nanotechnology

*Edited by*

*Ajay Kumar*

*Amit Kishore Singh*

*Krishna Kumar Choudhary*

Woodhead Publishing is an imprint of Elsevier
The Officers' Mess Business Centre, Royston Road, Duxford, CB22 4QH, United Kingdom
50 Hampshire Street, 5th Floor, Cambridge, MA 02139, United States
The Boulevard, Langford Lane, Kidlington, OX5 1GB, United Kingdom



**British Library Cataloguing-in-Publication Data**
A catalogue record for this book is available from the British Library

**Library of Congress Cataloging-in-Publication Data**
A catalog record for this book is available from the Library of Congress

ISBN: 978-0-12-817004-5 (print)

ISBN: 978-0-12-817005-2 (online)

For information on all Woodhead Publishing publications
visit our website at https://www.elsevier.com/books-and-journals

*Publisher:* Charlotte Cockle
*Acquisition Editor:* Nancy Maragioglio
*Editorial Project Manager:* Alexandra Romano
*Production Project Manager:* Debasish Ghosh
*Cover Designer:* Alan Studholme

Typeset by MPS Limited, Chennai, India

Transferred to Digital Printing in 2019

# Role of Plant Growth Promoting Microorganisms in Sustainable Agriculture and Nanotechnology

# Contents

# List of Contributors

**Waquar Akhter Ansari** ICAR-National Bureau of Agriculturally Important Microorganisms Kushmaur, Maunath Bhanjan, India

**Shubhi Avasthi** Department of Botany, Abhilashi Post Graduate Institute of Sciences, Mandi, India

**Raina Bajpai** Department of Mycology and Plant Pathology, Institute of Agricultural Sciences, Banaras Hindu University, Varanasi, India

**Anand Kumar Chaudhari** Department of Botany, Institute of Science, Banaras Hindu University, Varanasi, India

**Krishna Kumar Choudhary** Centre for Plant Sciences, School of Basic and Applied Sciences, Central University of Punjab, Bathinda, India

**Somenath Das** Department of Botany, Institute of Science, Banaras Hindu University, Varanasi, India

**Anupam Dikshit** Biological Product Lab, Department of Botany, University of Allahabad, Allahabad, India

**Ajay Kumar Gautam** Faculty of Science, School of Agriculture, Abhilashi University, Mandi, India

**Surendra K. Gond** Department of Botany, MMV, Banaras Hindu University, Varanasi, India

**Akanksha Gupta** Center of Advanced Study in Botany Institute of Science, Banaras Hindu University, Varanasi, India

**Amandeep Hora** Department of Biotechnology, Guru Nanak College, Chennai, India

**Ram Krishna** Institute of Environment and Sustainable Development, Banaras Hindu University, Varanasi, India; Division of Vegetable Improvement, ICAR-Indian Institute of Vegetable Research, Varanasi, India

**Ajay Kumar** Center of Advanced Study in Botany Institute of Science, Banaras Hindu University, Varanasi, India

**Akshay Kumar** Department of Botany, Institute of Science, Banaras Hindu University, Varanasi, India

**Baby Kumari** Department of Biotechnology, Vinoba Bhave University, Hazaribagh, India

**Rusi Lata** Department of Botany, MMV, Banaras Hindu University, Varanasi, India

**M.A. Mallick** Faculty of Science, Vinoba Bhave University, Hazaribagh, India

**Mahendra Mani** Department of Biotechnology, Guru Nanak College, Chennai, India

**Jhumishree Meher** Department of Mycology and Plant Pathology, Institute of Agricultural Sciences, Banaras Hindu University, Varanasi, India

**Pooja Mishra** Microbial Technology Department, CSIR-Central Institute of Medicinal and Aromatic Plants (CSIR-CIMAP), Lucknow, India

**Arpan Mukherjee** Institute of Environment and Sustainable Development, Banaras Hindu University, Varanasi, India

**Anand Pandey** Biological Product Lab, Department of Botany, University of Allahabad, Allahabad, India

**K.D. Pandey** Centre of Advanced Study in Botany, Banaras Hindu University, Varanasi, India

**Jaisingh Patel** Department of Plant food and Environmental Sciences, Dalhousie University, Halifax, Nova Scotia, Canada

**Bhanu Prakash** Department of Botany, Institute of Science, Banaras Hindu University, Varanasi, India

**Snigdha Rai** Molecular Biology Section, Centre for Advanced Study in Botany, Department of Botany, Banaras Hindu University, Varanasi, India

**Md. Mahtab Rashid** Department of Mycology and Plant Pathology, Institute of Agricultural Sciences, Banaras Hindu University, Varanasi, India

**Alok Kumar Shrivastava** Department of Botany, Mahatma Gandhi Central University, Motihari, India

**Amit Kishore Singh** Botany Department, Kamla Nehru Post Graduate College, Raebareli, India

**Anand Vikram Singh** Center of Advanced Study in Botany Institute of Science, Banaras Hindu University, Varanasi, India

**Divya Singh** Department of Chemical Engineering & Technology, Indian Institute of Technology, Banaras Hindu University, Varanasi, India

**Kshitij Kumar Singh** Campus Law Centre, Faculty of Law, University of Delhi, Delhi, India

**Major Singh** ICAR-Directorate of Onion and Garlic Research, Pune, India

**Monika Singh** Centre of Advanced Study in Botany, Banaras Hindu University, Varanasi, India

**Prashant Kumar Singh** Department of Vegetable and Fruit Science, Institute of Plant Science, Agriculture Research Organization (ARO), The Volcani Center, Rishon LeZion, Israel; Department of Vegetables and Field Crops, Institute of Plant Sciences, Agricultural Research Organization, The Volcani Center, Rishon LeZion, Israel

**Prem Pratap Singh** Center of Advanced Study in Botany, Institute of Science, Banaras Hindu University, Varanasi, India; Department of Botany, Institute of Science, Banaras Hindu University, Varanasi, India

**Sandeep Kumar Singh** Center of Advanced Study in Botany, Institute of Science, Banaras Hindu University, Varanasi, India

**Vipin Kumar Singh** Center of Advanced Study in Botany, Institute of Science, Banaras Hindu University, Varanasi, India

**Manoj K. Solanki** Department of Food Quality and Safety, Institute for Post-Harvest and Food Sciences, The Volcani Center, Agricultural Research Organization, Rishon LeZion, Israel

**Akhileshwar Kumar Srivastava** The National Institute for Biotechnology, Ben-Gurion University of the Negev, Negev, Israel

**Meenakshi Srivastava** Centre of Advanced Study in Botany, Banaras Hindu University, Varanasi, India

**Piush Srivastava** Biological Product Lab, Department of Botany, University of Allahabad, Allahabad, India

**Basavaraj Teli** Department of Mycology and Plant Pathology, Institute of Agricultural Sciences, Banaras Hindu University, Varanasi, India

**Arpita Tripathi** Biological Product Lab, Department of Botany, University of Allahabad, Allahabad, India

**Hariom Verma** B.R.D. Government Degree College, Dudhi Nagar, India

**Jay Prakash Verma** Institute of Environment and Sustainable Development, Banaras Hindu University, Varanasi, India

**Wang Wenjing** State Key Laboratory of Cotton Biology, Henan Key Laboratory of Plant Stress Biology, School of Life Science, Henan University, Kaifeng, P.R. China; Department of Life Science, School of Biology and Food Science, Shangqiu Normal University, Shangqiu, P. R. China

**Sudheer Kumar Yadav** Department of Botany, Institute of Science, Banaras Hindu University, Varanasi, India

# Plant growth-promoting microorganisms in sustainable agriculture

1

*Anand Pandey*[1], *Arpita Tripathi*[1], *Piush Srivastava*[1], *Krishna Kumar Choudhary*[2] *and Anupam Dikshit*[1]
[1]Biological Product Lab, Department of Botany, University of Allahabad, Allahabad, India, [2]Centre for Plant Sciences, School of Basic and Applied Sciences, Central University of Punjab, Bathinda, India

## 1.1 Introduction

The increasing human population is creating too much stress on existing resources of food, fuel, and raw materials. To meet the demand, agriculture practices are using intensive amounts of chemical-based fertilizers and pesticides that ultimately lead to land degradation and biodiversity loss (Carsten and Mathis, 2014). In many developing countries, the economic and self-employment sectors, agriculture plays a pivotal role (Gindling and Newhouse, 2012). According to one estimate, the average loss in crop production limits to USD 10 billion annually in developing Asian countries, the primary cause being land degradation (Lal et al., 2010). Several factors have been found to be responsible for degradation of soil. Among them, consequences due to anthropogenic activities are among the most cited in land degradation. Land degradation results in exhaustion, salinization, and desertification of soil (Kesavan and Swaminathan, 2008). The ever-advancing fields of science have provided many technological inventions for the improvement of crop productivity, but there is still global demand for food to meet the world's population. Microbes associated with plants can be used to overcome problems related to soil salinity, fertility, degradation, and habitat loss. Living among the wide variety of living fauna present in active soil microbial entities are unique as they are directly involved in enhancing soil fertility and growth and promotion of plants while lowering biotic and abiotic stresses (Glick, 2010). According to one estimate, thousands of bacterial species and millions of fungal species can be found in one gram of soil (Blackwell, 2011), of which the majority are considered to be beneficial for both plants and soil. This chapter is a summary of the role of beneficial soil microbes (BSMs) including mycorrhizae and cyanobacteria in sustainable agriculture. The BSMs that can be utilized in the sustainability of the environment are also discussed.

**Role of Plant Growth Promoting Microorganisms in Sustainable Agriculture and Nanotechnology.**
**DOI: https://doi.org/10.1016/B978-0-12-817004-5.00001-4**

## 1.2 Beneficial soil microbes

The soil is an active part of the lithospheric zone on Earth. It acts like a physical covering that naturally surrounds the earth surface and represents the border between all living and dead geological matter, water resources, and the cavities—in the form of gaseous pores—holding the latter two matters, thereby laying the foundation of all terrestrial ecosystems (Aislabie Jand and Deslippe, 2013). BSMs enhance plant growth by improving nutrient uptake, forming complex soil matrices, and helping in plant defense response through secretions of various metabolites. In addition, BSMs can improve tolerance of adverse environmental conditions such as salt stress, drought stress, weed infestation, nutrient deficiency, and heavy metal contamination. Recently, researchers have found that soil microbes play both beneficial and harmful roles in the soil ecosystem. However, BSMs have gained attention not only for their plant growth-promoting attributes, but also for their role in the decomposition of organic waste and detoxification of toxic substances such as pesticides and in alleviation of soil stressors (Aislabie Jand and Deslippe, 2013; Ma et al., 2016). Their direct interaction with the plant's root system results in nutrient and mineral uptake from the decomposed soil organic matter and also helps in plant-growth promotion as well as in suppression of phytopathogens (Nihorimbere et al., 2011). Soil also contains harmful microbes that invade or parasitize plants and reduce productivity. The presence of BSMs makes soil healthier and suppresses growth of unhealthy soil microflora.

### 1.2.1 Cyanobacteria

Cyanobacteria are photosynthetic prokaryotes and are ubiquitous in nature. They are found commonly in lakes, ponds, springs, wetlands, streams, and rivers. Additionally, cyanobacteria are also an important component of soil (Singh et al., 2016a,b). Fritsch (1907) first noticed the loads of cyanobacteria in rice fields. Cyanobacteria is important in the maintenance of rice-field fertility due to $N_2$ fixation, making them a suitable natural source for soil fertility enhancement (Song et al., 2005). Free-living or symbiotic blue green algae (BGA) have a long-documented history in sustainable agriculture. According to Smil (1999), free-living cyanobacteria or symbiotic cyanobacteria (like with the water fern, *Azolla*) fix 4–6 billion kilograms of $N_2$ annually. Efficient $N_2$-fixing cyanobacteria like *Nostoc, Anabaena, Aulosira, Calothrix* sp., etc., have been recognized in different diverse agroecological regions and applied to enhance rice production (Prasad and Prasad, 2001). The use of cyanobacteria in rice fields has been in practice for a long time, but recently their use with other crops has also been tested. Cyanobacteria not only contribute in global $N_2$ supply, but studies also find them to be involved in phytohormone production in free-living and symbiotic associations. For example, *Nostoc, Chlorogloeopsis, Calothrix, Plectonema, Anabaena,*

*Cylindrospermum*, and *Anabaenopsis* have been shown to be associated with the production of indole acetic acid (IAA) in the soil rhizospheric zone in paddies and wheat fields (Karthikeyan et al., 2007; Manjunath et al., 2011). Cyanobacteria also help in the soil formation process, as during their growth they excrete certain biomolecules. (Kaushik, 2014; Rosa and Philippis, 2015). Some reports indicate that cyanobacteria can grow successfully in saline soil where most plants (except halophytes) fail to grow and help to increase the fertility of such soils (Singh and Dhar, 2010).

### *1.2.2 Plant growth-promoting rhizobacteria*

Plant growth is facilitated by microbes present in the rhizosphere (Kloepper and Schroth, 1978; Lugtenberg and Kamilova, 2009; Mishra and Arora, 2012; Ahemad and Kibret, 2014; Goswami et al., 2016). PGPR species like *Azospirillum, Rhizobium, Azotobacter, Arthrobacter, Bacillus, Pseudomonas*, etc., are some bacterial varieties that have the capability to enhance the fertility of soil and the growth of plants (Cheng, 2009; Mishra and Arora, 2012; Arora, 2015). Recently, additional research has been done to reveal the mechanisms of plant–microbe interactions (Beneduzi et al., 2012). The growth promotion route involves the phytohormone production, nitrogen fixation, siderophore production, solubilization of inorganic substances (P, K, Zn etc.) and making it conveniently available to the plants. From another point of view, the indirect route can be seen in the form of growth-inhibition activity against phytopathogens by one of several mechanisms, such as antibiotic or antifungal metabolite(s) production, iron depletion from the rhizosphere (as a result of chelation), induced systemic resistance (ISR), and production of wall-degrading enzymes such as chitinase (Vejan et al., 2016a,b). In addition, PGPR are also recognized as potential microbes that can protect plants from various environmental stresses in normal as well as stressed environments (Khare and Arora, 2011; Kang et al., 2014). There is strong evidence that supports the efficacy of PGPRs in sustainable agriculture, though initially the role of PGPR was explored only for enhancing the crop productivity. Recent studies have suggested it plays a key role in proper functioning of the agroecosystem (Ahemad and Kibret, 2014; Arora and Mishra, 2016; Cheng, 2009; Choudhary et al., 2013; Choudhary and Agrawal, 2017). Many researchers have also shown them to be helpful for degraded land restoration, soil quality improvement, reduction of environmental soil pollutants, and in combating climate change.

## 1.3 Role of plant growth-promoting bacteria in soil fertility

The presence of a wide range of insects, ants, termites, earthworms, and most importantly, microorganisms take part in soil mineralization and nutrient cycling, processes that directly influence the productivity of plants (Meena et al., 2014a,b, 2015a,b). PGPR are the class of BSMs involved in all the necessary actions of soil. PGPR colonization in the rhizosphere depends on the moisture content of the soil (Shrivastava et al., 2014). Root-associated rhizobacteria are mainly responsible for synthesizing the numerous biomolecules that improve soil health. Organic compounds, as they act on the residues of the plant, undergo decomposition and mineralization. These mineralized chemicals add value to the soil. In addition to the above mentioned processes, chemicals responsible for plant growth are synthesized, affecting the root morphology of plant. The soils can be enhanced by rhizospheric bacteria, making nutrients more readily available. (Choudhary et al., 2011; Meena et al., 2013a,b, 2014a,b, 2015c). Nitrogen in plants is essential for amino acid and protein synthesis. It is accumulated from the atmosphere, soil, and rhizosphere through biological $N_2$ fixation. The increased level of phosphate available to plants is due to the P solubilization activity possessed by the PGPRs. PGPRs are also known to produce various volatile compounds and metabolites that improve plant and soil health. In the rhizosphere, a variety of enzymes are released that restrain the pathogen growth and thereby contribute to biocontrol actions (Vejan et al., 2016a,b). In general, for best yield, productivity and improved soil health, all of these mechanisms are important more fertile soil. (Meena et al., 2016a,b,c) (Fig. 1.1).

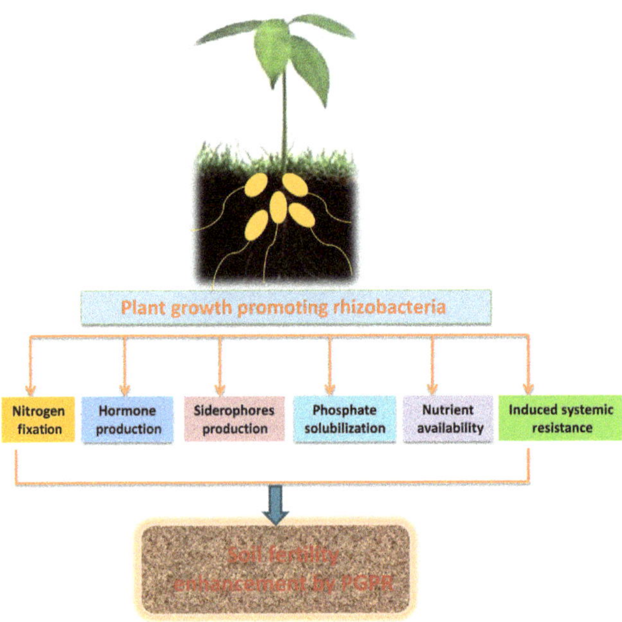


**Figure 1.1** Pictorial depiction of growth promotion by PGPR.

## 1.4 Mechanism of growth promotion by plant growth-promoting bacteria

PGPRs promote growth of plants either directly or indirectly (Castro et al., 2009). Direct involvement is through $N_2$ fixation, P and K solubilization along with the production of siderophores and phytohormones. All these processes help to increase as well as maintain soil fertility. Indirect mechanisms are additional support provided by these BSMs that help in the maintenance of a healthy environment in the soil rhizosphere but do not affect the soil fertility directly. These involve production of antibiotics, polysaccharides, various hydrolytic enzymes, and cyanide compounds leading to ISR (Fig. 1.2).

Growth of a plant can be promoted by PGPRs directly through easy availability of N, P, Fe, etc., along with enhanced levels of phytohormones. Through nutrient mineralization and solubilization processes, these bacteria provide the available nutrients to plants. PGPR also produces IAA, the most common one among all the phytohormones (Barazani and Friedman 1999). From an economic as well as environmental safety point of view, PGPRs are also influential disease suppressors (Mazurier et al., 2009; Duffy et al., 2003; Ramatte et al., 2003) through the production of pyoluteorinare and 2,4-diacetylphloroglucinol (2,4-DAPG), etc. (de Souza et al., 2003; Beneduzi et al., 2012).

Thus, PGPRs provide resistance to plants against phytopathogens and pests, which might be through systemic acquired resistance (SAR) or ISR. The feedback mechanism through defense signals is involved in SAR, which is activated immediately after an outbreak of pathogens (Pieterse et al., 2009). Barriers that help in providing resistance are phenylalanine ammonia lyase (PAL), lipoxygenase (LOX), chitinase, β-1,3-glucanase, polyphenol oxidase (PPO), peroxidase (PO), superoxide dismutase (SOD), catalase (CAT), and ascorbate peroxidase (APX) that provide ISR (Meena et al., 2016a; Sharma et al., 2016; Saha et al., 2016; Sindhu et al., 2016; Shrivastava et al., 2016).

### *1.4.1 Mineral solubilization by soil microbes*

The phosphate solubilizing bacteria (PSB) solubilizes the inorganic soil phosphates of Ca, Fe, and Al via production of siderophores, several acids (organic), and hydroxyl and carboxyl groups, and chelating them to the bound phosphates and the available calcium (Sharma et al., 2013). But some isolates of *Enterobacter* sp., *Pantoea* sp., and *Klebsiella* sp. solubilize more calcium phosphates, as compared to iron and aluminum phosphates (Chung et al., 2005). Consequently, the role of microorganisms is pivotal in cycling of soil phosphorus by channelizing it (Richardson and Simpson, 2011).

The availability of potassium in soil, though sufficient in India, still requires a lot in the form of K fertilizers. It is worth mentioning here that utilization of native sources available such as K-feldspar, muscovite, biotite, phlogopite, and green

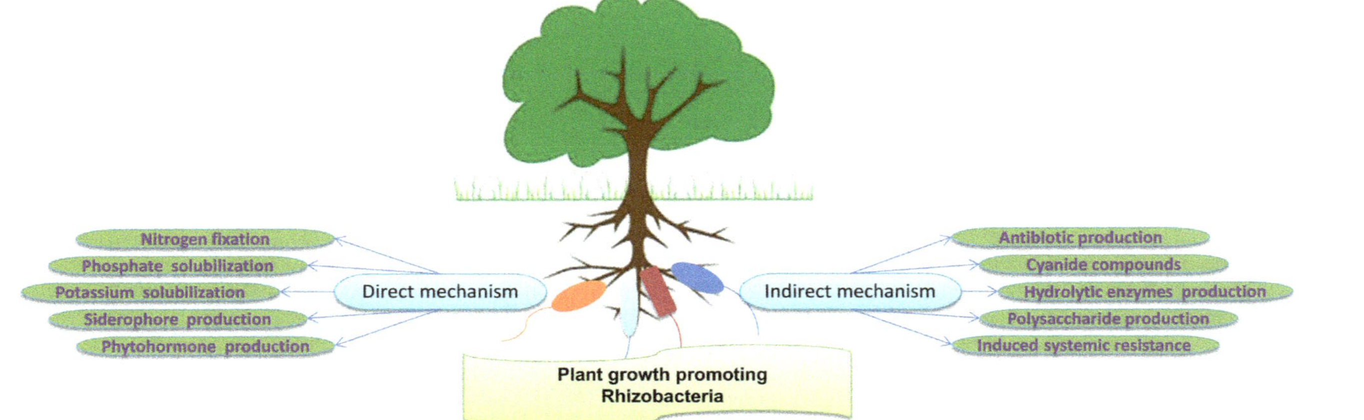


**Figure 1.2** Mechanisms of growth promotion by PGPR.
*Source*: Premachandra, D, Hudek, L, Brau, L (2016) Bacterial modes of action for enhancing of plant growth. J. Biotechnol. Biomater. 6 (3), 236. doi:10.4172/2155- 952X.1000236.

sands along with waste mica can be of great use (Nishanth and Biswas, 2008) and biointervention is helpful in increasing the K pool. But at the same time, limited access to K-bearing minerals retards their usefulness as K fertilizers in agriculture. Potassium-solubilizing bacteria (KSB), potassium-dissolving bacteria (KDB), and potassium-solubilizing rhizobacteria (KSR) have significant potential for K solubilization (Reitemeir, 1951). This class of soils microbes mainly includes, *Pseudomonas* sp., *Bacillus circulans*, *Acidithiobacillus ferroxidans*, *B. mucilaginosus*, *Burkholderia* sp., and *Paenibacillus* sp. are some efficient microbes (Bagyalakshmi et al., 2012; Liu et al., 2012; Verma et al., 2015; Saiyad et al., 2015). Potassium-solubilizing microorganisms as bioinoculants play beneficial roles in agricultural sustainability (Meena et al., 2013a,b, 2015c; Zhang and Kong, 2014; Singh et al., 2015, 2016a,b).

### 1.4.2 Siderophore production

Iron is one of the most fundamental mineral elements required to sustain all forms of life. In its oxidized state, $Fe^{3+}$ is the abundant ionic form, capable of forming insoluble oxyhydroxides and hydroxides, leading its unavailability to plants and microbes whereas $Fe^{2+}$ ionic form with low pH is freely available and is more readily absorbed by the plants (Rajkumar et al., 2009). Uptake of iron by bacteria and fungi is due to the presence of siderophores, which have specificity and affinity to chelate with iron. Siderophores are amino acids (nonproteinous) of less than 1000 Da, connected with rhizospheric microorganisms (Krewulak and Vogel 2008; Lemanceau et al., 2009). Similarly plants have phytosiderophores, which are helpful for acquiring micronutrients (cationic) from rhizosphere (Marschner et al., 1986; Zhang et al., 1991). Bacteria like *Burkholderia*, *Grimontella*, and *Enterobacter* have high siderophore production (Dimkpa et al., 2009).

### 1.4.3 Heavy metal toxicity

The increasing infiltration of industrial effluents into water bodies has led to heavy metal pollution of soils and is one of the worst environmental hazards today, with serious impacts on human health and agriculture. This situation is aggravated by the heavy use of various agrochemicals (herbicides, insecticides, pesticides, fungicides, etc), along with the overuse of nitrogen, phosphorous and potassium (NPK) fertilizers, thereby making situation more grievious (Saberi and Hassan, 2014; Lu et al., 1992). In particular, the trace elements and Cd in phosphate are especially concerning. Remediation for such metal-stressed soil has thus become key as contaminated soils are inappropriate for agriculture. Pollution of soil with heavy metals has long-lasting negative effects on all living organisms, including humans. Among the heavy metal pollutants, lead and cadmium are the most widespread and have the highest potential for toxicity.

### 1.4.4 Microbe-induced bioremediation for plant-growth promotion

The success of phytoremediation of any soil depends largely on the degree of contamination as well as the contaminating factors. Moreover, the rate of remediation in the defined soil vicinity is determined by how much the plants or microbes take up the contaminants. In ambient conditions, the plants with high metal accumulating efficiency, tends to grow at their minimal rate in the presence of excessive rhizoshperic metal concentration. To purge heavy metal contaminants from soil, physicochemical and biological techniques can be used (Glick, 2010; Sheoran et al., 2011; Ali et al., 2013; Ullah et al., 2015). Biological remediation is considered as the most suitable for removal of toxic metals, because of its ecofriendly nature and low cost (Doble and Kumar, 2005). Application of BSMs, in particular PGPRs, for soil bioremediation under climate-changing scenarios and excessive fertilizer use is beneficial (Nautiyal et al., 2013; Tiwari et al., 2016). Increased plant growth and high survival rate under stressed conditions is common for BSMs, indicating their remediation potential by conversion of complex substances to simple and nontoxic compounds. Resistance potential against toxic heavy metals makes PGPRs a suitable bioremediating agent (Thassitou and Arvanitoyannis, 2001; Mustapha and Halimoon, 2015). Metal sequestration by soil microbes is due to their ability to produce nontoxic organic acids, which boost metal chelation by lowering the pH of the soil (Mishra et al., 2017).

### 1.4.5 Remediation of heavy metals by bacteria

Soil bacteria form the vital microflora in the rhizosphere and are responsible for mitigating heavy metal toxicity (Hassan et al., 2017). Additionally, growth promotion and enhanced plant productivity is achieved by secretion of growth regulators by soil microbes, which also helps in nutrient uptake by plants (Nadeem et al., 2014). Formation of metal complexes from siderophore complexes, metabolites from bacteria, and transporter proteins limits metal contaminants (Rajkumar et al., 2010; Ahemad, 2012). These BSMs have special metabolic functions to overcome metal stress, making them of prime importance in agronomy. Primarily these methods include meal transport throughout the cytoplasmic membrane, cell-wall adherence and accumulation, and caging of metal-ion associations with other organisms (Ullah et al., 2015). Such evidence has been already observed, in which elimination of toxic metals was accomplished by a novel gene modulation technique to create a whole new species (Ullah et al., 2015). Overexpression of target genes or by some means of gene insertion to achieve a transgenic can serve as novel alternative for getting the desired microorganisms with higher throughput capacity of metal sequestration, translocation, etc.(Singh et al., 2011).

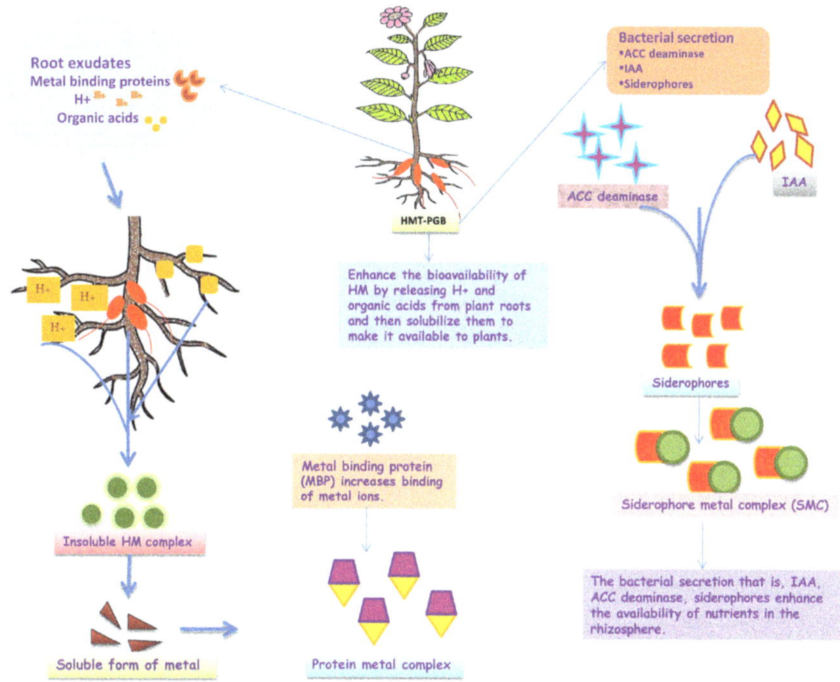


**Figure 1.3** BSMs mediated Plant growth promotion by different methods.
*Sources*: Tiwari, S., Lata, C. (2018). Heavy metal stress, signaling, and tolerance due to plant-associated microbes: an overview. Front. Plant Sci. https://doi.org/10.3389/fpls.2018.00452 and Mishra, J., Singh, R., Arora, N.K. (2017). Alleviation of heavy metal stress in plants and remediation of soil by rhizosphere microorganisms. Front. Microbiol. https://doi.org/10.3389/fmicb.2017.01706.

### 1.4.6 Remediation of heavy metals by fungi

The rhizosphere is colonized by a large number of microorganisms including mycorrhizal fungi. The root-colonizing mycorrhizae have the potential to increase the uptake of available diverse essential elements, along with heavy metal ions significantly (Shukla et al., 2015; Kumar et al., 2015). The metal bioavailability of soil can be rapidly transformed by the bacteria. The accumulation of heavy metals can be alleviated at various ecological trophic level, by the evolved strategies of mycorrhizae in the presence of mixed systems of microbial culture (Hall, 2002; Kramer, 2005). Arbuscular mycorrhizal fungus-inhabiting soil is generally used for its growth promotion of soil microbes in close vicinity to the inhabiting plant rhizosphere. Thus, vescicular arbuscular mycorrhiza (VAM) association in plants is recognized for its growth promotion—because of their extensive mycelial mats that not only acquires nutrients and provides them to plants, particularly phosphorus—but also binds the soil (Smith and Read, 1997). It has been demonstrated in various studies that VAM fungi can potentially shield heavy metal stress (Kumar et al., 2015). Hence, the VAM-mediated nutritional effects along with the heavy metal dislocation from the available nutrient pool at the interface of soil and plant is responsible for heavy metal removal.

Filamentous fungi like *Trichoderma, Penicillium*, and *Aspergillus* have the capability of tolerating metal stress (Ezzouhri et al., 2009; Oladipo et al., 2017). The cell-wall chemistry of fungus plays an important role in its metal-binding properties, which is attributed to its negatively charged functional groups like carboxylic, sulfhydryl, and phosphate (Ong et al., 2017). Evidence has been found for effective reduction in the metal concentration in soil with the help of the filamentous spore-forming fungi *Aspergillus niger* var. (Coreño-Alonso et al., 2014), and reductions through *Trichoderma* sp. in chickpea plants (Tripathi et al., 2013; Tripathi et al., 2017). On the other hand, arbuscular mycorrhizal fungi (AMF) also forms important microflora in soil and institutes one-to-one linkups between soil plants roots, facilitating nutrient uptake by enhancing root surface area as well as binding of soil (Saxena et al., 2017). Arbuscular mycorrhizae are also useful in alleviation of toxic metals from plants (Leyval et al., 1997; Meharg, 2003). A variety of ecological processes and organisms are involved in heavy metal remediation (Pawlowska and Charvat, 2004). In connection with AMF, many similar mechanisms have also been proposed as for PGPR and for remediation of metals and their allocation in plants. These include binding of metals to the fungal cell wall along with vacuolar deposition, chemical binding of metal and its sequestration through root apoplasm; metallothioneins or phytochelatins formation inside the fungus; and exchange of metal ions via tonoplast of both symbionts (Jan and Parray, 2016).

### 1.4.7 Remediation of heavy metals by plant growth-promoting bacteria

Increased accumulation of heavy metals in some plants growing in the contaminated soil atmosphere can be attributed to the local soil flora, which facilitates the ambient conditions for the plant (Dikshit et al., 2013). This increased efficacy can be visualized as an essential tool for reuse of nutrients, detoxification of noxious waste, and alternation and soil rebuild. For remediation, the application of green plants is now considered an antiquated practice that can be supplemented or replaced, if needed, by application of rhizospheric microflora, which enables as well as magnifies the detoxification process by both elimination and immobilization.

The mechanism for sequestration of heavy metals from contaminated soil remains in question, though much research has been done. The prerequisite conditions required for such hyperaccumulator plants is a high rate of growth for the biomass that can withstand high stress conditions (Nie et al., 2002).

The soil microbial communities are affected by the heavy metal contamination in many ways. First, a reduction in microbial biomass (Brookes and

McGrath, 1984; Fliessbach et al., 1994); second, by reduction of the microflora inhabiting a particular soil (Chaudri et al., 1993; Koomen et al., 1990); and lastly, through hindrance of the whole microbial community, which may result in an inaccurate soil profile (Frostegård et al., 1993, 1996; Gray and Smith, 2005; Sandaa et al., 1999).

## 1.5 Induced systemic resistance

ISR occurs when resistance to the pathogenic form of microbes, particularly filamentous fungi, gram-negative soil bacteria, and some viruses, is achieved with rhizospheric bacteria (Lugtenberg and Kamilova, 2009). PGPRs like *Bacillus, Pseudomonas*, and *Azospirillum* are well known for their ISR. Additionally, stress response in the form of ethylene and jasmonic acid signaling within plants has been observed, which triggers resistance capability of plants against various pathogens (Glick, 2012). Lipopolysaccharides, cyclic lipopeptides, siderophores, and 2,4-diacetylphloroglucinol and volatile substances (2, 3-butanediol and acetoin) are some common ISR-inducing agents (Lugtenberg and Kamilova, 2009).

## 1.6 New paradigms in sustainable agriculture

Contaminated and barren soil is an urgent problem in agriculture today. The soil rhizosphere is an important region for microbial bioremediation, but many rhizospheric bacteria that are competent in degrading specific kinds of organic pollutants do not flourish easily in some soil atmospheres. The efficacy of these BSMs in bioremediation of soil lies with the other native soil microbial species though vigorous but not stable and competent to carry forward. Advancements in molecular biology and genetics can be used to develop microorganisms that are competent enough to not only enhance the fertility of the soil, but simultaneously remediate the negative factors as well. For soil pollutants like trichloroethylene and some PCBs, the molecular mechanism of degradation has long been known (Bradford, 1976; Brazil et al., 1995). Apart from this, cloning of desirable genes into a competent vector bacterial BSM will provide soil remediation at a higher rate. Essential factors for such gene targeting include suitable strain identification, overexpression of a desired gene, and tolerance factors of the bacteria. Other important approaches include the application of nanoparticle adhesion to BSMs and increasing the surface activity. Nanoparticles can also be used in bioinoculants, which have a rate of growth and survival in the rhizosphere. Thus, a sustainable approach could be achieved to conserve the fertility of soil, and at the same time, enhance productivity.

## Acknowledgments

The authors are thankful to the Head, Department of Botany, for providing laboratory facilities; and to CSIR and UGC for financial assistance. Krishna Kumar Chaudhary is thankful to Research Seed Money Grant, Central University of Punjab and UGC Start-Up Grant, New Delhi.

## References

Ahemad, M., 2012. Implication of bacterial resistance against heavy metals in bioremediation: a review. J. Inst. Integr. Omics Appl. Biotechnol. 3, 39–46.

Ahemad, M., Kibret, M., 2014. Mechanisms and applications of plant growth promoting rhizobacteria: current perspective. J. King Saud. Univ. Sci. 26, 1–20.

Aislabie Jand, A., Deslippe, J.R., 2013. Soil microbes and their contribution to soil services. In: Dymond, J.R. (Ed.), Ecosystem Services in New Zealand-Conditions and Trends. Manaaki Whenua Press, New Zealand, pp. 143–161.

Ali, H., Khan, E., Sajad, M.A., 2013. Phytoremediation of heavy metals—concepts and applications. Chemosphere 91, 869–881.

Arora, N.K. (Ed.), 2015. Plant Microbes Symbiosis: Applied Facets. Springer, Dordrecht, The Netherlands.

Arora, N.K., Mishra, J., 2016. Prospecting the roles of metabolites and additives in future bioformulations for sustainable agriculture. Appl. Soil Ecol. Available from: https://doi.org/10.1016/j.apsoil.2016.05.020.

Bagyalakshmi, B., Ponmurugan, P., Marimuthu, S., 2012. Influence of potassium solubilizing bacteria on crop productivity.

Barazani, O., Friedman, J., 1999. Is IAA the major root growth factor secreted from plant-growth-mediating bacteria?. J. Chem. Ecol. 25 (10), 2397–2406.

Beneduzi, A., Ambrosini, A., Luciane, M.P., Passaglia, L.M.P., 2012. Plant growth promoting rhizobacteria (PGPR): their potential as antagonists and biocontrol agents. Genet. Mol. Biol. 35, 1044–1051.

Blackwell, M., 2011. The fungi: 1, 2, 3 ... 5.1 million species? Am. J. Bot. 3, 426–438.

Bradford, M.M., 1976. A rapid and sensitive method for the quantitation of microgram quantities of protein utilizing the principle of protein-dye binding. Anal. Biochem. 72, 248–254.

Brazil, G.M., Kenefick, L., Callanan, M., Haro, A., de Lorenzo, V., Dowling, D.N., 1995. Construction of a rhizosphere pseudomonad with potential to degrade polychlorinated biphenyls and detection of bph gene expression in the rhizosphere. Appl. Environ. Microbiol. 61, 1946–1952.

Brookes, P.C., McGrath, S.P., 1984. Effects of metal toxicity on the size of the soil microbial biomass. J. Soil Sci. 35, 341–346.

Carsten, S.J., Mathis, H.H., 2014. Agricultural soils, pesticides and microbial diversity. Curr. Opin. Biotechnol. 27, 15–20.

Castro, R.O., Cornejo, H.A.C., Rodriguez, L.M., Bucio, J.L., 2009. The role of microbial signals in plant growth and development. Plant Signal. Behav. 4 (8), 701–712.

Chaudri, A.M., McGrath, S.P., Giller, K.E., Rietz, E., Sauerbeck, D.R., 1993. Enumeration of indigenous *Rhizobium leguminosarum* biovar trifolii in soils previously treated with metal-contaminated sewage sludge. Soil Biol. Biochem. 25, 301–309.

Cheng, W., 2009. Rhizosphere priming effect: its functional relationships with microbial turnover, evapotranspiration, and C-N budgets. Soil Biol. Biochem. 41, 1795–1801.

Choudhary, K.K., Agrawal, S.B., 2017. Effect of UV-B radiation on leguminous plants. In: Lichtfouse, E. (Ed.), Sustainable Agricultural Reviews. Springer International Publishing, Switzerland, pp. 115–162.

Choudhary, D.K., Sharma, K.P., Gaur, R.K., 2011. Biotechnological perspectives of microbes in agro-ecosystems. Biotechnol. Lett. 33, 1905–1910.

Choudhary, K.K., Pandey, D., Agrawal, S.B., 2013. Deterioration of rhizospheric soil health due to elevated ultraviolet-B. Arch. Agron. Soil Sci. 59, 1419–1437.

Chung, H., Park, M., Madhaiyana, M., Seshadri, S., Song, J., Cho, H., et al., 2005. Isolation and characterization of phosphate solubilizing bacteria from the rhizosphere of crop plants of Korea. Soil Biol. Biochem. 3, 1970–1974.

Coreño-Alonso, A., Solé, A., Diestra, E., Esteve, I., Gutiérrez-Corona, J.F., López, G.R., et al., 2014. Mechanisms of interaction of chromium with *Aspergillus niger* var. tubingensis strain Ed8. Bioresour. Technol. 158, 188–192.

de Souza, J.T.A., Arnould, C., Deulvot, C., Lemanceau, P., Gianinazzi-Pearson, V., Raaijmakers, J.M., 2003. Effect of 2,4-diacetylphloroglucinol on Pythium: cellular responses and variation in sensitivity among propagules and species. Phytopathology 93, 966–975.

Dikshit, A., Shukla, S.K., Mishra, R.K., 2013. Plant Growth Promotion by Plant Growth Promoting Rhizobacteria (PGPR) with Nanomaterials in Current Agricultural Scenario. Lap Lambert Academic Publishing, ISBN 978- 3-659-36774-8.

Dimkpa, C.O., Merten, D., Svatos, A., Büchel, G., Kothe, E., 2009. Siderophores mediate reduced and increased uptake of cadmium by *Streptomyces tendae* F4 and sunflower (*Helianthus annuus*), respectively. J. Appl. Microbiol. 5, 687–1696.

Doble, M., Kumar, A., 2005. Biotreatment of Industrial Effluents. Butterworth-Heinemann, Oxford, pp. 19–38.

Duffy, B., Schouten, A., Raajimakers, J., 2003. Pathogen selfdefense: mechanisms to counteract microbial antagonism. Annu. Rev. Phytopathol. 45, 501–538.

Ezzouhri, L., Castro, E., Moya, M., Espinola, F., Lairini, K., 2009. Heavy metal tolerance of filamentous fungi isolated from polluted sites in Tangier, Morocco. Afr. J. Microbiol. Res. 3, 35–48.

Fliessbach, A., Martens, R., Reber, H.H., 1994. Soil microbial biomass and microbial activity in soils treated with heavy metal contaminated sewage sludge. Soil Biol. Biochem. 26, 1201–1205.

Fritsch, F.E., 1907. A general consideration of aerial and fresh water algal flora of Ceylon. Proc. R. Soc. London 11, 79–197.

Frostegård, A., Tunlid, A., Bååth, E., 1993. Phospholipid fatty acid composition, biomass and activity of microbial communities from two soil types experimentally exposed to different heavy metals. Appl. Environ. Microbiol. 59, 3605–3617.

Frostegård, A., Tunlid, A., Bååth, E., 1996. Changes in microbial community structure during long-term incubation in two soils experimentally contaminated with metals. Soil Biol. Biochem. 28, 55–63.

Gindling, T.H., Newhouse, D., 2012. Self-employment in the developing world. Policy Res Working Paper, 1813, 9450–6201.

Glick, B.R., 2010. Using soil bacteria to facilitate phytoremediation. Biotechnol. Adv. 28, 367–374.

Glick, B.R., 2012. Plant growth-promoting bacteria: mechanisms and applications. Hindawi Publishing Corporation, Scientifica 2012, 1–15.

Goswami, D., Thakker, J.N., Dhandhukia, P.C., 2016. Portraying mechanics of plant growth promoting rhizobacteria (PGPR): a review. Cogent. Food Agric. 2, 1127500.

Gray, E.J., Smith, D.L., 2005. Intracellular and extracellular PGPR: commonalities and distinctions in the plant-bacterium signaling processes. Soil Biol. Biochem. 37, 395–412.

Hall, J.L., 2002. Cellular mechanism of heavy metal detoxification and tolerance. J. Exp. Bot. 53, 1–11.

Hassan, T.U., Bano, A., Naz, I., 2017. Alleviation of heavy metals toxicity by the application of plant growth promoting rhizobacteria and effects on wheat grown in saline sodic field. Int. J. Phytoremed. 19, 522–529.

Use of mycorrhiza as metal tolerance strategy in plants. In: Jan, S., Parray, J.A. (Eds.), Approaches to Heavy Metal Tolerance in Plants. Springer, Singapore.

Kang, S.M., Khan, A.L., Waqas, M., You, Y., Kim, J., Hamayun, M., et al., 2014. Plant growth promoting rhizobacteria. Annu. Rev. Microbiol. 63, 541–556.

Karthikeyan, N., Prasanna, R.L., Nain, L., Kaushik, B.D., 2007. Evaluating the potential of plant growth promoting cyanobacteria as inoculants for wheat. Eur. J. Soil Biol. 43, 23–30.

Kaushik, B.D., 2014. Developments in cyanobacterial biofertilizer. Proc. Indian Nat. Sci. Acad. 80, 379–388.

Kesavan, P.C., Swaminathan, M.S., 2008. Strategies and models for agricultural sustainability in developing Asian countries. Philos. Trans. R. Soc. B 363, 877–891.

Khare, E., Arora, N.K., 2011. Physiologically stressed cells of fluorescent Pseudomonas EKi as better option for bioformulation development for management of charcoal rot caused by *Macrophomina phaseolina* in field conditions. Curr. Microbiol. 62, 1789–1793.

Kloepper, J.W., Schroth, M.N., 1978. Plant growth-promoting rhizobacteria on radishes. Proc. Fourth Int. Conf. Plant Pathol. Bacteriol. 2, 879–882.

Koomen, I., McGrath, S.P., Giller, K.E., 1990. Mycorrhizal infection of clover is delayed in soils contaminated with heavy metals from past sewage sludge applications. Soil Biol. Biochem. 22, 871–873.

Kramer, U., 2005. Phytoremediation: novel approaches to cleaning up polluted soils. Curr. Opin. Biotechnol. 16, 133–141.

Krewulak, H.D., Vogel, H.J., 2008. Structural biology of bacterial iron uptake. Biochim. Biophys. Acta 1778, 1781–1804.

Kumar, R., Mishra, R.K., Qidwai, A., Shukla, S.K., Mishra, V., Pandey, A., et al., 2015. Detoxification and tolerance of heavy metals in plants. Plant Metal Interaction: Emerging Remediation Techniques, Chapter 13. Elsevier Inc, pp. 335–360, ISBN: 978-0-12-803158-2.

Lal, R., Sivakumar, M.V.K., Faiz, S.M.A., Mustafizur Rahman, A.H.M., Islam, K.R. (Eds.), 2010. Climate Change and Food Security in South Asia. Springer, Netherlands.

Lemanceau, P., Bauer, P., Kraemer, S., Briat, J.F., 2009. Iron dynamics in the rhizosphere as a case study for analyzing interactions between soils, plants and microbes. Plant Soil 321, 513–535.

Leyval, C., Turnau, K., Haselwandter, K., 1997. Effect of heavy metal pollution on mycorrhizal colonization and function-physiological, ecological and applied aspects. Mycorrhiza 7, 139–153.

Liu, D., Lian, B., Dong, H., 2012. Isolation of Paenibacillus sp. and assessment of its potential for enhancing mineral weathering. Geophys. J. R. Astron. Soc. 29, 413–421.

Lu, R.K., Shi, Z.Y., Xio, L.M., 1992. Cadmium contents of rock phosphates and phosphate fertilizers of China and their effects on ecological environment. Acta Pedol. Sin. 29, 150–157.

Lugtenberg, B., Kamilova, F., 2009. Plant-growth-promoting rhizobacteria. Annu. Rev. Microbiol. 63, 541–556.

Ma, Y., Oliveira, R.S., Freitas, H., Zhang, C., 2016. Biochemical and molecular mechanisms of plant-microbe-metal interactions: relevance for phytoremediation. Front. Plant Sci. 10.3389, 00918.

Manjunath, M., Prasanna, R., Shama, P., Nain, L., Singh, R., 2011. Developing PGPR consortia using novel genera Providencia and Alcaligenes along with cyanobacteria for wheat. Arch. Agron. Soil Sci. 57, 873–887.

Marschner, H., Römheld, V., Kissel, M., 1986. Different strategies in higher-plants in mobilization and uptake of iron. J. Plant Nutr. 9, 695–713.

Mazurier, S., Corberand, T., Lemanceau, P., Raaijmakers, J.M., 2009. Phenazine antibiotics produced by fluorescent pseudomonads contribute to natural soil suppressiveness to Fusarium wilt. ISME J. 3, 977–991.

Meena, V.S., Maurya, B.R., Bohra, J.S., Verma, R., Meena, M.D., 2013a. Effect of concentrate manure and nutrient levels on enzymatic activities and microbial population under submerged rice in alluvium soil of Varanasi. Crop Res. 45 (1, 2 & 3), 6–12.

Meena, V.S., Maurya, B.R., Verma, R., Meena, R.S., Jatav, G.K., Meena, S.K., 2013b. Soil microbial population and selected enzyme activities as influenced by concentrate manure and inorganic fertilizer in alluvium soil of Varanasi. Bioscan 8 (3), 931–935.

Meena, V.S., Maurya, B.R., Bahadur, I., 2014a. Potassium solubilization by bacterial strain in waste mica. Bangladesh J. Bot. 43, 235–237.

Meena, V.S., Maurya, B.R., Verma, J.P., 2014b. Does a rhizospheric microorganism enhance $K^+$ availability in agricultural soils?. Microbiol. Res. 169, 337–347.

Meena, V.S., Maurya, B.R., Meena, R.S., 2015a. Residual impact of wellgrow formulation and NPK on growth and yield of wheat (*Triticum aestivum* L.). Bangladesh J. Bot. 44 (1), 143–146.

Meena, V.S., Maurya, B.R., Verma, J.P., Aeron, A., Kumar, A., Kim, K., et al., 2015b. Potassium solubilizing rhizobacteria (KSR): isolation, identification, and K-release dynamics from waste mica. Ecol. Eng. 81, 340–347.

Meena, V.S., Verma, J.P., Meena, S.K., 2015c. Towards the current scenario of nutrient use efficiency in crop species. J. Clean Prod. 102, 556–557.

Meena, R.S., Bohra, J.S., Singh, S.P., Meena, V.S., Verma, J.P., Verma, S.K., et al., 2016a. Towards the prime response of manure to enhance nutrient use efficiency and soil sustainability a current need: a book review. J. Clean Prod. 112 (1), 1258–1260.

Meena, S.K., Rakshit, A., Meena, V.S., 2016b. Effect of seed bio-priming and N doses under varied soil type on nitrogen use efficiency (NUE) of wheat (*Triticum aestivum* L.) under greenhouse conditions. Biocatal. Agric. Biotechnol. 6, 68–75.

Meena, M.K., Gupta, S., Datta, S., 2016c. Antifungal potential of PGPR, their growth promoting activity on seed germination and seedling growth of winter wheat and genetic variabilities among bacterial isolates. Int. J. Curr. Microbiol. Appl. Sci. 5 (1), 235–243.

Meharg, A.A., 2003. The mechanistic basis of interactions between mycorrhizal associations and toxic metal cations. Mycol. Res. 107, 1253–1265.

Mishra, S., Arora, N.K., 2012. Evaluation of rhizospheric *Pseudomonas* and *Bacillus* as biocontrol tool for *Xanthomonas campestris* pv. campestris. World J. Microbiol. Biotechnol. 28, 693–702.

Mishra, J., Singh, R., Arora, N.K., 2017. Alleviation of heavy metal stress in plants and remediation of soil by rhizosphere microorganisms. Front. Microbiol. 8, 1706.

Mustapha, M.U., Halimoon, N., 2015. Screening and isolation of heavy metal tolerant bacteria in industrial effluent. Procedia Environ. Sci. 30, 33–37.

Nadeem, S.M., Ahmad, M., Zahir, Z.A., Javaid, A., Ashraf, M., 2014. The role of mycorrhizae and plant growth promoting rhizobacteria (PGPR) in improving crop productivity under stressful environments. Biotechnol. Adv. 32, 429–448.

Nautiyal, C.S., Srivastava, S., Chauhan, P.S., Seem, K., Mishra, A., Sopory, S.K., 2013. Plant growth-promoting bacteria *Bacillus amyloliquefaciens* NBRISN13 modulates gene expression profile of leaf and rhizosphere community in rice during salt stress. Plant Physiol. Biochem. 66, 1–9.

Nie, L., Shah, S., Burd, G.I., Dixon, D.G., Glick, B.R., 2002. Phytoremediation of arsenate contaminated soil by transgenic canola and the plant growth-promoting bacterium *Enterobacter cloacae* CAL2. Plant Physiol. Biochem. 40, 355–361.

Nihorimbere, V., Ongena, M., Smargiassi, M., Thonart, P., 2011. Beneficial effect of the rhizosphere microbial community for plant growth and health. Biotechnol. Agron. Soc. Environ. 2, 327–337.

Nishanth, D., Biswas, D.R., 2008. Kinetics of phosphorus and potassium release from rock-phosphate and waste mica enriched compost and their effect on yield and nutrient uptake by wheat (*Triticum aestivum* L.). Bioresour. Technol. 99, 3342–3354.

Oladipo, O.G., Awotoye, O.O., Olayinka, A., Bezuidenhout, C.C., Maboeta, M.S., 2017. Heavy metal tolerance traits of filamentous fungi isolated from gold and gemstone mining sites. Braz. J. Microbiol. 49, 29–37.

Ong, G.H., Ho, X.H., Shamkeeva, S., Fernando, M.S., Shimen, A., Wong, L.S., 2017. Biosorption study of potential fungi for copper remediation from Peninsular Malaysia. Remediat. J. 27, 59–63.

Pawlowska, T.E., Charvat, I., 2004. Heavy-metal stress and developmental patterns of arbuscular mycorrhizal fungi. Appl. Environ. Microbiol. 70, 6643–6649.

Pieterse, C.M., Leon-Reyes, A., Van der Ent, S., Van Wees, S.C., 2009. Networking by small-molecule hormones in plant immunity. Nat. Chem. Biol. 5, 308–316.

Prasad, R.C., Prasad, B.N., 2001. Cyanobacteria as a source biofertilizer for sustainable agriculture in Nepal. J. Plant Sci. Bot. Orientalis 8, 127–133.

Rajkumar, M., Ae, N., Prasad, M.N.V., Freitas, H., 2010. Potential of siderophore-producing bacteria for improving heavy metal phytoextraction. Trends Biotechnol. 28, 142–149.

Rajkumar, M., Vara Prasad, M.N., Freitas, H., Ae, N., 2009. Biotechnological applications of serpentine soil bacteria for phytoremediation of trace metals. Crit. Rev. Biotechnol. 29 (2), 120–130.

Ramatte, A., Frapolli, M., Defago, G., Moenne-Loccoz, Y., 2003. Phylogeny of HCN synthase encoding hcnBC genes in biocontrol fluorescent pseudomonads and its relationship with host plant species and HCN synthesis ability. Mol. Biol. Pl Microbe Interact 16, 525–535.

Reitemeir, R.F., 1951. Soil potassium. In: Norman, A.G. (Ed.), American Society of Agronomy, vol 3. Academic Press, Int Publ, New York, pp. 113–164. , Advances in agronomy.

Richardson, A.E., Simpson, R.J., 2011. Soil microorganisms mediating phosphorus availability update on microbial phosphorus. Plant Physiol. 156 (3), 989–996.
Rosa, F., Philippis, R.D., 2015. Role of cyanobacterial exopolysaccharides in phototrophic biofilms and in complex microbial mats. Life 5, 1218–1238.
Saberi, A.R., Hassan, S.A., 2014. The effects of nitrogen fertilizer and plant density on mustard (*Brassica juncea*); an overview. Glob. Adv. Res. J. Agric. Sci. 3 (8), 205–210.
Saha, M., Maurya, B.R., Meena, V.S., Bahadur, I., Kumar, A., 2016. Identification and characterization of potassium solubilizing bacteria (KSB) from indo-Gangetic Plains of India. Biocatal. Agric. Biotechnol. 7, 202–209.
Saiyad, S.A., Jhala, Y.K., Vyas, R.V., 2015. Comparative efficiency of five potash and phosphate solubilizing bacteria and their key enzymes useful for enhancing and improvement of soil fertility. Int. J. Sci. Res. Pub. 5, 1–6.
Sandaa, R.A., Torsvik, V., Enger, Ø., 1999. Analysis of bacterial communities in heavy metal-contaminated soils at different levels of resolution. FEMS Microbiol. Ecol. 30, 237–251.
Saxena, B., Shukla, K., Giri, B., 2017. Arbuscular mycorrhizal fungi and tolerance of salt stress in plants. In: Wu, Q.S. (Ed.), Arbuscular Mycorrhizas and Stress Tolerance of Plants. Springer, Singapore, pp. 67–97.
Sharma, S.B., Sayyed, R.Z., Trivedi, M.H., Gobi, T.A., 2013. Phosphate solubilizing microbes: sustainable approach for managing phosphorus deficiency in agricultural soils. Spring 2, 1–14.
Sharma, A., Shankhdhar, D., Shankhdhar, S.C., 2016. Potassium-solubilizing microorganisms: mechanism and their role in potassium solubilization and uptake. In: Meena, V.S., Maurya, B.R., Verma, J.P., Meena, R.S. (Eds.), Potassium Solubilizing Microorganisms for Sustainable Agriculture. Springer, India, pp. 203–219.
Sheoran, V., Sheoran, A., Poonia, P., 2011. Role of hyperaccumulators in phytoextraction of metals from contaminated mining sites: a review. Crit. Rev. Environ. Sci. Technol. 41, 168–214.
Shrivastava, S., Prasad, R., Varma, A., 2014. Anatomy of root from eyes of a microbiologist. In: Morte, A., Varma, A. (Eds.), Root Engineering. Springer, Berlin, pp. 3–22.
Shrivastava, M., Srivastava, P.C., D'Souza, S.F., 2016. KSM soil diversity and mineral solubilization, in relation to crop production and molecular mechanism. In: Meena, V.S., Maurya, B.R., Verma, J.P., Meena, R.S. (Eds.), Potassium Solubilizing Microorganisms for Sustainable Agriculture. Springer, India, pp. 221–234.
Shukla, S.K., Mishra, R.K., Pandey, M., Mishra, V., Pathak, A., Pandey, A., et al., 2015. Land reformation using plant growth promoting rhizobacteria with context to heavy metal contamination, Plant Metal Interaction: Emerging Remediation Techniques, Chapter, 21. Elsevier Inc, pp. 499–530, ISBN: 978-0-12-803158-2.
Sindhu, S.S., Parmar, P., Phour, M., Sehrawat, A., 2016. Potassium-solubilizing microorganisms (KSMs) and its effect on plant growth improvement. In: Meena, V.S., Maurya, B. R., Verma, J.P., Meena, R.S. (Eds.), Potassium Solubilizing Microorganisms for Sustainable Agriculture. Springer, India, pp. 171–185.
Singh, N.K., Dhar, D.W., 2010. Cyanobacterial reclamation of saltaffected soil. In: Lichtfouse, E. (Ed.), Genetic Engineering, Biofertilisation, Soil Quality and Organic Farming. Springer, Netherlands, 245-24.
Singh, J.S., Abhilash, P.C., Singh, H.B., Singh, R.P., Singh, D.P., 2011. Genetically engineered bacteria: an emerging tool for environmental remediation and future research perspectives. Gene 480, 1–9.

Singh, N.P., Singh, R.K., Meena, V.S., Meena, R.K., 2015. Can we use maize (*Zea mays*) rhizobacteria as plant growth promoter?. Vegetos 28 (1), 86–99.

Singh, J.S., Kumar, A., Rai, A.N., Singh, D.P., 2016a. Cyanobacteria: a precious bio-resource in agriculture, ecosystem, and environmental sustainability. Front. Microbiol. 7, 529.

Singh, M., Dotaniya, M.L., Mishra, A., Dotaniya, C.K., Regar, K.L., Lata, M., 2016b. Role of biofertilizers in conservation agriculture. In: Bisht, J.K., Meena, V.S., Mishra, P.K., Pattanayak, A. (Eds.), Conservation Agriculture: An Approach to Combat Climate Change in Indian Himalaya. Springer, Singapore, pp. 113–134.

Smil, V., 1999. Nitrogen in crop production: an account of global flows. Glob. Biogeochem. Cycl. 13, 647–662.

Smith, S.E., Read, D.J., 1997. Mycorrhizal Symbiosis. Academic Press, London, pp. 563–568. Available from: http://doi.org/10.1046/j.1469-8137.1997.00851-2.

Song, T., Martensson, L., Eriksson, T., Zheng, W., Rasmussen, U., 2005. Biodiversity and seasonal variation of the cyanobacterial assemblage in a rice paddy field in Fujian, China. FEMS Microbiol. Ecol. 123, 54131-1400.

Thassitou, P., Arvanitoyannis, I., 2001. Bioremediation: a novel approach to food waste management. Trends Food Sci. Technol. 12, 185–196.

Tiwari, S., Lata, C., Chauhan, P.S., Nautiyal, C.S., 2016. *Pseudomonas putida* attunes morphophysiological, biochemical and molecular responses in *Cicer arietinum* L. during drought stress and recovery. Plant Physiol. Biochem. 99, 108–117.

Tripathi, P., Singh, P.C., Mishra, A., Chaudhry, V., Mishra, S., Tripathi, R.D., et al., 2013. Trichoderma inoculation ameliorates arsenic induced phytotoxic changes in gene expression and stem anatomy of chickpea (*Cicer arietinum*). Ecotoxicol. Environ. Saf. 89, 8–14.

Tripathi, P., Singh, P.C., Mishra, A., Srivastava, S., Chauhan, R., Awasthi, S., et al., 2017. Arsenic tolerant Trichoderma sp. reduces arsenic induced stress in chickpea (*Cicer arietinum*). Environ. Pollut. 223, 137–145.

Ullah, A., Heng, S., Munis, M.F.H., Fahad, S., Yang, X., 2015. Phytoremediation of heavy metals assisted by plant growth promoting (PGP) bacteria: a review. Environ. Exp. Bot. 117, 28–40.

Verma, P., Yadav, A.N., Khannam, K.S., Panjiar, N., Kumar, S., Saxena, A.K., et al., 2015. Assessment of genetic diversity and plant growth promoting attributes of psychrotolerant bacteria allied with wheat (*Triticum aestivum* L.) from the northern hills zone of India. Ann. Microbiol. Available from: https://doi.org/10.1007/s13213-014-1027-4.

Vejan, P., Abdullah, R., Khadiran, T., Ismail, S., Nasrulhaq Boyce, A., 2016a. Role of plant growth promoting rhizobacteria in agricultural sustainability—a review. Molecules 21 (5), 573.

Vejan, P., Abdullah, R., Khadiran, T., Ismail, S., Boyce, A.N., 2016b. Role of plant growth promoting rhizobacteria in agricultural sustainability—a review. Molecules 21, 1–17.

Zhang, C., Kong, F., 2014. Isolation and identification of potassium-solubilizing bacteria from tobacco rhizo- spheric soil and their effect on tobacco plants. Appl. Soil Ecol. 82, 18–25.

Zhang, F.S., Römheld, V., Marschner, H., 1991. Diurnal rhythm of release of phytosiderophores and uptake rate of zinc in iron-deficient wheat. Soil Sci. Plant Nutr. 37 (4), 671–678.

## Further reading

Afzal, M., Shabir, G., Tahseen, R., Ejazul, I., Iqbal, S., Khan, Q.M., 2014. Endophytic Burkholderia sp. Strain PsJN improves plant growth and phytoremediation of soil irrigated with textile effluent. Clean Soil Air Water 42, 1304–1310.

Ahmad, M., Nadeem, S.M., Naveed, M., Zahir, Z.A., 2016. Potassium-solubilizing bacteria and their application in agriculture. In: Meena, V.S., Maurya, B.R., Verma, J.P., Meena, R.S. (Eds.), Potassium Solubilizing Microorganisms for Sustainable Agriculture. Springer, India, pp. 293–313.

Chen, M., Xu, P., Zeng, G., Yang, C., Huang, D., Zhang, J., 2015. Bioremediation of soils contaminated with polycyclic aromatic hydrocarbons, petroleum, pesticides, chlorophenols and heavy metals by composting: applications, microbes and future research needs. Biotechnol. Adv. 33, 745–755.

De, P.K., 1939. The role of blue-green algae in nitrogen fixation in rice fields. Proc. R. Soc. London 127, 121–139.

Glick, B.R., 2001. Phytoremediation: synergistic use of plants and bacteria to clean up the environment. Biotechnol. Adv. 21, 383–393.

Lugtenberg, B.J.J., Dekkers, L., Bloemberg, C.V., 2001. Molecular determinants of rhizosphere colonization by *Pseudomonas*. Annu. Rev. Phytopathol. 39, 461–490.

Verma, R., Maurya, B.R., Meena, V.S., 2014. Integrated effect of bio-organics with chemical fertilizer on growth, yield and quality of cabbage (*Brassica oleracea* var capitata). Indian J. Agric. Sci. 84 (8), 914–919.

# Microbes as a novel source of secondary metabolite products of industrial significance

2

*Bhanu Prakash, Prem Pratap Singh, Akshay Kumar, Somenath Das and Anand Kumar Chaudhari*
Department of Botany, Institute of Science, Banaras Hindu University, Varanasi, India

## 2.1 Introduction

Natural products such as plants, animals, marine organisms, and microorganisms have been widely used as sources of novel bioactive compounds of industrial significance. Bioactive compounds such as alkaloid, terpenes, phenolic, and flavonoids possess strong pesticidal and therapeutic potential have been generated from natural products especially by higher plants as secondary metabolites. However, in view of the low concentration of bioactive molecules in plants, threat of biodiversity loss, chemotypic variation, and scarcity of land and water required for large-scale cultivation, industries are looking for alternative source of bioactive molecules. Today, the use of microorganisms and their metabolite products are considered as candidates to control agricultural pests and provide novel therapeutic agents (antibiotics, immunosuppressants, and lipid-lowering agents). Secondary metabolites are organic compounds that are not directly involved in the normal growth but play significant role in defense systems of organisms (Stamp, 2003). Secondary metabolite products of microbial sources have been widely used in agricultural and pharmaceutical sectors because of their advantages over other natural resources such as large-scale cultivation without seasonal variation, sufficient availability, and low cost. Recent reports have revealed that microorganisms, especially plant growth-promoting rhizobacteria (PGPR), have made significant contributions to agricultural industries. PGPR are widely used as bioinoculants in organic farming as a biofertilizer agent to improve soil fertility and also to prevent pest (Kumar et al., 2015a,b). *Bacillus* sp., *Pseudomonas*, *Streptomyces* and *Trichoderma*, and *Gliocladium* and *Fusarium* are widely used for the preparation of different consortia that have the potential to reduce pests in agriculture (Kumar et al., 2015b, 2017; Kloepper et al., 2004). Thus, the sustainable use of microbes and their products could reduce the use of chemical fertilizers and pesticides, which often pollute the environment. Endophytes (bacteria, fungi, and actinomycetes) colonize internal living tissues without causing any immediate negative symptoms to host plants. Most of them have the capability to produce bioactive secondary metabolites with unique structural and functional properties that could be successfully used as therapeutic agents

**Role of Plant Growth Promoting Microorganisms in Sustainable Agriculture and Nanotechnology.**
**DOI: https://doi.org/10.1016/B978-0-12-817004-5.00002-6**

after proper characterization. In addition, the endophytic microbes associated with the medicinal plants have the potential to produce active principle compounds of host plants. A range of plant secondary metabolites including alkaloids, benzopyranones, flavonoids, phenolic acids, quinones, steroids, terpenoids, and xanthones, have already been extracted from the diverse group of endophytes. The anticancerous compounds vincristine, vinblastine, and paclitaxel (Taxol) have been successfully extracted from the endophytic microbes such as *Alternaria* sp., *Fusarium oxysporum*, and *Taxomyces andreanae* (Guo et al., 1998; Zhang et al., 2000; Stierle et al., 1993). Phomoxanthone A and B are the antituberculosis compounds isolated from the endophytic fungi *Phomopsis* sp. (Isaka et al., 2001).

In this chapter, we provide an overview of the potential role of microbes and their metabolite products in the agriculture and pharma industries. In addition, existing limitations and the role of modern scientific and technological innovation to enhance the effectiveness and commercial application of microbial products are also discussed.

## 2.2 Role of microbes in agriculture

Chemical fertilizers have been used increasingly in recent decades to manage pests and boost agricultural production. However, most chemical fertilizers have deleterious effects to ecological components including soil fertility (Babalola, 2010). Therefore, considerable attention has been given to the use of biopesticides and other organic agricultural practices as alterantives to chemical pesticides. Higher plant products such as nicotine (nightshade plants), strychnine (*Strychnos* spp.), pyrethrum (*Chrysenthumum cinerariaefolium*), sabadilla (*Schoenocaulon officinale*) rotenone (*Derris elliptica*), ryanodine (*Ryania speciosa*), and neem extract (*Azadirachta indica*), have been used as biopesticides and in various crop protection strategies (Dubey et al., 2011). However, in view of their low yield, inadequate availability of raw materials, chemical variability, and biodegradability microbial products are currently used as alternatives of plant secondary metabolites in sustainable agriculture. In nature, a wide range of bacterial groups are associated with plants that promote plant growth by managing/interfering with neighboring pests, these beneficial rhizobacteria are referred to as PGPR (Reichenbach and Höfle, 2008; Weller et al., 2002). The PGPR associated with rhizosphere or inside the nodular structure of root can induce plant growth by solubilization of nutrients, efficient nitrogen fixation and growth regulators production or as biocontrol agent by competitive exclusion of pathogenic microbes (Bashan and de-Bashan, 2010). *Agrobacterium*, *Arthrobacter*, *Azotobacter*, *Azospirillum*, *Burkholderia*, *Caulobacter*, *Chromobacterium*, *Erwinia*, *Flavobacterium*, and *Pseudomonas* are generally associated with the rhizospheric region while the endophytes and Frankia are associated with the nodular region (Gray and Smith, 2005; Verma et al., 2010). *Trichoderma* spp. and their consortium have been sold as microbial products to enhance plant growth and production via inducing plant systemic resistance against phytopathogens

(Shoresh et al., 2010). Today, a wide range of microorganisms such as bacteria, actinomycetes, and fungi has been used to control various phytopathogenic fungi (Herrera-Estrella and Chet, 2003). The volatile organic compounds produced by bacterial species such as *Bacillus subtilis* strain GB03 and *Bacillus amyloliquefaciens* strain IN937a play a significant role in disease prevention in plants as well as induce plant host resistance (Heil and Karban, 2010). One recent study revealed that bacterial volatiles also provide resistance to abiotic factors such as salt and drought (Zhang et al., 2008). Tajpoor et al. (2013) reported that the combined effect of organic and biofertilizer treatments enhance the growth and essential oil yield of dill plant (*Anethum graveolens* L.). Banchio et al. (2009) studied the effect of PGPR such as *Pseudomonas fluorescens*, *B. subtilis*, and *Azospirillum brasilense* on the growth of *Origanum majoricum* and its volatile secondary metabolites and reported that the inoculants significantly increase the fresh weight with enhance metabolite products. Kapoor et al. (2007) studied the effects of *Glomus macrocarpum* and *Glomus fasciculatum* alone or supplemented with P fertilizer on artemisinin and reported that the inoculants effectively enhanced the concentration of artemisinin in *Artimisia annua.* Santoro et al. (2011) studied the effects of PGPR such as *P. fluorescens*, *B. subtilis*, and *A. brasilense* on volatile compounds of *Mentha piperita* (peppermint). They reported that the volatile organic compounds of rhizobacteria significantly induce the biosynthesis of secondary metabolites and also affect the pathway of monoterpene metabolism. Thus, microbes and their associated products would have better prospect in sustainable agriculture as a preferred alternative to synthetic chemical to boost the agricultural production and to meet the needs of the growing population. The role of microbes and their bioactivity to boost the plant growth against biotic and abiotic factors has been summarized in Table 2.1.

## 2.3 Role of microbes and their products in pharma industries

From millennia natural products and their bioactive secondary metabolites have been used in traditional system of medicine in India, China, Egypt, and other countries. Although secondary metabolite products of plant products have always been used for the treatment of disease, their continuous use is in question due to limited availability of raw material and high cost. In the modern era of medicine, secondary metabolite products of microbial origin and their semisynthetic formulation have been used extensively as alternatives to plant products to treat infectious diseases, cancer, hypertension, and inflammation. The discovery of the antibiotic penicillin (*Penicillium notatum*) and its broad therapeutic potential use in medicine led to intensive investigation of microbial products as a source of novel bioactive agents. Most of the currently available antifungal and antibacterial products were derived from microbial sources. The major group of antibiotics (i.e., β-lactams, aminoglycosides, lipopeptides, glycopeptides, nucleosides, peptides, tetracyclines, and macrolides) was produced from actinomycetes (Adegboye and Babalola, 2013). Cyclosporins

**Table 2.1** The role of microbes and their bioactivity to boost the plant growth against biotic and abiotic factors.

| PGPR | Plant | Attribute | References |
|---|---|---|---|
| *Bacillus megaterium, B. subtilis, B. subtilis* subsp. *subtilis*, and *Pseudomonas* sp. | *Arachis hypogaea* | *Biocontrol and plant- growth promotion:* Inhibit *Aspergillus niger* growth that causes root rot diseases, produces auxin (indole-3-acetic acid, IAA), and increases nitrogen fixing activities in plant. | Yuttavanichakul et al. (2012) |
| *Bacillus cereus* (Sneb 560), *B. subtilis* (Sneb 815), *Pseudomonas putida* (Sneb 821), *P. fluorescens* (Sneb 825), and *Serratia proteamaculans* (Sneb 851) | *Solanum lycopersicum* | *Biocontrol:* Incorporating these five bacterial isolated in *S. lycopersicum*, promotes the plant biomass and inhibits root-knot nematode *Meloidogyne incognita.* | Zhao et al. (2018) |
| *Bacillus subtilis* LHS11 and FX2 | *Brassica napus* | *Biocontrol and plant-growth promotion:* Antagonistic activity against *Sclerotinia sclerotiorum* (Lib.) de Bary and showed phosphate solubilization, nitrogen fixation, and IAA production activities. | Sun et al. (2017) |
| *Bacillus velezensis* strains 5YN8 and DSN012 | *Piper nigrum* | *Biocontrol:* Control the gray mold disease caused by necrotrophic pathogen *Botrytis cinerea.* | Jiang et al. (2018) |
| *B. subtilis*, *Bacillus pumilus*, *Burkholderia cepacia*, *P. putida*, *Bacillus amyloliquefaciens*, *Bacillus atrophaeus*, *Bacillus macerans*, and *Flavobacter balastinium* | *Solanum tuberosum* | *Biocontrol:* Inhibit the dry rot disease of *S. tuberosum* caused by *Fusarium sambucinum*, *Fusarium oxysporum*, and *Fusarium culmorum under* storage condition. | Recep et al. (2009) |
| *Sphingomonas* sp. LK11 | *S. lycopersicum* | *Plant growth promotion:* Production of gibberellins and IAA. | Khan et al. (2014) |
| *B. amyloliquefaciens* | *Capsicum annum* | *Biocontrol and plant growth-promotion:* Induces resistance against anthracnose disease (*Colletotrichum truncatum*) and enhances germination of seed along with vegetative growth. | Gowtham et al. (2018) |

| | | | |
|---|---|---|---|
| *B. amyloliquefaciens* strain CEIZ-11 | *S. lycopersicum* | *Biocontrol:* Control damping-off disease caused by *Pythium aphanidermatum.* | Zouari et al. (2016) |
| *Pseudomonas aeruginosa* | *Launaea nudicaulis* | *Biocontrol:* Activity against *Macrophomina phaseolina*, *Fusarium solani*, and *F. oxysporum, inhibits the maximum infection of M. phaseolina* on mungbean roots. | Mansoor et al. (2007) |
| *P. aeruginosa* | *Triticum aestivum* | *Bioremediation and plant growth-promotion:* Causes oxidative stress tolerance under Zn stress and enhances nutrient availability, antioxidant defense system and reduction in Zn uptake for wheat plant growth promotion. | Islam et al. (2014) |
| *Enterobacter aerogenes* strain K6 | *Oryza sativa* | *Heavy metal tolerance and plant growth-promotion:* Provide resistance against $Cd^{2+}$, $Pb^{2+}$, and $As^{3+}$, and PGP traits like also associated with IAA) production, nitrogen fixation, phosphate solubilization, and 1-aminocyclopropane-1-carboxylate (ACC) deaminase activity for plant-growth promotion. | Pramanik et al. (2018) |
| *Enterobacter cloacae* strain *(HSNJ4)* | *B. napus* | *Stress tolerance and plant-growth promotion:* Balances the relative content of IAA and ethylene to enhance the salt tolerance. In addition, proline content and antioxidant enzyme activity were also enhanced. | Li et al. (2017) |
| *Pseudomonas koreensis* | *Helianthus annuus* | *Stress tolerance*: Provides tolerance to plant in drought conditions. | Macleod et al. (2015) |
| *Pseudomonas* sp. | *Eleusine coracana* | *Stress tolerance*: Provides tolerance against oxidative stress inducing by drought condition and induces plant fitness. | Chandra et al. (2018) |

(*Continued*)

**Table 2.1** (Continued)

| PGPR | Plant | Attribute | References |
|---|---|---|---|
| *P. aeruginosa* BHU B13-398 and *B. subtilis* BHU M | *Vigna radiata* | *Plant-growth promotion and biocontrol:* Exhibits PGP traits for plant-growth promotion such as phosphate solubilization, ammonia, siderophore, and hydrogen cyanide (HCN) production along with resistance against root rot pathogen *Rhizoctonia solani.* | Kumari et al. (2018) |
| *B. cereus* and *B. safensis* | *Lens culinaris* | *Plant-growth promotion and biocontrol*: PGP traits showed were production of siderophores, indole acetic acid, and phosphate solubilization along with biocontrol activity against *Alternaria* sp. | Roy et al. (2018) |
| *Bacillus* sp. strain WU-5, WU-9, and WU-1 | *P. nigrum* | *Stress tolerance and plant-growth promotion:* Provides salinity tolerance via increasing the proline content and antioxidant enzyme activity and induces plant growth such as phosphate solubilization, ACC deaminase, ammonia, and siderophore production. | Wang et al. (2018) |
| *B. amyloliquefaciens* isolate WE15 and *B. firmus* isolate WD19 | *Brassica oleracea var. alboglabra* | *Phytoremediation and plant-growth promotion: phytoremediation* of soil mediated iron ion contamination and induces plant fitness via phosphate solubilization, auxin phytohormone, and siderophore production. | Sarawaneeyaruk et al. (2018) |
| *Streptomyces* sp. PM1 and PM5 | *S. lycopersicum* | *Biocontrol:* Showed resistance against *Pectobacterium carotovorum* subsp. *Brasiliensis* causing soft rot disease. | Dias et al. (2017) |
| *Bacillus aryabhattai* | *O. sativa* | *Heavy metal tolerance and plant growth promotion:* Arsenic tolerant along with PGP traits such as ACC deaminase activity, IAA production, nitrogen fixation, and siderophore production. | Ghosh et al. (2018) |

and *rapamycin* (immunosuppressive agents), mevastatin and lovastatin (cholesterol-lowering compounds), phomoenamide, sansanmycin, trichoderin (antituberculosis), and ivermectins (antihelmintics) are some of the prominent microbial products with strong industrial significance in the pharmaceutical industries (Cragg and Newman, 2013; Alvin et al., 2014). Pepstatin A produced by the *Streptomyces* species has exhibited inhibitory activity against HIV-1 protease (Cragg and Newman, 2001; Yang et al., 2001). The genus *Agaricus* have inherent bioactive compounds such as ganoderan, heteroglycan, mannoglycan, glycoprotein, glucan, mannoglucan, proteoglycan, and proteoglycan and have exhibited immunomodulatory potential (El Enshasy and Hatti-Kaul, 2013). Kulanthaivel et al. (2004) isolated novel lipoglycopeptides (potent inhibitors from a *Streptomyces* sp.) and reported that it has strong potential as a broad-spectrum antibiotic. Banskota et al. (2006) isolated a glycosidic polyketide ECO-0501 from *Amycolatopsis orientalis* ATCC 43491 strain possessed strong antibacterial activity against methicillin-resistant *Staphylococcus aureus* and vancomycin-resistant *Enterococci*. McDonald et al. (2002) extracted and purified muraymycins, a novel nucleoside-lipopeptide antibiotic from *Streptomyces* sp. LL-AA896, and reported inhibition of peptidoglycan biosynthesis enzyme activity as possible mechanism of action. Platensimycin, a bioactive compound extracted from *Streptomyces* sp., exhibited broad-spectrum antibacterial activity by inhibiting lipid biosynthesis (Wang et al., 2006). OPT-80, isolated from the actinomycete *Dactylosporangium aurantiacum*, exhibited strong antibiotic potential against *Clostridium difficile*—associated diarrhea and vancomycin-resistant *Enterococcus* infection (Johnson, 2007). Table 2.2 summarizes the bioactive compounds of microbial origin, along with their therapeutic potential and mode of action.

## 2.4 Current challenges for microbial products and unexplored areas of research

Indeed, microorganisms and their metabolite products have been extensively used as novel candidates for the management of agricultural pests (PGPR, bacterial inoculants, endophytes, and actinomycetes) and as therapeutic agents (antibiotics, immunosuppressants, and cholesterol-lowering agents). However, most of the world's microbial diversity is still unexplored. At present, approximately 6000 bacterial species have been identified and classified under various taxonomic groups of microbes over the 1.5 million species of prokaryote (Harvey, 2000). Lack of systematic exploitation of the microbial community of diverse ecosystems such as polar ice, geothermal vents, dark caves, and deep-sea sites are the major avenues available for the discovery of novel microbial bioactive compounds of industrial significance. In addition, the isolation and in vitro cultivation of less-culturable microorganisms with strong bioactivity limits their commercial production. The discrepancy between the morphological and molecular methods of identification of microbial strains in culture is another controversial issue in terms of identification of microbial strains in all ecosystems. In this context, the need for proper

**Table 2.2** The bioactive compounds of microbial origin, their therapeutic potential and mode of action.

| Name | Classification | Lead compound | Property | Mode of action | References |
|---|---|---|---|---|---|
| *Streptomyces toxytricini* | Bacterium | Lipstatin | Antiobesity | Suppresses the activity of gastrointestinal lipases causing significant cessation in the absorption of dietary fats. | Borgström (1988) |
| *Pestalotiopsis microspora* | Endophytic Fungus | Isopestacin | Antioxidant | Provide oxidative stress tolerance by neutralizing free radical ions such as superoxide- and hydroxide-free ions. | Strobel et al. (2002) |
| *Streptomyces hygroscopicus var. limoneus* | Bacteria | Validamycin | Antifungal | Suppresses the energy mechanism of fungi by inhibiting the carbohydrate storage and utilization enzyme, trehalase. | Iwasa et al. (1970), El Nemr and El Sayed (2011) |
| *Streptomyces peucetius var. caesius* | Bacteria | Doxorubicin | Oncology | Damage cellular components and compartments by free radical's generation along with DNA intercalation and disruption. | Thorn et al. (2011) |
| *Glarea lozoyensis* | Fungus | Pneumocandin B | Antifungal | Competitively inhibits the β(1,3)-D-glucan synthase causing disruption of fungal cell wall synthesis. | Letscher-Bru and Herbrecht (2003) |
| *Micromonospora* strain SANK62390 | Bacteria | Trehazolin | Antifungal | Trehalase inhibitor, shuts off the energy mechanism of fungi. | El Nemr and El Sayed (2011) |
| *Coleophoma empetri* | Fungus | FR901379 | Antifungal | Dismantles the biosynthesis of 1,3-β-glucan, the main fungal cell wall component. | Boyer-Joubert et al. (2003) |
| *Bacillus* sp. | Bacteria | 1-deoxynojirimycin | Type 1 Gaucher disease | Directly impedes the glucose machinery by suppressing its absorption in intestine and increasing its hepatic metabolism. | Parenti et al. (2007), Li et al. (2013), Butler (2008) |

| | | | | | |
|---|---|---|---|---|---|
| *Micromonospora* strain DPJ12 | Bacteria | Diazepinomicin | Oncology | Effectively inhibits the cancer cell proliferation and its migration (Ras-MAPK pathway), also, induces the cell apoptotic machinery. | Abdelmohsen et al. (2012) |
| *Penicillium citrinum* | Mold | Mevastatin | Cardiovascular diseases | Cease the activity of 5-hydroxy-3-methylglutarylcoenzyme A (HMG-CoA) reductase, thus, inhibits the cholesterol synthesis and lowered lipid activity. | Tobert (2003), Liwa et al. (2017) |
| *Aspergillus terreus* | Fungus | Lovastatin | Cardiovascular diseases | Cut-off in the activity of 5-hydroxy-3-methylglutarylcoenzyme A (HMG-CoA) reductase, thus, lowered lipid activity. | Liwa et al. (2017) |
| *Streptomyces bottropensis* DO-45 | Bacteria | Trioxacarcin C | Oncology | Inhibits DNA and RNA synthesis of tumor cell. | Fujimoto and Morimoto (1983) |
| *Aspergillus fumigatus* | Fungus | Fumagillin | Antiparasitic; Oncology | Restraining blood flow around tumors through angiogenesis. | van den Heever et al. (2014) |
| *Streptomyces cattleya* | Fungus | Thienamycin | Antibacterial | Binds with penicillin-binding proteins (PBPs) causing cessation in cell wall biosynthesis. | Doi and Chambers (2015) |
| *Actinomadura* sp. 007 | Bacteria | ZHD-0501 | Oncology | Restricts the tumor cell proliferation by inhibiting the cell cycle at the $G_2$/M phase. | Olano et al. (2009) |
| *S. hygroscopicus* | Bacterium | Sirolimus | Cardiovascular surgery, Oncology, Immuno suppression | Prevents signal flow in downstream-signaling pathway by binding to its receptor and stop the T-cell proliferation, causing inhibition of the immune responses. | Rath et al. (2010) |

(*Continued*)

**Table 2.2** (Continued)

| Name | Classification | Lead compound | Property | Mode of action | References |
|---|---|---|---|---|---|
| *A. nidulans var. echinulatus* | Fungus | Echinocandin B | Antifungal | Disrupting fungal cell wall integrity by inhibiting of β(1,3)-D-glucan synthase. | Butler (2005), Grover (2010) |
| *Streptomyces* sp. KORDI-3973 | Bacteria | Streptopyrrolidine | Antiangiogenesis activity | Inhibited cell proliferation due to cessation of biochemical activity of endothelial cell. | Shin et al. (2008) |
| *Sorangium cellulosum* | Myxo-bacterium | Epothilone B | Oncology | Showed the strong cytotoxicity activity by inhibiting $G_2$/M phase in cell cycle. The dismantles cytoskeleton dynamics through its bind-n-cease activity over microtubules, hence, impeding proper cell functioning and its maintenance. | Reichenbach and Höfle (2008) |
| *Streptomyces* sp. KS3 | Bacteria | Komodoquinone A | Oncology | Causes cell arrest $G_1$ and $G_2$/M phase and inhibit tumor cells. | Itoh et al. (2003), Olano et al. (2009) |

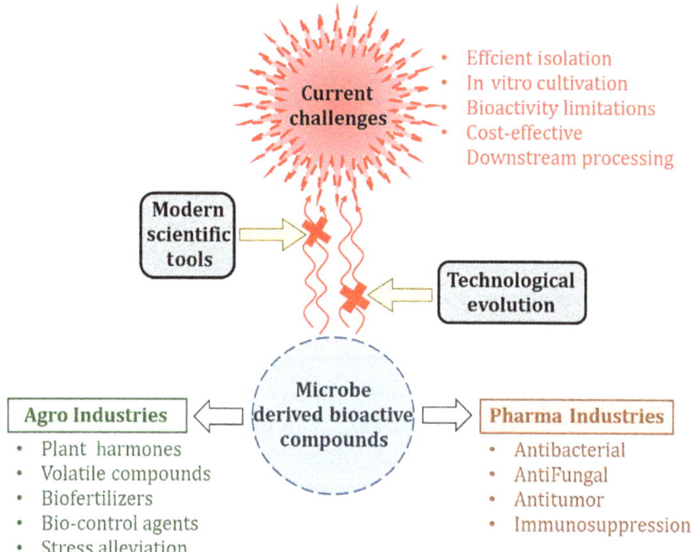


**Figure 2.1** Systematic representation of the bioactive compounds of microbial origin and prospect of recent scientific and technological innovation to overcome the existing lacuna.

characterization and isolation of active compounds from microbes requires expertise. Furthermore, there has been little advancement in the adequate production of bioactive compounds and their optimization and scale-up processes. In the case of therapeutic agents, the preclinical validation of a novel drug candidate is a prerequisite step for using animal models for their commercialization. Therefore the in vitro efficacy of bioactive compounds must be validated using animal models. There is also the need for a diversified approach to unveil the unexplored microbes and their metabolite products using collaborative research between industries and research centers and governmental agencies to facilitate the development, manufacture, and production of microbial products of industrial significance in both agriculture and pharma industries. Fig. 2.1 represents the bioactive compounds of microbial origin and recent scientific and technological innovation.

## 2.5 Role of modern scientific and technological innovation

Indeed, before the advent of modern scientific and technological innovation botanical products were the only sources used for the management of agricultural pests and therapeutic agents since ancient time. However, today the demand for microbial products has increased significantly as a result of large-scale worldwide applications. Although current microbial products exhibit play key roles as alternatives to conventional pesticides and drugs of plant origin, there is still a shortage of novel lead compounds, especially in the field of pharmaceuticals as most of the world's biodiversity still remains unexplored. Therefore, the use of recent advancement in science (e.g., synthetic biology, metabolic engineering, biotechnology, and recombinant DNA technologies) and technology (e.g., small- and large-scale bioreactors, bioreactor engineering, smart screening methods, robotic separation with structural analysis, advance extraction techniques, and nanotechnology) could increase the large-scale production, effectiveness, and commercial applications of microbial products. Today cloning and genetic engineering have been extensively used to regulate the biosynthesis pathway of secondary metabolites of microbial sources and to design novel rational compounds in the agricultural and pharmaceutical industries. Encapsulation of microbial products may enhance bioavailability with targeted delivery. Only a small part of the world's microbial biodiversity has been cultured in the laboratory. The microbial products of marine sources have strong bioactivity but unfortunately the culture of these microorganisms at laboratory condition is tedious and difficult. The use of combinatorial genetics offers an alternative solution by selecting the genetic material responsible for synthesis of the metabolic pathway of the bioactive agent and its incorporation into a more convenient culturable microorganism such as *Streptomyces* that can be cultured in the laboratory. This has been widely used by Galilaeus Oy (Kaarina, Finland), Terragen (Vancouver, BC, Canada), and Kosan Biosciences (Hayward, CA, United States) for novel anthracyclines, metabolites from lichens and marine organisms, and large quantities of epothilone, respectively (Harvey, 2000). The use of combinatorial genetics could offer continuous supply of microbial metabolites with more diverse chemicals structure in nonnatural environments. A range of novel bioactive metabolite products has been synthesized by introducing combinations of genes into a productive strain of *Streptomyces* (McDaniel et al., 1995; Cane et al., 1998; Gokhale et al., 1999; Seow et al., 1997/>).

## 2.6 Conclusion and future prospects

Microorganisms have tremendous efficacy in the production of secondary metabolites that could be used in the agriculture and pharmaceutical industries. Recent advancements in science could help boost the commercialization of microbial products. However, further research is needed to elucidate the mechanistic pathway of the gene regulation responsible for secondary metabolite production and to understand the influence of the role of environmental factors as adjuvants. The extent of such knowledge could provide an alternative way for the in vitro synthesis of desired secondary metabolites with enhanced production at the industrial level for their worldwide application. To meet this aim, collaborative research among research centers and various industries as well as regulatory agencies to synthesize novel microbial products with enhanced effectiveness for commercial application is needed.

## References


Abdelmohsen, U.R., Szesny, M., Othman, E.M., Schirmeister, T., Grond, S., Stopper, H., et al., 2012. Antioxidant and anti-protease activities of diazepinomicin from the sponge-associated *Micromonospora* strain RV115. Mar. Drugs 10 (10), 2208–2221.

Adegboye, M.F., Babalola, O.O., 2013. Actinomycetes: a yet inexhaustive source of bioactive secondary metabolites. Microbial Pathogens and Strategies for Combating Them: Science, Technology and Education. Formatex, Badajoz, pp. 786–795.

Alvin, A., Miller, K.I., Neilan, B.A., 2014. Exploring the potential of endophytes from medicinal plants as sources of antimycobacterial compounds. Microbiol. Res. 169 (7–8), 483–495.

Babalola, O.O., 2010. Beneficial bacteria of agricultural importance. Biotechnol. Lett. 32 (11), 1559–1570.

Banchio, E., Bogino, P.C., Santoro, M., Torres, L., Zygadlo, J., Giordano, W., 2009. Systemic induction of monoterpene biosynthesis in *Origanum majoricum* by soil bacteria. J. Agric. Food Chem. 58 (1), 650–654.

Banskota, A.H., McAlpine, J.B., Sørensen, D., Ibrahim, A., Aouidate, M., Piraee, M., et al., 2006. Genomic analyses lead to novel secondary metabolites. J. Antibiot. 59 (9), 533.

Bashan, Y., de-Bashan, L.E., 2010. How the plant growth-promoting bacterium *Azospirillum* promotes plant growth—a critical assessment. Advances in Agronomy. Academic Press, pp. 77–136.

Borgström, B., 1988. Mode of action of tetrahydrolipstatin: a derivative of the naturally occurring lipase inhibitor lipstatin. Biochim. Biophys. Acta, Lipids Lipid Metab. 962 (3), 308–316.

Boyer-Joubert, C., Lorthiois, E., Moreau, F., 2003. Annual Reports in Medicinal Chemistry, Vol. 38. Elsevier Inc.

Butler, M.S., 2005. Natural products to drugs: natural product derived compounds in clinical trials. Nat. Prod. Rep. 22 (2), 162–195.

Butler, M.S., 2008. Natural products to drugs: natural product-derived compounds in clinical trials. Nat. Prod. Rep. 25 (3), 475–516.

Cane, D.E., Walsh, C.T., Khosla, C., 1998. Harnessing the biosynthetic code: combinations, permutations, and mutations. Science 282 (5386), 63–68.

Chandra, D., Srivastava, R., Glick, B.R., Sharma, A.K., 2018. Drought-tolerant *Pseudomonas* spp. improve the growth performance of finger millet (*Eleusine coracana* (L.) Gaertn.) under non-stressed and drought-stressed conditions. Pedosphere 28 (2), 227–240.

Cragg, G.M., Newman, D.J., 2001. Natural product drug discovery in the next millennium. Pharm. Biol. 39, 8–17.

Cragg, G.M., Newman, D.J., 2013. Natural products: a continuing source of novel drug leads. Biochim. Biophys. Acta Gen. Subj. 1830 (6), 3670–3695.

Dias, M.P., Bastos, M.S., Xavier, V.B., Cassel, E., Astarita, L.V., Santarém, E.R., 2017. Plant growth and resistance promoted by *Streptomyces* spp. in tomato. Plant Physiol. Biochem. 118, 479–493.

Doi, Y., Chambers, H.F., 2015. Other β-lactam antibiotics. In: Bennett, J.E., Dolin, R., Blaser, M.J. (Eds.), Mandell, Douglas, and Bennett's Principles and Practice of Infectious Diseases. Elsevier Inc, pp. 293–297.

Dubey, N.K., Shukla, R., Kumar, A., Singh, P., Prakash, B., 2011. Global scenario on the application of natural products in integrated pest management programmes. Nat. Prod. Plant Pest Manage. 1, 1–20.

El Enshasy, H.A., Hatti-Kaul, R., 2013. Mushroom immunomodulators: unique molecules with unlimited applications. Trends Biotechnol. 31 (12), 668–677.

El Nemr, A., El Sayed, H., 2011. Potential trehalase inhibitors: syntheses of trehazolin and its analogues. Advances in Carbohydrate Chemistry and Biochemistry. Academic Press, pp. 45–114.

Fujimoto, K., Morimoto, M., 1983. Antitumor activity of trioxacarcin C. J. Antibiot. 36 (9), 1216–1221.

Ghosh, P.K., Maiti, T.K., Pramanik, K., Ghosh, S.K., Mitra, S., De, T.K., 2018. The role of arsenic resistant *Bacillus aryabhattai* MCC3374 in promotion of rice seedlings growth and alleviation of arsenic phytotoxicity. Chemosphere 211, 407–419.

Gokhale, R.S., Tsuji, S.Y., Cane, D.E., Khosla, C., 1999. Dissecting and exploiting intermodular communication in polyketide synthases. Science 284 (5413), 482–485.

Gowtham, H.G., Murali, M., Singh, S.B., Lakshmeesha, T.R., Murthy, K.N., Amruthesh, K. N., et al., 2018. Plant growth promoting rhizobacteria *Bacillus amyloliquefaciens* improves plant growth and induces resistance in chilli against anthracnose disease. Biol. Contr. 126, 209–217.

Gray, E.J., Smith, D.L., 2005. Intracellular and extracellular PGPR: commonalities and distinctions in the plant–bacterium signalling processes. Soil Biol. Biochem. 37 (3), 395–412.

Grover, N.D., 2010. Echinocandins: a ray of hope in antifungal drug therapy. Indian J. Pharmacol. 42 (1), 9.

Guo, B., Li, H., Zhang, L., 1998. Isolation of fungus producing vinblastine. J. Yunnan Univ. (Nat. Sci.) 20 (3), 214–215.

Harvey, A., 2000. Strategies for discovering drugs from previously unexplored natural products. Drug Discov. Today 5 (7), 294–300.

Heil, M., Karban, R., 2010. Explaining evolution of plant communication by airborne signals. Trends Ecol. Evol. 25 (3), 137–144.

Herrera-Estrella, A., Chet, I., 2003. The biological control agent *Trichoderma*: from fundamentals to applications. In: Arora, D. (Ed.), Handbook of Fungal Biotechnology. Dekker, New York, pp. 1000–10020.

Isaka, M., Jaturapat, A., Rukseree, K., Danwisetkanjana, K., Tanticharoen, M., Thebtaranonth, Y., 2001. Phomoxanthones A and B, novel xanthone dimers from the endophytic fungus *Phomopsis* species. J. Nat. Prod. 64 (8), 1015–1018.

Islam, F., Yasmeen, T., Ali, Q., Ali, S., Arif, M.S., Hussain, S., et al., 2014. Influence of *Pseudomonas aeruginosa* as PGPR on oxidative stress tolerance in wheat under Zn stress. Ecotoxicol. Environ. Saf. 104, 285–293.

Itoh, T., Kinoshita, M., Aoki, S., Kobayashi, M., 2003. Komodoquinone A, a novel neuritogenic anthracycline, from marine *Streptomyces* sp. KS3. J. Nat. Prod. 66 (10), 1373–1377.

Iwasa, T., Yamamoto, H., Shibata, M., 1970. Studies on validamycins, new antibiotics. J. Antibiot. 23 (12), 595–602.

Jiang, C.H., Liao, M.J., Wang, H.K., Zheng, M.Z., Xu, J.J., Guo, J.H., 2018. *Bacillus velezensis*, a potential and efficient biocontrol agent in control of pepper gray mold caused by *Botrytis cinerea*. Biol. Contr. 126, 147–157.

Johnson, A.P., 2007. Drug evaluation: OPT-80, a narrow-spectrum macrocyclic antibiotic. Curr. Opin. Investig. Drugs 8 (2), 168–173.

Kapoor, R., Chaudhary, V., Bhatnagar, A.K., 2007. Effects of arbuscular mycorrhiza and phosphorus application on artemisinin concentration in *Artemisia annua* L. Mycorrhiza 17 (7), 581.

Khan, A.L., Waqas, M., Kang, S.M., Al-Harrasi, A., Hussain, J., Al-Rawahi, A., et al., 2014. Bacterial endophyte *Sphingomonas* sp. LK11 produces gibberellins and IAA and promotes tomato plant growth. J. Microbiol. 52 (8), 689–695.

Kloepper, J.W., Ryu, C.M., Zhang, S., 2004. Induced systemic resistance and promotion of plant growth by *Bacillus* spp. Phytopathology 94 (11), 1259–1266.

Kulanthaivel, P., Kreuzman, A.J., Strege, M.A., Belvo, M.D., Smitka, T.A., Clemens, M., et al., 2004. Novel lipoglycopeptides as inhibitors of bacterial signal peptidase I. J. Biol. Chem. 279 (35), 36250–36258.

Kumar, A., Vandana, R.S., Singh, M., Pandey, K.D., 2015a. Plant growth promoting rhizobacteria (PGPR). A promising approach for disease management. Microbes and Environmental Management.. Studium Press, New Delhi, pp. 195–209.

Kumar, A., Vandana, R.S., Yadav, A., Giri, D.D., Singh, P.K., Pandey, K.D., 2015b. Rhizosphere and their role in plant–microbe interaction. Microbes in Soil and Their Agricultural Prospects.. Nova Science Publisher, Inc., New York, pp. 83–97.

Kumar, A., Verma, H., Singh, V.K., Singh, P.P., Singh, S.K., Ansari, W.A., et al., 2017. Role of *Pseudomonas* sp. in sustainable agriculture and disease management. Agriculturally Important Microbes for Sustainable Agriculture.. Springer, Singapore, pp. 195–215.

Kumari, P., Meena, M., Gupta, P., Dubey, M.K., Nath, G., Upadhyay, R.S., 2018. Plant growth promoting rhizobacteria and their biopriming for growth promotion in mung bean (*Vigna radiata* (L.) R. Wilczek). Biocatal. Agric. Biotechnol. 16, 163–171.

Letscher-Bru, V., Herbrecht, R., 2003. Caspofungin: the first representative of a new antifungal class. J. Antimicrob. Chemother. 51 (3), 513–521.

Li, Y.G., Ji, D.F., Zhong, S., Lin, T.B., Lv, Z.Q., Hu, G.Y., et al., 2013. 1-Deoxynojirimycin inhibits glucose absorption and accelerates glucose metabolism in streptozotocin-induced diabetic mice. Sci. Rep. 3, 1377.

Li, H., Lei, P., Pang, X., Li, S., Xu, H., Xu, Z., et al., 2017. Enhanced tolerance to salt stress in canola (*Brassica napus* L.) seedlings inoculated with the halotolerant *Enterobacter cloacae* HSNJ4. Appl. Soil Ecol. 119, 26–34.

Liwa, A.C., Barton, E.N., Cole, W.C., Nwokocha, C.R., 2017. Bioactive plant molecules, sources and mechanism of action in the treatment of cardiovascular disease. Pharmacognosy 315–336.

Macleod, K., Rumbold, K., Padayachee, K., 2015. A systems approach to uncover the effects of the PGPR *Pseudomonas koreensis* on the level of drought stress tolerance in *Helianthus annuus*. Proc. Environ. Sci. 29, 262–263.

Mansoor, F., Sultana, V., Ehteshamul-Haque, S., 2007. Enhancement of biocontrol potential of *Pseudomonas aeruginosa* and *Paecilomyces lilacinus* against root rot of mungbean by a medicinal plant *Launaea nudicaulis* L. Pak. J. Bot. 39 (6), 2113–2119.

McDaniel, R., Ebert-Khosla, S., Hopwood, D.A., Khosla, C., 1995. Rational design of aromatic polyketide natural products by recombinant assembly of enzymatic subunits. Nature 375 (6532), 549.

McDonald, L.A., Barbieri, L.R., Carter, G.T., Lenoy, E., Lotvin, J., Petersen, P.J., et al., 2002. Structures of the muraymycins, novel peptidoglycan biosynthesis inhibitors. J. Am. Chem. Soc. 124 (35), 10260–10261.

Olano, C., Méndez, C., Salas, J.A., 2009. Antitumor compounds from marine actinomycetes. Mar. Drugs 7 (2), 210–248.

Parenti, G., Zuppaldi, A., Pittis, M.G., Tuzzi, M.R., Annunziata, I., Meroni, G., et al., 2007. Pharmacological enhancement of mutated α-glucosidase activity in fibroblasts from patients with Pompe disease. Mol. Ther. 15 (3), 508–514.

Pramanik, K., Mitra, S., Sarkar, A., Maiti, T.K., 2018. Alleviation of phytotoxic effects of cadmium on rice seedlings by cadmium resistant PGPR strain *Enterobacter aerogenes* MCC 3092. J. Hazard Mater. 351, 317–329.

Rath, C.M., Scaglione, J.B., Kittendorf, J.D., Sherman, D.H., 2010. NRPS/PKS hybrid enzymes and their natural products. Comprehensive Natural Products II. Elsevier, pp. 453–492.

Recep, K., Fikrettin, S., Erkol, D., Cafer, E., 2009. Biological control of the potato dry rot caused by *Fusarium* species using PGPR strains. Biol. Contr. 50 (2), 194–198.

Reichenbach, H., Höfle, G., 2008. Discovery and development of the epothilones. Drugs R D 9 (1), 1–10.

Roy, T., Bandopadhyay, A., Sonawane, P.J., Majumdar, S., Mahapatra, N.R., Alam, S., et al., 2018. Bio-effective disease control and plant growth promotion in lentil by two pesticide degrading strains of *Bacillus* sp. Biol. Contr. 127, 55–63.

Santoro, M.V., Zygadlo, J., Giordano, W., Banchio, E., 2011. Volatile organic compounds from rhizobacteria increase biosynthesis of essential oils and growth parameters in peppermint (*Mentha piperita*). Plant Physiol. Biochem. 49 (10), 1177–1182.

Sarawaneeyaruk, S., Lorliam, W., Krajangsang, S., Pringsulaka, O., 2018. Enhancing plant growth under municipal wastewater irrigation by plant growth promoting rhizospheric *Bacillus* spp. J. King Saud Univ .

Seow, K.T., Meurer, G., Gerlitz, M., Wendt-Pienkowski, E., Hutchinson, C.R., Davies, J., 1997. A study of iterative type II polyketide synthases, using bacterial genes cloned from soil DNA: a means to access and use genes from uncultured microorganisms. J. Bacteriol. 179 (23), 7360–7368.

Shin, H.J., Kim, T.S., Lee, H.S., Park, J.Y., Choi, I.K., Kwon, H.J., 2008. Streptopyrrolidine, an angiogenesis inhibitor from a marine-derived *Streptomyces* sp. KORDI-3973. Phytochemistry 69 (12), 2363–2366.

Shoresh, M., Harman, G.E., Mastoury, F., 2010. Induced systemic resistance and plant responses to fungal biocontrol agents. Ann. Rev. Phytopathol. 48, 21–43.

Stamp, N., 2003. Out of the quagmire of plant defense hypotheses. Q. Rev. Biol. 78 (1), 23–55.

Stierle, A., Strobel, G., Stierle, D., 1993. Taxol and taxane production by *Taxomyces andreanae*, an endophytic fungus of Pacific yew. Science 260 (5105), 214–216.

Strobel, G., Ford, E., Worapong, J., Harper, J.K., Arif, A.M., Grant, D.M., et al., 2002. Isopestacin, an isobenzofuranone from *Pestalotiopsis microspora*, possessing antifungal and antioxidant activities. Phytochemistry 60 (2), 179–183.

Sun, G., Yao, T., Feng, C., Chen, L., Li, J., Wang, L., 2017. Identification and biocontrol potential of antagonistic bacteria strains against *Sclerotinia sclerotiorum* and their growth-promoting effects on *Brassica napus*. Biol. Contr. 104, 35–43.

Tajpoor, N., Moradi, R., Zaeim, A.N., 2013. Effects of various fertilizers on quantity and quality of dill (*Anethum graveolens* L.) essential oil. Int. J. Agric. Crop Sci. 6 (19), 1334.

Thorn, C.F., Oshiro, C., Marsh, S., Hernandez-Boussard, T., McLeod, H., Klein, T.E., et al., 2011. Doxorubicin pathways: pharmacodynamics and adverse effects. Pharmacogenet. Genom. 21 (7), 440.

Tobert, J.A., 2003. Lovastatin and beyond: the history of the HMG-CoA reductase inhibitors. Nat. Rev. Drug Discov. 2 (7), 517.

van den Heever, J.P., Thompson, T.S., Curtis, J.M., Ibrahim, A., Pernal, S.F., 2014. Fumagillin: an overview of recent scientific advances and their significance for apiculture. J. Agric. Food Chem. 62 (13), 2728–2737.

Verma, J.P., Yadav, J., Tiwari, K.N., Lavakush, S.V., 2010. Impact of plant growth promoting rhizobacteria on crop production. Int. J. Agric. Res. 5 (11), 954–983.

Wang, J., Soisson, S.M., Young, K., Shoop, W., Kodali, S., Galgoci, A., et al., 2006. Platensimycin is a selective FabF inhibitor with potent antibiotic properties. Nature 441 (7091), 358.

Wang, W., Wu, Z., He, Y., Huang, Y., Li, X., Ye, B.C., 2018. Plant growth promotion and alleviation of salinity stress in *Capsicum annuum* L. by *Bacillus* isolated from saline soil in Xinjiang. Ecotoxicol. Environ. Saf. 164, 520–529.

Weller, D.M., Raaijmakers, J.M., Gardener, B.B.M., Thomashow, L.S., 2002. Microbial populations responsible for specific soil suppressiveness to plant pathogens. Annu. Rev. Phytopathol. 40 (1), 309–348.

Yang, S.S., Cragg, G.M., Newman, D.J., Bader, J.P., 2001. Natural product-based anti-HIV drug discovery and development facilitated by the NCI developmental therapeutics program. J. Nat. Prod. 64 (2), 265–277.

Yuttavanichakul, W., Lawongsa, P., Wongkaew, S., Teaumroong, N., Boonkerd, N., Nomura, N., et al., 2012. Improvement of peanut rhizobial inoculant by incorporation of plant growth promoting rhizobacteria (PGPR) as biocontrol against the seed borne fungus, *Aspergillus niger*. Biol. Contr. 63 (2), 87–97.

Zhang, L., Guo, B., Li, H., Zeng, S., Shao, H., Gu, S., et al., 2000. Preliminary study on the isolation of endophytic fungus of *Catharanthus roseus* and its fermentation to produce products of therapeutic value. Chin. Trad. Herbal Drugs 31 (11), 805–807.

Zhang, H., Kim, M.S., Sun, Y., Dowd, S.E., Shi, H., Paré, P.W., 2008. Soil bacteria confer plant salt tolerance by tissue-specific regulation of the sodium transporter HKT1. Mol. Plant Microbe. Interact. 21 (6), 737–744.

Zhao, D., Zhao, H., Zhao, D., Zhu, X., Wang, Y., Duan, Y., et al., 2018. Isolation and identification of bacteria from rhizosphere soil and their effect on plant growth promotion and root-knot nematode disease. Biol. Contr. 119, 12–19.

Zouari, I., Jlaiel, L., Tounsi, S., Trigui, M., 2016. Biocontrol activity of the endophytic *Bacillus amyloliquefaciens* strain CEIZ-11 against *Pythium aphanidermatum* and purification of its bioactive compounds. Biol. Contr. 100, 54–62.

# Further reading

Rodrguez-Daz, M., Rodelas-Gonzals, B., Pozo-Clemente, C., Martnez-Toledo, M.V., & Gonzlez-Lpez, J. (n.d.). A Review on the Taxonomy and Possible Screening Traits of Plant Growth Promoting Rhizobacteria. Plant-Bacteria Interactions, pp. 55–80.

Ruiz, B., Chávez, A., Forero, A., García-Huante, Y., Romero, A., Sánchez, M., et al., 2010. Production of microbial secondary metabolites: regulation by the carbon source. Crit. Rev. Microbiol. 36 (2), 146–167.

Ryu, C.M., Farag, M.A., Hu, C.H., Reddy, M.S., Wei, H.X., Paré, P.W., et al., 2003. Bacterial volatiles promote growth in *Arabidopsis*. Proc. Natl. Acad. Sci. 100 (8), 4927–4932.

Ryu, C.M., Farag, M.A., Hu, C.H., Reddy, M.S., Kloepper, J.W., Paré, P.W., 2004. Bacterial volatiles induce systemic resistance in *Arabidopsis*. Plant Physiol. 134 (3), 1017–1026.

Tarbell, D.S., Carman, R.M., Chapman, D.D., Cremer, S.E., Cross, A.D., Huffman, K.R., et al., 1961. The chemistry of Fumagillin1. J. Am. Chem. Soc. 83 (14), 3096–3113.

# Modern molecular and omics tools for understanding the plant growth-promoting rhizobacteria

3

*Ram Krishna*[1,2], *Waquar Akhter Ansari*[3], *Jay Prakash Verma*[1] *and Major Singh*[4]

[1]Institute of Environment and Sustainable Development, Banaras Hindu University, Varanasi, India, [2]Division of Vegetable Improvement, ICAR-Indian Institute of Vegetable Research, Varanasi, India, [3]ICAR-National Bureau of Agriculturally Important Microorganisms Kushmaur, Mau Nath Bhanjan, India, [4]ICAR-Directorate of Onion and Garlic Research, Pune, India

## 3.1 Introduction

The plant growth-promoting rhizobacteria (PGPR) found in plant rhizosphere have the ability to promote plant growth by various mechanisms like biological nitrogen fixation, engineering of rhizosphere, production of siderophore, phytohormone, volatile organic compounds (VOCs), 1-aminocyclopropane-1-carboxylate deaminase (ACC), phosphate solubilization, quorum sensing and antifungal activity, stimulation of systemic resistance, promotion of beneficial plant-microbe symbioses, etc. (Verma et al., 2018). The utilization of PGPR in agriculture is gradually increasing as it offers an alternative to chemical fertilizers, pesticides, and other chemical-based products. PGPR are the group of free-living microbial strains residing in the rhizosphere to provide nutrient source for enhancing plant growth attributes (Verma et al., 2014). Though these microorganisms are microscopic in size, their impact on plants is enormous. While these microorganisms are found everywhere in nature and the most important sources are soil and water. A teaspoon full of productive soil contains nearly 1 billion bacteria. Our lack of understanding of soil microorganisms in their natural environments is largely due to the undesirable field conditions for studying soil microorganisms.

Traditional techniques like culturing do not provide satisfactory sampling as they do not reflect natural sampling of the PGPR world. A natural environment has a very large number of PGPR; however, many thousands of microbe species with PGPR activity have not yet been scientifically described due to lack of suitable modern tools and techniques to understand these microbes. Conventional methods for detection and identification of various kinds of PGPR samples are based on culturing, enumeration, and colony isolation. While there are various culture and microscopy techniques in wide use today, these techniques provide

**Role of Plant Growth Promoting Microorganisms in Sustainable Agriculture and Nanotechnology.**
**DOI: https://doi.org/10.1016/B978-0-12-817004-5.00003-8**

inadequate information about the PGPR. Many microbes appear alike under the microscope, and many will not grow beyond natural environmental conditions. Because of a lack of modern tools and techniques, less than 1% of microbial species have been recognized to date (Brenner et al., 2015). Hence, over time only a relatively small number of microbes have been studied by the microbiologists, mostly grown in a laboratory (Brehm-Stecher and Johnson, 2004). Microbes are the most primitive organisms and readily adjust to wide ranging, adverse environmental conditions, like the extremes between desert and polar regions. The microbes are able to perform varying biological processes including decomposition of chemical pesticides and other compounds produced by different organisms (Jaiswal et al., 2017; Verma et al., 2016). In conclusion, it is important to study modern PGPR techniques in the microorganisms' natural habitat to identify their functions and their qualitative variations in various environmental conditions.

Current tools and techniques together with molecular analysis are allowing scientists to discriminate between PGPR and qualitative analysis. Combined, these techniques will be very advantageous for improving agricultural practices and for revealing novel biofertilizers helpful to agriculture.

## 3.2 Need for modern molecular tools and techniques

Many research literature have proposed that culturable plant-beneficial microbes are present in the environment, however, a huge number of genetic information are still unknown due to the viable but nonculturable microorganisms living in natural environmental conditions that are not yet culturable (Giovannoni et al., 1990). It is also necessary to reveal the hidden properties of both culturable and viable but nonculturable microbes with the help of modern molecular tools and techniques. Molecular techniques will also provide a glimpse into existing and extinct microbes, as all examples of PGPR are not known to science, nor their capabilities, effectiveness, and the rate of extinction or emergence (Fakruddin and Mannan, 2013).

Modern molecular tools and techniques are also important for:

1. Enhancing knowledge of genetic resources diversity
2. Understanding microbial distribution and determining functional roles
3. Understanding biodiversity regulation and magnitude

## 3.3 Modern molecular techniques

Conventional techniques for PGPR characterization and categorization as earlier described are primarily based on culturable PGPR analysis. However, because of the viable but nonculturable nature of PGPR and majority of them from natural habitat, analysis of whole community structure is enormously problematic (Dokic et al., 2010). Modern molecular techniques include both phylogenetic and molecular features and are used to differentiate PGPR range based on genetic material diversity and do not require cultivation of PGPR. PGPR can be explored using various molecular techniques including DNA–DNA and DNA–RNA hybridization, cloning and sequencing of DNA, as well as PCR-oriented techniques like temperature gradient gel electrophoresis (TGGE), automated ribosomal intergenic spacer analysis (ARISA), denaturing gradient gel electrophoresis (DGGE), ribosomal intergenic spacer analysis (RISA), genomics, transcriptomics, proteomics, and metabolomics.

### 3.3.1 Polymerase chain reaction

Polymerase chain reaction (PCR) is the targeted amplification of nucleic acid sequence, a specific gene, conserved, or arbitrary sequences can be targeted. The 16S rRNA gene sequence is a suitable PCR target by species–strain-specific or universal primers for phylogenic purposes and identification as it is distributed among PGPR universally, and has significant variations among species and strains DNA sequences. Nonspecific DNA binding dyes like SYBER Green I and SYBER Gold that bind to the double-stranded PCR products (Lidder and Sonnino, 2012). The numerous amplicons within a single reaction mixture can be detected by fluorescent probes tagged with various dye, whereas the double-stranded DNA dyes are restricted to a single product per reaction. The post-PCR detection techniques include electrophoresis, nucleic acid probing, and hybridization analysis.

Based on bacterial sample types, different versions of PCR have been derived for fulfilling simultaneous identification of multiple bacteria, differentiation and quantification of live bacterial cells. For the detection of multiple species belong to single genera the multiplex PCR has been evolved (Touron et al., 2005). Traditional PCR is unable to indicate the viability or unviability of bacterial cells, and for that reason reverse transcriptase PCR was evolved for viability detection of cells. This technique is based on the messenger RNA-dependent cDNA synthesis by reverse transcriptase enzyme (Rodriguez-Lázaro et al., 2007). The technique is time efficient as one does not need any preenrichment steps. Reverse transcriptase PCR also detects the viable number of cells not detectable during culturing. The technique of quantitative PCR (Couillerot et al., 2010) is based on the successive PCR amplicons monitoring with the reaction proceeding by using either fluorescent probes or dyes which are sequence specific or nonspecific. The PCR protocols require coordination with laboratories to make PCR results reproducible and reliable when used in various locations or times. Though it is a significantly accurate technique in template DNA quality, microbial and chemical cleanliness, humidity, temperature, equipment, individual expertise, reaction materials, and the reaction conditions (Fricker et al., 2007), occasional samples may have materials that can degrade the target DNA sequence or restrain activity of enzyme in PCR, resulting in false negative results (Glynn et al., 2006).

### 3.3.2 G + C mole% content

The techniques of DNA reassociation and kinetics reproducing the mole percent of G + C is the most primitive molecular technique used for microbial characterization, especially for taxonomic purposes. The G + C mole % of DNA can be determined by thermal denaturation. The mole% G + C value varies from 25% to 75% based on microbial species and remains stable for a specific organism. It is concluded that closely interrelated microbes have very similar mole% G + C patterns, and there is only a 3%–5% difference in taxonomically linked groups. It is a quantitative technique and unaffected by PCR biases and includes whole extracted DNA. However, similarities in base composition is not a verification of relationship. Although the base composition difference is strong evidence of a missing relationship, this technique needs comparatively large amount of DNA (Clegg et al., 2000).

### 3.3.3 Reassociation of DNA

Variation of DNA sequences with the help of genetic complication measurement leads to genetic diversity of a microbial community's contemporary plant rhizosphere (Torsvik et al., 1996). In this technique, total genomic DNA is isolated from microbial samples then purified, denatured, and permitted to reanneal. The reassociation rate is directly proportional to sequence similarity. As the diversity of DNA sequences increases, the rate of DNA reassociation will decrease (Theron and Cloete, 2000). Many factors influence the DNA reassociation technique, such as DNA product concentration (Co) and incubation time (t).

### 3.3.4 Nucleic acid hybridization

The dot blot hybridization technique is a powerful tool used in bacterial molecular biology for DNA analysis as well as RNA qualitative and quantitative with the help of specially designed oligonucleotide probes from known sequences, varying in specifications from domain to species and are 5′-end tagged with markers (Goris et al., 2007). The FISH (fluorescent in situ hybridization) technique is one of the most famous DNA hybridizing techniques. Using this technique, spatial distribution of microbial communities in various ecosystems can be determined. However, due to low hybridization sensitivity of nucleic acids isolated directly from soil samples is the major important limitation of this technique.

### 3.3.5 Restriction fragment-length polymorphism

Restriction fragment-length polymorphism (RFLP) is based on polymorphisms of DNA and is used analyze various communities of microbes (Moyer et al., 1996). A very easy and potent technique for bacterial strain's identification of species and below species level. In RFLP techniques, digested DNA is electrophoresed and then blotted onto nitrocellulose or nylon membranes from agarose gels and hybridized with suitable probes prepared from cloned DNA fragments of linked microbes. This technique has proven to be suitable especially in DNA–DNA hybridization and enzyme electrophoresis combinations for discriminating closely related strains and in intraspecies variation determination. However, occasionally the same banding pattern does not denote a close linkage between the microbes to be compared.

### 3.3.6 Terminal restriction fragment-length polymorphism

This technique is a modified version of RFLP and addresses some of its limitations. It is an alternate technique for rapid microbe diversity analysis in different ecosystems (Thies, 2007). This technique has similar principle as RFLP except the fluorescent dye labeling of one PCR primer, such as TET (4,7,20,70-tetrachloro-6-carboxyfluorescein) or 6-FAM (phosphoramiditefluorochrome 5-carboxyfluorescein). The PCR of sample DNA is performed using 16S rDNA universal primers, one of which is labeled with fluorescent dye. The fluorescent labeled terminal RFLP (FLT−RFLP) patterns can then be prepared by digesting the labeled PCR product with the help of restriction enzymes. The fragments are separated with gel electrophores in an automated sequence analyzer. The operational taxonomic unit (OTU) is prepared, counted with the help of unique fragment, and then each OTU frequency analyzed. The banding pattern can be utilized to measure species diversity and similarities as well as evenness between samples. While this technique may misjudge the real diversity of microbes as only dominant population species are detected because of availability of large amount of DNA template (Liu et al., 1997), the incomplete or partial digestion of DNA by restriction enzymes may result in diversity overestimation. It has also been noted that universal primers are not able to amplify whole sequences from bacterial and archaeal domains, and that the primers are designed from existing 16S rRNA and internal transcribed spacer (ITS) sequence data bases, which have sequences from culturable microbes and hence may not work be representative of the true diversity of the microbial sample. Additionally, different restriction enzymes will generate dissimilar fingerprints of community. T−RFLP is a unique technique for comparing relationships among different microbial samples and has also been used for the measurement of spatial as well as temporal changes in communities of bacteria to analyze complicated microbial communities, detect and monitor populations, and assess arbuscular mycorrhizal fungi diversity in rhizosphere.

### 3.3.7 Ribosomal intergenic spacer analysis/automated ribosomal intergenic spacer analysis/amplified ribosomal DNA restriction analysis

These techniques are principally similar to RFLP and T−RFLP and offer microbial fingerprinting based on ribosomes. InRISA and ARISA, the 16S and 23S ribosomal subunits intergenic spacer region is amplified by PCR followed by denaturation and separation in denaturing conditions on a polyacrylamide gel. The encoded tRNA is helpful for differentiation of closely related species and strains of bacteria due to intergenic spacer heterogeneity of the sequence and length. The polymorphisms in this technique are detected by silver staining. In ARISA, forward primer labeled with fluorescent dye, through which detection automatically performed. Both techniques (RISA and ARISA) can provide highly reproducible profiles of microbial communities. However, some of the limitations of this technique are the large amount of genomic DNA required, longer time to complete, silver staining intensity in a few cases, and low resolution (Kirk et al., 2004). ARISA is more sensitive, requiring less time than RISA, although the traditional PCR limitations also apply to RISA (Kirk et al., 2004). The diversity of microbes in soil and rhizosphere has been studied with the help of RISA (Borneman and Triplett, 1997).

### 3.3.8 DNA microarrays

This technique is very important for carrying out bacterial studies with high accuracy due to the fact that a single array can comprise thousands of DNA sequences (DeSantis et al., 2007). The target-specific genes encoded for enzymes like naphthalene dioxygenase, nitrate reductase and nitrogenase, etc., can be used to explain the functional diversity of bacterial communities in microarrays. The environmental sample "standards" (DNA fragments having below 70% hybridization) representing various species likely to be found in any environment can also be utilized in microarray. The reverse sample genome probing is a technique that utilizes genomic microarrays for microbial composition analysis by:

1. Genomic DNA isolation from pure bacterial cultures
2. Testing of cross-hybridization to achieve less than 70% cross-hybridized DNA fragments
3. Solid support-based genomic array preparation
4. Random labeling of a specific mixture of entire community DNA and internal standard

This technique may have large number of target gene sequences but detects only the most abundant species. In addition, culturing of bacterial species is required, but in this technique cloned DNA sequences of nonculturables can also be used. Experimentally it has also been proven that use of DNA fragments or genes in place of genomes on the microarray can remove the need of for live microbial culture, as cloning of genes can be done in plasmids or DNA fragments can be continuously used for PCR amplification. Furthermore, DNA fragments enhance the hybridization specificity over the genome utilization, and assessment of functional genes can be easily performed in the microbial community (Greene and Voordouw, 2003).

### 3.3.9 Denaturant gradient gel electrophoresis/temperature gradient gel electrophoresis

In DGGE or TGGE, DNA fragments of similar length but with dissimilar sequence base-pairing can be separated. From the testing samples DNA is isolated and amplified in PCR using 16S or 18S rRNA sequence-specific universal primers. The amplicons are separated based on mobility differences of DNA fragments partially melted in acrylamide gels with a linear DNA denaturant gradient like formamide and urea. The sequence variation in DNA fragments results in a dissimilarities in melting behavior and thus separation in denaturing gradient gels. The product melting takes place at various melting points that are nucleotide stretches with similar melting temperatures. The variation of sequence fragments will stop migration at various points in the gel as per the denaturant concentration. Theoretically, DGGE can separate even a single base pair difference in DNA sequences (Miller et al., 1999). The TGGE technique is based on a similar principle but has a temperature gradient in place of chemical denaturants. Both techniques are rapid, consistent, reproducible, and less expensive than other techniques. Analysis of numerous samples can be performed simultaneously and tracking of microbial population changes against any adversity or stimuli is possible with these techniques. However, biases of PCR, laborious sample preparation and handling, and uneven DNA isolation productivity are some of its limitations.

### 3.3.10 Single-strand conformation polymorphism

The single-strand conformation polymorphism (SSCP) technique works on electrophoresis-based discrimination of DNA sequences and allows differentiation of DNA fragments with equal length but variation in sequences of nucleotide. The SSCP technique was initially developed to distinguish point mutations or novel polymorphisms in DNA. In this technique, separation of single-stranded DNA is carried out in polyacrylamide gel because of the variation in mobility caused from its secondary structure (i.e., heteroduplex). This technique is helpful in analyzing the genetic diversity of microbes. However, due to the existence of multiple stable conformations of some single-stranded DNA, multiple bands may be produced on gel of the same DNA sequence. But this technique does not need a GC clamp or gradient gel construction and has been used to study rhizosphere community diversity of bacteria and fungi (Stach et al., 2001).

## 3.4 Genomics

The recent development of sequencing technologies as well as efficient bioinformatics tools displayed the unknown metabolic potential of PGPR. In fact, the presence of gene clusters within genomes was revealed by genomic analysis of species like *Bacillus atrophaeus*, *Bacillus halotolerans*, and *Pseudomonas* (Ma et al., 2018; Zhang et al., 2018; Kuzmanović et al., 2018). The genome mining has been accomplished as potential technique to evaluate the genetic potential of a strain by the survey of genome such as secondary metabolites analysis SHell (anti-SMASH) (Blin et al., 2017), prediction informatics for secondary metabolomes (PRISM) (Skinnider et al., 2017), global alignment for natural-products chemInformatics (GARLIC), generalized retrobiosynthetic assembly prediction engine (GRAPE) platform (Dejong et al., 2016) and Integrated Microbial Genomes Atlas of Biosynthetic gene Clusters (IMG-ABC) (Hadjithomas et al., 2017) are used bacterial genome mining results in identification of novel plant growth-promoter substance producers. Metabolic engineering of gifted strains (novel bacterial strains having PGPR activities) have been prioritized due to advancement in genome mining and comparative genomics. Primary and secondary metabolic pathway reconstruction uncovers important metabolic genes for metabolic engineering. For example, whole genome sequence data analysis of the *Micromonospora* genus provided insight into their ability to produce new natural products (Carro et al., 2018).

## 3.5 Transcriptomics

Recently, transcriptome techniques have been used to increase our understanding of the interactions between plants and microbes like *Pseudomonas fluorescens* FPT9601-T5 (Wang et al., 2005). DNA sequencing has unveiled the high abundance of biosynthetic gene clusters (BGCs) of PGPR; however, the majority of their actual products have not been characterized in laboratory condition because of complex regulation at transcriptional, translational, and post-translational levels. To disclose the mechanisms of controlling the metabolic switch, developmental differentiation and biosynthesis of natural products, transcriptional alteration in gene expression levels have been significantly studied (Palazzotto et al., 2015). A widely used approach to stimulate the expression of PGPR BGCs and thus biosynthesis of novel natural products, is the utilization of biological (co-cultivation), molecular, and chemical elicitors. Three indole-3-acetic acid (IAA) biosynthesis pathways have been identified in *Azospirillum brasilense*—two tryptophan (Trp)-dependent ones and one independent one. The production of 90% of IAA in *A. brasilense* was synthesized by the indole-3-pyruvic acid (IPA) pathway (Trp- > IPA- > indole-3-acetaldehyde- > IAA) in the presence of Trp. Recently the transcriptome analysis of *Bacillus subtilis* OKB105 in seedlings of rice was performed (vanPuyvelde et al., 2011). Researchers found 176 genes, about 3.8% of the total transcriptome of *B. subtilis* strain OKB105, and there was significant alteration in the expression pattern in response to rice seedlings. Out of this, 52 genes upregulated and most participated in metabolism and nutrient transport and stress responses, including *araA*, *ywkA*, *yfls*, etc. All 124 downregulated genes, which included *cheV*, *fliL*, *spmA*, and *tua*, participated in motility sporulation, chemotaxis, and teichuronic acid biosynthesis.

## 3.6 Proteomics

Associations among various metabolic pathways and production of natural products have been explored broadly by using various proteomics approaches. By comparing protein expression intensities, proteomics provides details on distinct pathways, emphasizing major factors in natural product biosynthesis. Many proteomic studies of crops and PGPR have been carried out to understand the protein expression profiling based plant and PGPR interactions identification like *Paenibacillus polymyxa* SQR and watermelon growth, *A. brasilense* Sp7 and maize and tomato seedlings and *P. polymyxa* and *Arabidopsis thaliana* studies thoroughly (Yaoyao et al., 2017) and explore the proteins and mechanism behind the enhanced performance of plants. Using differential proteomics it was possible to reveal the means of regulating morphological variation and metabolite production during different phases of growth (Palazzotto and Weber, 2018). The proteomics play a similar role to transcriptomic and genomics in strain classification. Analysis based on genomic, transcriptomic, and proteomic parameters was carried out on the thiopeptide GE2270 producing unusual actinomycetes *Planobispora rosea*, and diverse "omics" data revealed the physiology during GE2270 biosynthesis (Tocchetti et al., 2015).

## 3.7 Metabolomics

The analysis of metabolomic approaches has been effectively applied, it gives important information on plant and PGPR interactions. Metabolomics also helps to understand the impact on plant metabolism with PGPR association and identification of metabolic compounds which play an essential role in crop improvement. The inaccessible nature of the biosynthetic ability of cultured and uncultured microorganisms has been revealed by genomics-based methods. Distinctive approaches have been used to understand such "silent" gene clusters (Liu et al., 2015). Metabolomics permits a metabolic examination of various biological samples. Various new natural products have been found using nuclear magnetic resonance (NMR)-based metabolomic and mass spectrometry (MS). The MS/MS database Global Natural Product Social (GNPS) molecular networking platform has significantly improved the evaluation of mass spectrometry data, allowing a rapid dereplication of identified molecules and strain prioritization (Crüsemann et al., 2016). Furthermore, VarQuest, an innovative algorithm for recognition of peptidic natural product alternatives via database search of mass spectra, was established to extend GNPS by Gurevich et al. (2018). An example of metabolomics approach to find novel natural products is the recognition of a previously undescribed prenylated isatin antibiotic of *Streptomyces* sp. MBT28. This molecule has antibiotic activity against *B. subtilis* was discovered by NMR based metabolomics approach (Wu et al., 2015).

## 3.8 Metaomics

Because of the fact that only a minute percent of existing prokaryotes is amenable for cultivation utilizing standard techniques, a plenty of natural products remain undiscovered. Metaomics techniques (metagenomics, metatranscriptomics, and metaproteomics) are used to deal with the chemistry of uncultivated bacteria. The analysis of DNA samples isolated from natural environment allows to explore "dark matter", biosynthetic pathways of marine unculturable bacteria such as obligate symbiont of sponges and other marine invertebrates. These approaches have shown symbiont bacteria to be "skilled" makers of new polyketides and peptides with antitumor action (Wilson et al., 2014). A big contribution to discovery of novel effective antibiotics has been gifted by the development of new microbiome screening technique to explore previously unculturable bacteria. In 2015, Ling and coworkers used the situ development strategy iChip to study teixobactin, from the previously uncultured microorganism *Eleftheria terrae*. This flow of gene expression in a complex microbial community can be explored utilizing a metatranscriptomics methodology. However, while metagenomics analysis can be used to study the chemical capability of various obscure species by mining natural DNA, metatranscriptomics can be used to study the transient reactions to environmental conditions by analyzing the aggregate arrangements of messenger RNAs (Jiang et al., 2016). The combination of metagenomics, metatranscriptomics, and metaproteomics allows researchers to study the physiological movements of complex natural microbial communities (Jansson and Baker, 2016). Diverse metaproteomic instruments are also used to explore BGCs and pathways. With high-throughput metaproteome data, it is also feasible to explore posttranslational modifications in in situ environments and hence disclose the mechanisms utilized by bacteria to regulate physiological processes as well as natural products fabrication (Fan et al., 2017).

## 3.9 Collective omics approach

The appearance of increasingly influential metabolomics methods combined with genomics have allowed the discovery of previously unknown natural products and exploration of their significance in nature. Metabologenomics combines genome sequencing and robotized gene cluster prediction with MS-based metabolomics (Maansson et al., 2016). The data of genome are utilized to interrogate the chemical data and vice versa to explore molecules families with interested phenotypes in large strain collections (Paulus et al., 2017). To find novel phosphonic acids, a large-scale genome mining investigation of special phosphonyl-pyruvate mutase quality (pepM) was conducted on 10,000 actinomycetes (Ju et al., 2015). Out of 278 strains, 64 diverse groups of phosphonate BGCs were recognized. By characterization of strains within these clusters another prototype pathway for phosphonate biosynthesis and 11 previously undescribed phosphonic acid natural products were found. A broad study of lanthipeptide-related BGCs in *Actinobacteria* showed that lanthipeptide synthetases can produce natural products other than lanthipeptides. In fact, genomics and MS data suggest cross-talk between lanthipeptide biosynthetic enzymes and polyketide synthases (PKSs) and nonribosomal peptide synthases (NRPSs) systems and consequently natural products with new platforms. A multiomics approach was applied successfully to understand how BGCs are exploited in animal and microbe symbiosis in aquatic and terrestrial ecosystems. One case is the broad study researching the natural product diversity generated by the nematode symbionts *Photorhabdus* and *Xenorhabdus*. Despite comparative genomics evaluation shows a high similarity at DNA level, high-resolution MS analyses reveal a huge chemical diversity. The utilization of genomic and metabolomic techniques in an integral way has permitted the rapid identification of bioactive items including xefoampeptides and tilivalline. A mix of science, chemistry, genetics, metagenomics, and metatranscriptomics was effectively used in 752 metagenomic tests from the NIH Human Microbiome Project (Donia et al., 2014).

## 3.10 Next generation sequencing

Next generation sequencing (NGS) techniques include high-throughput sequencing and pyrosequencing used to determine new microorganism(s) groups in complex soil ecosystems and to explore the complexities of microbial populations (Bartram et al., 2011). Using the NGS technique, extremely complex microbial communities can be analyzed (Fakruddin and Mannan, 2013). Recently Roche 454 Life Science developed a pyrosequencing technique by which large amounts of DNA reads can be generated. This technique was successfully used to analyze with having complex microbial communities like wastewater, marine sediments, and soil. Beside eradicating the utilization of cloning vectors and library construction, and their related biases, NGS can also read through secondary structures and generate large quantity of sequences of up to 100 Mb per run. In addition, different bioinformatics tools like the RDP Pyrosequencing Pipeline and Newbler Assembler have been used to analyze raw NGS data in silico to determine the complex composition of microbial communities in environmental samples (Van den Bogert et al., 2011).

## Acknowledgment

The authors express sincere thanks to the funding agencies DIC-BHU, DBT, and DST-NRDMS for providing funds for research work on climate-resilient or stress-tolerant bacterial strains for sustainable agriculture.

## References

Bartram, A.K., Lynch, M.D.J., Stearns, J.C., Moreno-Hagelsieb, G., Neufeld, J.D., 2011. Generation of multimillion-sequence 16S rRNA gene libraries from complex microbial communities by assembling paired-end illumina reads. Appl. Environ. Microbiol. 77, 3846–3852.

Blin, K., Medema, M.H., Kottmann, R., Lee, S.Y., Weber, T., 2017. The antiSMASH database, a comprehensive database of microbial secondary metabolite biosynthetic gene clusters. Nucl. Acids Res. 45, D555–D559.

Borneman, J., Triplett, E.W., 1997. Molecular microbial diversity in soils from eastern Amazonia: evidence for unusual microorganisms and microbial population shifts associated with deforestation. Appl. Environ. Microbiol. 63 (7), 2647–2653.

Brehm-Stecher, B.F., Johnson, E.A., 2004. Single-cell microbiology: tools, technologies, and applications. Microbiol. Mol. Biol. Rev. 68, 538–559.

Brenner, D.J., Staley, J.T., Krieg, N.R., 2015. Classification of procaryotic organisms and the concept of bacterial speciation. Bergey's Manual of Systematic Bacteriology. Wiley, Chichester, UK, pp. 27–32.

Carro, L., Nouioui, I., Sangal, V., Meier-Kolthoff, J.P., Trujillo, M.E., Montero-Calasanz, M. D.C., et al., 2018. Genome-based classification of micromonosporae with a focus on their biotechnological and ecological potential. Sci. Rep. 8, 525.

Clegg, C.D., Ritz, K., Griffiths, B.S., 2000. % G + C profiling and cross hybridisation of microbial DNA reveals great variation in below-ground community structure in UK upland grasslands. Appl. Soil Ecol. 14 (2), 125–134.

Couillerot, O., Bouffaud, M.L., Baudoin, E., Muller, D., Caballero-Mellado, J., Moënne-Loccoz, Y., 2010. Development of a real-time PCR method to quantify the PGPR strain *Azospirillum lipoferum* CRT1 on maize seedlings. Soil. Biol. Biochem. 42, 2298–2305.

Crüsemann, M., O'Neill, E.C., Larson, C.B., Melnik, A.V., Floros, D.J., Da Silva, R.R., et al., 2016. Prioritizing natural product diversity in a collection of 146 bacterial strains based on growth and extraction protocols. J. Nat. Prod. 80, 588–597.

Dejong, C.A., Chen, G.M., Li, H., Johnston, C.W., Edwards, M.R., Rees, P.N., et al., 2016. Polyketide and nonribosomal peptide retro-biosynthesis and global gene cluster matching. Nat. Chem. Biol. 12, 1007–1014.

DeSantis, T.Z., Brodie, E.L., Moberg, J.P., Zubieta, I.X., Piceno, Y.M., Andersen, G.L., 2007. High-density universal 16S rRNA microarray analysis reveals broader diversity than typical clone library when sampling the environment. Microb. Ecol. 53, 371–383.

Dokic, L., Savic, M., Narancic, T., Vasiljevic, B., 2010. Metagenomic analysis of soil microbial communities. Arch. Biol. Sci. Belgrade 62, 559–564.

Donia, M.S., Cimermancic, P., Schulze, C.J., Wieland Brown, L.C., Martin, J., Mitreva, M., et al., 2014. A systematic analysis of biosynthetic gene clusters in the human microbiome reveals a common family of antibiotics. Cell 158, 1402–1414.

Fakruddin, M., Mannan, K.S.B., 2013. Methods for analyzing diversity of microbial communities in natural environments. Ceylon J. Sci., (Biol. Sci.) 42, 19.

Fan, B., Li, Y.L., Li, L., Peng, X.J., Bu, C., Wu, X.Q., et al., 2017. Malonylome analysis of rhizobacterium Bacillus amyloliquefaciens FZB42 reveals involvement of lysine malonylation in polyketide synthesis and plant-bacteria interactions. J. Prot. 154, 1–12.

Fricker, M., Messelhäußer, U., Busch, U., Scherer, S., Ehling-Schulz, M., 2007. Diagnostic real-time PCR assays for the detection of emetic *Bacillus cereus* strains in foods and recent food-borne outbreaks. Appl. Environ. Microbiol. 73, 1892–1898.

Giovannoni, S.J., Britschgi, T.B., Moyer, C.L., Field, K.G., 1990. Genetic diversity in Sargassa sea bacterioplankton. Nature 345, 60–65.

Glynn, B., Lahiff, S., Wernecke, M., Barry, T., Smith, T.J., Maher, M., 2006. Current and emerging molecular diagnostic technologies applicable to bacterial food safety. Int. J. Dairy Technol. 59, 126–139.

Goris, J., Konstantinidis, K.T., Klappenbach, J.A., Coenye, T., Vandamme, P., Tiedje, J.M., 2007. DNA-DNA hybridization values and their relationship to whole-genome sequence similarities. Int. J. Syst. Evol. Microbiol. 57, 81–91.

Greene, E.A., Voordouw, G., 2003. Analysis of environmental microbial communities by reverse sample genome probing. J. Microbiol. Methods 53, 211–219.

Gurevich, A., Mikheenko, A., Shlemov, A., Korobeynikov, A., Mohimani, H., Pevzner, P.A., 2018. Increased diversity of peptidic natural products revealed by modification-tolerant database search of mass spectra. Nat. Microbiol. 3, 319–327.

Hadjithomas, M., Chen, I.M.A., Chu, K., Huang, J., Ratner, A., Palaniappan, K., et al., 2017. IMG-ABC: new features for bacterial secondary metabolism analysis and targeted biosynthetic gene cluster discovery in thousands of microbial genomes. Nucl. Acids Res. 45, D560–D565.

Jaiswal, D.K., Verma, J.P., Yadav, J., 2017. Microbe induced degradation of pesticides in agricultural soils. In: Environmental Science and Engineering (Subseries: Environmental Science), pp. 167–189.

Jansson, J.K., Baker, E.S., 2016. A multi-omic future for microbiome studies. Nat. Microbiol. 1.

Jiang, Y., Xiong, X., Danska, J., Parkinson, J., 2016. Metatranscriptomic analysis of diverse microbial communities reveals core metabolic pathways and microbiomespecific functionality. Microbiome 4, 2.

Ju, K.-S., Gao, J., Doroghazi, J.R., Wang, K.-K.A., Thibodeaux, C.J., Li, S., et al., 2015. Discovery of phosphonic acid natural products by mining the genomes of 10,000 actinomycetes. Proc. Natl. Acad. Sci. 112, 12175–12180.

Kirk, J.L., Beaudette, L.A., Hart, M., Moutoglis, P., Klironomos, J.N., Lee, H., et al., 2004. Methods of studying soil microbial diversity. J. Microbiol. Methods 58, 169–188.

Kuzmanović, N., Eltlbany, N., Ding, G., Baklawa, M., Min, L., Wei, L., et al., 2018. Analysis of the genome sequence of plant beneficial strain Pseudomonas sp. RU47. J. Biotechnol. 281, 183–192.

Lidder, P., Sonnino, A., 2012. Biotechnologies for the management of genetic resources for food and agriculture. Adv. Genet. 78, 1–167.

Ling, L., Schneder, T., Peoples, A., Spoering, A., Engels, I., Conlon, B., et al., 2015. A new antibiotic kills pathogen without detectable resistance. Nature 517, 455–459.

Liu, W.T., Marsh, T.L., Cheng, H., Forney, L.J., 1997. Characterization of microbial diversity by determining terminal restriction fragment length polymorphisms of genes encoding 16S rRNA. Appl. Environ. Microbiol. 63, 4516–4522.

Liu, J., Zhu, X., Seipke, R.F., Zhang, W., 2015. Biosynthesis of antimycins with a reconstituted 3-formamidosalicylate pharmacophore in *Escherichia coli*. ACS Syn. Biol. 4, 559–565.

Ma, J., Wang, C., Wang, H., Liu, K., Zhang, T., Yao, L., et al., 2018. Analysis of the complete genome sequence of *Bacillus atrophaeus* GQJK17 reveals its biocontrol characteristics as a plant growth-promoting rhizobacterium. Biomed. Res. Int. 2018.

Maansson, M., Vynne, N.G., Klitgaard, A., Nybo, J.L., Melchiorsen, J., Nguyen, D.D., et al., 2016. An integrated metabolomic and genomic mining workflow to uncover the biosynthetic potential of bacteria. mSystems 1, e00028–15.

Miller, K.M., Tobi, J., Schulze, A.D., Withler, R.E., 1999. Denaturing gradient gel electrophoresis (DGGE): a rapid and sensitive technique to screen nucleotide sequence variation in populations research report. Biotechniques 27, 1016–1030.

Moyer, C.L., Tiedje, J.M., Dobbs, F.C., Karl, D.M., 1996. A computer-simulated restriction fragment length polymorphism analysis of bacterial small-subunit rRNA genes: efficacy of selected tetrameric restriction enzymes for studies of microbial diversity in nature. Am. Soc. Microbiol. 62, 2501–2507.

Palazzotto, E., Renzone, G., Fontana, P., Botta, L., Scaloni, A., Puglia, A.M., et al., 2015. Tryptophan promotes morphological and physiological differentiation in *Streptomyces coelicolor*. Appl. Microbiol. Biotechnol. 99, 10177–10189.

Palazzotto, E., Weber, T., 2018. Omics and multi-omics approaches to study the biosynthesis of secondary metabolites in icroorganisms. Curr. Opin. Microbiol. 45, 109–116.

Paulus, C., Rebets, Y., Tokovenko, B., Nadmid, S., Terekhova, L.P., Myronovskyi, M., et al., 2017. New natural products identified by combined genomics-metabolomics profiling of marine Streptomyces sp. MP131-18. Sci. Rep. 7.

Rodriguez-Lázaro, D., Lombard, B., Smith, H., Rzezutka, A., D'Agostino, M., Helmuth, R., et al., 2007. Trends in analytical methodology in food safety and quality: monitoring microorganisms and genetically modified organisms. Trends Food Sci. Technol. 18, 306–319.

Skinnider, M.A., Merwin, N.J., Johnston, C.W., Magarvey, N.A., 2017. PRISM 3: expanded prediction of natural product chemical structures from microbial genomes. Nucl. Acids Res. 45, W49–W54.

Stach, J.E., Bathe, S., Clapp, J.P., Burns, R.G., 2001. PCR-SSCP comparison of 16S rDNA sequence diversity in soil DNA obtained using different isolation and purification methods. FEMS Microbiol. Ecol. 36 (2–3), 139–151.

Theron, J., Cloete, T.E., 2000. Molecular techniques for determining microbial diversity and community structure in natural environments. Crit. Rev. Microbiol. 26, 37–57.

Thies, J.E., 2007. Soil microbial community analysis using terminal restriction fragment length polymorphisms. Soil Sci. Soc. Am. J. 71, 579.

Tocchetti, A., Bordoni, R., Gallo, G., Petiti, L., Corti, G., Alt, S., et al., 2015. A genomic, transcriptomic and proteomic look at the GE2270 producer *Planobispora rosea*, an uncommon actinomycete. PLoS One. 10, e0133705.

Torsvik, V., Sørheim, R., Goksøyr, J., 1996. Total bacterial diversity in soil and sediment communities—a review. J. Ind. Microbiol. Biotechnol. 17, 170–178.

Touron, A., Berthe, T., Pawlak, B., Petit, F., 2005. Detection of Salmonella in environmental water and sediment by a nested-multiplex polymerase chain reaction assay. Res. Microbiol. 156, 541–553.

Van den Bogert, B., de Vos, W.M., Zoetendal, E.G., Kleerebezem, M., 2011. Microarray analysis and barcoded pyrosequencing provide consistent microbial profiles depending on the source of human intestinal samples. Appl. Environ. Microbiol. 77, 2071–2080.

vanPuyvelde, S., Cloots, L., Engelen, K., Das, F., Marchal, K., Vanderleyden, J., et al., 2011. Transcriptome analysis of the rhizosphere bacterium *Azospirillum brasilense* reveals an extensive auxin response. Microb. Ecol. 61, 723–728.

Verma, J.P., Yadav, J., Tiwari, K.N., Jaiswal, D.K., 2014. Evaluation of plant growth promoting activities of microbial strains and their effect on growth and yield of chickpea (*Cicer arietinum* L.) in India. Soil. Biol. Biochem. 70, 33–37.

Verma, J.P., Jaiswal, D.K., Maurya, P.K., 2016. Screening of bacterial strains for developing effective pesticide tolerant plnt growth-promoting microbial consortia from rhizosphere soils. Energy Ecol. Environ. 1, 408–418.

Verma, J.P., Jaiswal, D.K., Krishna, R., Prakash, S., Yadav, J., Singh, V., 2018. Characterization and screening of thermophilic Bacillus strains for developing plant growth promoting consortium from hot spring of Leh and Ladakh region of India. Front. Microbiol. 9.

Wang, Y., Ohara, Y., Nakayashiki, H., Tosa, Y., Mayama, S., 2005. Microarray analysis of the gene expression profile induced by the endophytic plant growth-promoting rhizobacteria, *Pseudomonas fluorescens* FPT9601-T5 in Arabidopsis. Mol. Plant Microb. Interact. 18, 385–396.

Wilson, M.C., Mori, T., Rückert, C., Uria, A.R., Helf, M.J., Takada, K., et al., 2014. An environmental bacterial taxon with a large and distinct metabolic repertoire. Nature 506, 58–62.

Wu, C., Du, C., Gubbens, J., Choi, Y.H., Van Wezel, G.P., 2015. Metabolomics-driven discovery of a prenylated isatin antibiotic produced by Streptomyces species MBT28. J. Nat. Prod. 78, 2355–2363.

Yaoyao, E., Yuan, J., Yang, F., Wang, L., Ma, J., Li, J., et al., 2017. PGPR strain *Paenibacillus polymyxa* SQR-21 potentially benefits watermelon growth by re-shaping root protein expression. AMB Expr. 7, 104.

Zhang, Z., Yin, L., Li, X., Zhang, C., Liu, C., Wu, Z., 2018. The complete genome sequence of *Bacillus halotolerans* ZB201702 isolated from a drought- and salt-stressed rhizosphere soil. Microb. Pathog. 123, 246–249.

## Further reading

Luis Royo, J., Hidalgo, M., Ruiz, A., 2007. Pyrosequencing protocol using a universal biotinylated primer for mutation detection and SNP genotyping. Nat. Protoc. 2, 1734–1739.

# Role of microbially synthesized nanoparticles in sustainable agriculture and environmental management

*Vipin Kumar Singh*[1] *and Amit Kishore Singh*[2]
[1]Center of Advanced Study in Botany, Institute of Science, Banaras Hindu University, Varanasi, India, [2]Botany Department, Kamla Nehru Post Graduate College, Raebareli, India

## 4.1 Introduction

The word "nano" refers to one billionth part of 1 m length (i.e., $10^{-9}$ m). The term nanotechnology was coined by Prof. Norio Taniguchi (Bulovic et al., 2004) to define those materials nanometer in size. With the advancement of newer technologies, the dimensions of nanotechnology have considerably grown. However, the technology has found greater application in materials and electronic engineering as compared to other disciplines (Prasad et al., 2017). Their very small size has given several important attributes to nanoparticles with respect to their application in food, textile, medicine, agricultural, and environmental sectors.

The more recent so-called green nanotechnology is based on the application of microbial processes for synthesis of nanoparticles. Production of nanoparticles by microorganisms involves the assemblage of different inorganic materials either inside or outside the cells. Although a large number of microbes including bacteria and fungi are known to synthesize metal nanoparticles, the microbial synthesis mechanisms are very much important. Microbe catalyzed nanoparticle synthesis is the emerging tool of green chemistry (Alghuthaymi et al., 2015). Synthesis of various metal nanoparticles by the activities of fungi, bacteria, actinomycetes, and yeasts has been widely demonstrated. Globally, the problem of environmental contamination has raised the publicity and interest of green nanotechnology among present day scientists and general people.

## 4.2 Microbial (green) synthesis of nanoparticles and advantages over nonbiological synthesis

Nanotechnology research and development programs have thus far focused mainly on the fabrication of molecules falling in the nanometer range, with the nanoparticle

**Role of Plant Growth Promoting Microorganisms in Sustainable Agriculture and Nanotechnology.**
**DOI: https://doi.org/10.1016/B978-0-12-817004-5.00004-X**

synthesis proceeding via physical, chemical, and biological processes. The methods employed for their synthesis are generally dependent on the characteristics of the bulk material used. Several techniques are currently used for the synthesis of metal nanoparticles with specific properties, but physicochemical methods have numerous drawbacks in contrast to biological processes such as use of the use of large amounts of toxic chemicals, high cost, generation of excessive secondary products, need for high thermal conditions, and high input of constant energy supply (Ghorbani, 2016). Very often chemically synthesized nanomaterials can be applied in health management because of their high toxicity (Khan, 2013). However, in a few cases, physicochemical methods for specific nanoparticle synthesis are indispensable because biological approaches for their synthesis are currently lacking.

Thus far, several scientific advances have occurred in the area of nanotechnology and several methods have been developed and suggested for nanoparticle synthesis (Sabri et al., 2016). Generally synthesis of nanoparticles is primarily based on three fundamental junctures.

1. Determination of most favorable conditions of the solvent.
2. The materials facilitating the nanoparticle synthesis should be prone to natural degradation pathways and reusable.
3. Exploitation of nontoxic substance as a capping agent to enhance the fabrication of nanoparticles for desired purposes (Khalil et al., 2014).

Nanotechnology is currently focused on diverse biological methods for fabrication of nontoxic and environmentally safe nanoparticles. Thus, resulting in the so-called bionanotechnology, which exploits a variety of microbes such as algae, bacteria, fungi, and plant extracts. Plant-derived biomolecules such as proteins, carbohydrates, and lipids have also been reported to catalyze the synthesis of nanoparticles. Green technology relies on basic rules for nanoparticle synthesis and is solely dependent on processes involving fewer byproducts, nontoxic substances, and cheaper materials with minimal energy consumption (Iravani et al., 2017).

## 4.3 Metal nanoparticle biosynthesis by bacteria

Bacterial species differing in size, shape, and mode of energy generation for vital cellular processes are one of the widely exploited single-cell microorganisms thriving in different habitats. Use of biological agents such as bacteria for fabrication of metal nanoparticles is a promising ecofriendly and nontoxic approach with numerous applications as mentioned previously. Microbial synthesis of nanoparticles has attracted researchers worldwide due to their biocompatibility and ease in synthesis on account of limited resources as well as multidimensional applications. Furthermore, limitations associated with physicochemical processes have intensified the research possibilities on metal nanoparticle biosynthesis by bacteria. However, biofabrication of nanoparticles is not always the best tool because of constraints such as instability and differing sizes and shapes, necessitating process optimization for better outcomes (Quester et al., 2013). The remediation of metal-contaminated sites through bioreduction of metals using microorganisms, especially bacteria to produce metal sulfide (Stephen and Macnaughtont, 1999), has already been extensively applied under natural environmental conditions and subsequently used for nanoparticle generation. Both extracellular and intracellular biosynthesis of metal nanoparticles have been described in the literature (Narayanan and Sakthivel, 2010). While the biosynthesis of nanoparticles is affected by conditions such as pH (He et al., 2007), salt content, temperature (Fatemi et al., 2018), nature of microorganisms used, compounds secreted by them, nutritional conditions, and the presence of electron donors and acceptors, their biological, chemical, and physical properties are largely determined by size and shape.

According to Nath and Banerjee (2013), the essential biological processes leading to the formation of metal nanoparticles is termed as biomineralization. This process can be categorized into two types (controlled and induced) based on the detoxification mechanism allowing the microorganism to with the heavy metals present in the environment (Narayanan and Sakthivel, 2010). The controlled mineralization process involves intracellular formation of nanoparticles while the induced process involves mineral formation outside the cells (Perez-Gonzalez et al., 2010). The intracellularly produced nanoparticles are challenging to get outside because of their localization to specific regions as well as their small quantity (Revati and Pandey, 2011). In contrast, the extracellularly generated nanoparticles are high in content and easier to procure for intended purposes. Hence, the microbes catalyzing the extracellular synthesis of nanoparticles are important (Moon et al., 2007). However, the major concern of extracellularly produced metal nanoparticles lies in their high polydispersity compared to intracellular ones, which must be reduced to the level of monodispersity in order to popularize the extracellular biological processes for nanoparticle biosynthesis (Bao et al., 2003). Some of the important bacterial species catalyzing the synthesis of metal nanoparticles are presented in Table 4.1.

## 4.4 Metal nanoparticle biosynthesis by fungi

Diverse fungal cells are equally important as bacterial cells in catalyzing the green synthesis of metallic nanoparticles under similar conditions. They are superior to other microbes used for fabrication of metal nanoparticles because of easy cultivation, fast growth, low-cost biomass production, larger surface area, excess biosynthesis of enzymes necessary for vital cellular processes, efficient and elevated synthesis of metal nanoparticles because of tolerance to higher metal concentrations, and comparatively simpler extraction of metals from biomass (Sastry et al., 2003; Castro-Longoria et al., 2011, 2012; Volesky and Holan, 1999). Some of the most notable fungal species responsible for metal nanoparticle biosynthesis are *Aspergillus* sp., *Trichoderma* sp., *Fusarium* sp., *Verticillium* sp., *Penicillium* sp., *Phomopsis* sp., and *Phanerochaete* sp. (Banerjee and Rai, 2018). Generally, the hazardous metals present in the vicinity are either reduced by the activity of fungal reductases ($\alpha$-NADPH and nitrate dependent) or adhere to their cell surfaces (biosorption) (Alghuthaymi et al., 2015). Thus, these nonessential toxic metals are

**Table 4.1** Metal nanoparticle biosynthesis by bacterial species.

| Bacterial species | Metal nanoparticles | References |
|---|---|---|
| *Lactobacillus fermentum* | Iron | Ghandehari et al. (2018) |
| *Shewanella loihica* PV-4 | Copper | Lv et al. (2018) |
| *Rhodococcus aetherivorans* BCP1 | Selenium | Presentato et al. (2018) |
| *Bacillus endophyticus* SCU-L | Silver | Gan et al. (2018) |
| *Bacillus* sp. GP | Palladium, gold | Zhang and Hu (2018) |
| *Deinococcus radiodurans* | Gold, silver, gold–silver complex | Li et al. (2018) |
| *Desulfotomacculum acetoxidans* | Iron | Das and Kerkar (2017) |
| *Acinetobacter* sp. SW 30 | Gold, selenium | Wadhwani et al. (2018) |
| *Shewanella oneidensis* | Copper | Kimber et al. (2018) |
| *S. loihica* PV-4 | Palladium | Wang et al. (2018) |
| *Bacillus* sp. FU4 | Copper | Taran et al. (2017) |
| *Streptomyces* sp. | Platinum | Sharma (2017) |
| *Shewanella* sp. CNZ-1 | Palladium | Zhang and Hu (2017) |
| *Rahnella aquatilis* HX2 | Selenium | Zhu et al. (2018) |
| *Alcaligenes* sp. CKCr-6A | Selenium | Mesbahi-Nowrouzi and Mollania (2018) |
| *Lactobacillus* sp. | Magnesium | Mohanasrinivasan et al. (2018) |
| *Delftia* sp. SFG | Bismuth | Shakibaie et al. (2018) |
| *Alishewanella* sp. WH16-1 | Selenium, chromium | Xia et al. (2018) |

eliminated from the environment in order to escape from cellular toxicity. Many toxic metals exert excess oxidative stress in fungal cells, which favors the elevated synthesis of enzyme-catalyzing metal reduction in order to alleviate the toxic impacts. Due to catalytic efficiency and easy isolation procedures, fungi are preferred for such large-scale biological activities. Moreover, fungi have been demonstrated to synthesize a variety of enzymes, protein/peptide molecules, and reducing agents of quinine types (naphthoquinones and anthaquinones) responsible for transformation of bulk materials into their nanoparticle form. Controlled synthesis of these biological molecules may be helpful in designing the synthesis of nanoparticles with desired shape and size. Although the intricate mechanism of nanoparticle biosynthesis still has to be explored in much detail, there is continuously rising interest among researchers to use cellular metabolites for nanoparticle production. However, some of the fungi under certain conditions may be specialized for the synthesis of a particular metal nanoparticle. The fungi-induced generation of metal nanoparticles may be extracellular or intracellular in nature depending on the fungal species involved. In extracellular mode, enzymes are secreted externally into the media leading to the conversion of bull metals into their respective nanoparticulate form. Under some conditions, the positively charged metal ions may interact with the negatively charged cell wall components under the influence of cell wall enzymes resulting in the intracellular synthesis of metal nanoparticles (Thakkar et al., 2010). Some of the important fungi directing the synthesis of metal nanoparticles are presented in Table 4.2.

**Table 4.2** Metal nanoparticle biosynthesis by selected fungal species.

| Fungal species | Metal nanoparticles | References |
|---|---|---|
| *Lentinus edodes*, *Pleurotus ostreatus*, *Ganoderma lucidum*, and *Grifola frondosa* | Gold, silver, silicon, selenium | Vetchinkina et al. (2017) |
| *Aspergillus* sp. WL-Au | Gold | Shen et al. (2017) |
| *Duddingtonia flagrans* | Silver | Silva et al. (2017) |
| *Neurospora crassa*, *Pestalotiopsis* sp., and *Myrothecium gramineum* | Copper | Li and Gadd (2017) |
| *Pleurotus* sp. | Iron | Mazumdar and Haloi (2017) |
| *Punctularia atropurpurascens*, *Botrytis cinerea*, *Penicillium expansum*, *P. ostreatus*, *Phanerochaete chrysosporium*, *Rhizopus stolonifer*, *Gymnopilus spectabilis*, *G. frondosa* | Silver | Sanguiñedo et al. (2018) |
| *Aspergillus nidulans* | Cobalt | Vijayanandan and Balakrishnan (2018) |
| *Trametes trogii* | Silver | Kobashigawa et al. (2018) |
| *Pichia pastoris* | Silver and selenium | Elahian et al. (2017) |
| *Alternaria* sp. | Silver | Singh et al. (2017) |
| *Pleurotus ostreatus* | Gold | El Domany et al. (2018) |
| *Aspergillus aculateus* | Iron | Bedi et al. (2018) |
| *Rhizomucor pusillus*, *Sporotrichum thermophile*, *Termoascus thermophilus*, *Termomyces lanuginosus* | Gold | Molnár et al. (2018) |
| *Aspergillus niger* | Zinc | Kalpana et al. (2018) |
| *Penicillium chrysogenum* | Platinum | Subramaniyan et al. (2018) |
| *Alternaria alternate* | Platinum | Sarkar and Acharya (2017) |
| *Nigrospora oryzae* | Cadmium | Gowri et al. (2018) |
| *Pichia kudriavzevii* | Zinc | Moghaddam et al. (2017) |
| *Aspergillus flavus* | Lead | Priyanka et al. (2017) |
| *Aspergillus flavus* | Zinc | Uddandarao and Balakrishnan (2017) |
| *Saccharomyces cerevisiae* | Titanium | Chaturvedi and Kumar (2018) |

## 4.5 Mechanism of nanoparticle synthesis

Generally, nanoparticle synthesis involves two basic approaches: "bottom-up" and "bottom down" approaches. Biological synthesis is a "bottom-up" approach where formation of nanoparticles occurs due to reduction/oxidation of metals, and the agents mainly responsible for the process are different enzymes secreted by microbial systems and various metabolites from plants (Prabhu and Poulose, 2012). Nanoparticles are biosynthesized by culturing microorganisms into specific nutrient media containing corresponding metal ions. Then, metal ions are converted into element metals through the action of enzymes. Indeed, depending on the location, synthesis of nanoparticles from microbial cells (particularly from bacteria, fungi, actinomycetes, yeasts, and even viruses) can be categorized into extracellular and intracellular synthesis according to Shankar et al. (2016). In the intracellular synthesis mechanism, the first step involves the transport of metal ions mediated by the microbial cell wall. The enzymes present within the cell reduce the metal ions to nanoparticles, which get diffused off through the cell wall. Overall, the intracellular synthesis process involves trapping, bioreduction, and capping of various nanoparticles such as, gold, silver, or other nanoparticles synthesized from bacteria, fungi, and actinomycetes (Li et al., 2011). For example, *Verticillum* (fungus) follows the above steps for nanoparticle synthesis via the intracellular method. However, extracellular synthesis is composed of secretion of enzymes, bioreduction, and capping of particles. One commonly used enzyme is nitrate reductase, which might be responsible for the synthesis of nanoparticles like silver and gold nanoparticles. The majority of the published reports (Singh, 2015) have claimed that extracellular synthesis of nanoparticles is preferable because the downstreaming and purification processes are easier as compared to intracellular methods. Furthermore, the extracellular method of synthesis is more cost effective and convenient than intracellular biosynthesis of nanoparticles (Banerjee and Rai, 2018). The main mechanisms of the intracellular and extracellular methods of nanoparticles utilized by microbes are shown in Fig. 4.1.

Both the intracellular and extracellular approaches as reported from diverse bacteria and fungi have been well exploited and extensively reviewed by several authors (Golinska et al., 2014; Hulkoti and Taranath, 2014; Hasan et al., 2018). However, some reports claim fungi-derived nanoparticles are more stable and efficient in comparison to those derived by bacterial processes due to following reasons:

- Fungi possess unique high cell wall binding and high intracellular metal uptake capacities (Moghaddam et al., 2017).
- Specific enzymes such as nitrate reductase secreted by the fungi facilitate the reduction of metal ions to nanosized particles (Ahmad et al., 2003).
- Fungi grow over the surface of inorganic substrate, which results in the synthesis of metal nanoparticles serving as catalyst (Kitching et al., 2015).
- Fastidious fungal growth to generate large biomass in a short time.
- Large biomass of fungus facilitates the production of a large amount of desired enzymes responsible for nanoparticle biosynthesis.

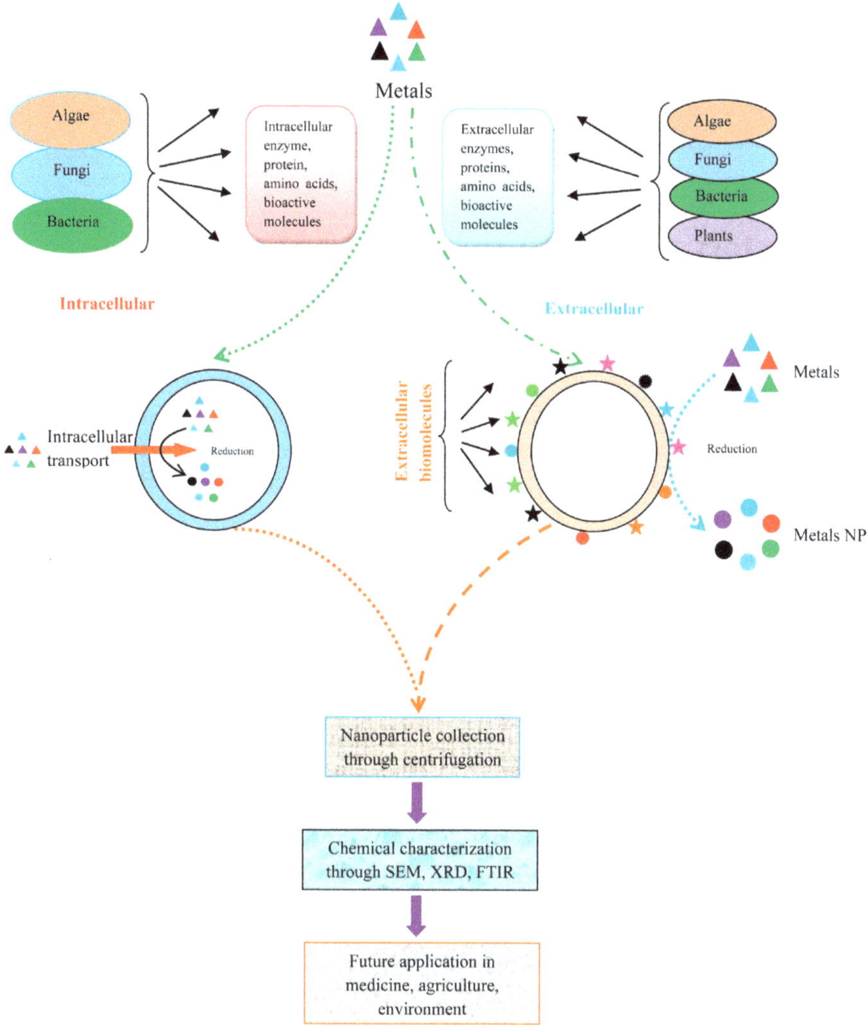


**Figure 4.1** Generalized mechanism of extracellular and intracellular biosynthesis of metal nanoparticles with characterization and application in different sectors.

However, in bacterially catalyzed nanoparticle production, the separation of secreted enzymes requires sophisticated equipment (centrifugation and sonicator) and chemical agents like methanol, while fungal filtrate can be easily excluded from the mycelia through simple filtration techniques. Overall, this makes the process more rapid and effective for fungal-derived synthesis of nanoparticles (Gade et al., 2008). Once the particles have been synthesized in the colloidal solution, they are subjected for freeze-drying followed by purification for the characterization process. Characterization of nanoparticles are carried out by different instruments working on different principles such as UV–visible spectrophotometer, X-ray diffractometer, dynamic light scattering, scanning and transmission electron microscopes, energy-dispersive spectroscopy, and Fourier transform infrared spectroscopy. These instruments are used to analyze the particles based on their size, shape, surface plasmon resonance, the charge on them, side groups present on them, and metallic characteristics (Kulkarni and Muddapur, 2014).

## 4.6 Factors affecting nanoparticle synthesis by microorganisms

Indeed, biosynthesis and size of nanoparticles are fundamental and important parameters that are critically affected by several factors like pH, temperature, redox conditions, irradiation (Mokhtari et al., 2009), incubation time (Ogi et al., 2010; Sinha and Khare, 2011), and effect of reaction time (Malik et al. 2014). Some dominant factors affecting nanoparticle biosynthesis are described in the following sections.

### *4.6.1 pH*

pH is one of the most critical factors that affects the potential of reaction medium and strongly influences the synthesis of nanoparticles. The size of gold nanoparticles formed by *Escherichia coli* and *Desulfovibrio desulfuricans* was smaller at acidic pH as compared to neutral or alkaline conditions (Deplanche and Macaskie, 2008). Monodispersed and spherical mercury nanoparticles were formed by *Enterobacter* at pH 7, while particles of irregular shape and size were formed at pH 6 and higher numbers of extremely smaller particles were found at higher pH (Sinha and Khare, 2011). In *E. coli*, the highest number of Ag nanoparticles, with the fastest rate of formation, was found at pH 10 (Gurunathan et al., 2009). Similar results were observed by Phanjom and Ahmed (2017) where pH range (4–10) favored the production AgNP using cell filtrate of *Aspergillus oryzae* (MTCC 1846), whereas no synthesis was observed in pH range 4–5, indicating that alkaline condition is necessary for mycosynthesis of nanoparticles.

### *4.6.2 Temperature*

Temperature is another important factor affecting the reaction rate and crystalline nature of nanoparticles. High temperature prefers the synthesis of nanoparticle either from bacterial or fungal resource. For example, increasing the temperature up to 60°C favors the optimal growth of *E. coli* that not only increased the rate of Ag nanoparticle synthesis but also led to the synthesis of a number of smaller size of the particles (Gurunathan et al., 2009). A similar case was seen with fungi (*Aspergillus, Fusarium*, etc.) derived nanoparticles in which increasing temperature (30°C–90°C) fastened the rate of reaction and thus the production of nanoparticles. In addition, most of the studies have shown that high temperature not only reduces the size but also tends to produce nanoparticles of uniform size (Phanjom and Ahmed, 2017). Further studies have also revealed that microorganisms growing at maximum possible temperature may secrete the enzymes responsible for controlling the size of nanoparticles (Iravani, 2014).

### 4.6.3 Redox condition

Most of the silver nanoparticles discussed so far from various studies have used aerobic synthesis (Bai et al., 2011; Kannan et al., 2011; Prakash et al., 2010). This is also the case for Pd nanoparticles, which are also synthesized aerobically (Bunge et al., 2010; Hennebel et al., 2011). However, other nanoparticles such as platinum (Konishi et al., 2007), technetium (Marshall et al., 2008), and tellurium (Baesman et al., 2007) have been synthesized anaerobically except for microaerobic synthesis of tellurium by *Magnetospirillum magneticum* AMB-1 (Tanaka et al., 2010). Interestingly, gold nanoparticles are synthesized both aerobically (Ahmad et al., 2003) and anaerobically (Deplanche and Macaskie, 2008; Konishi et al., 2007).

### 4.6.4 Irradiation

Irradiation (gamma radiation or visible slight)-induced synthesis of nanoparticles is one of the most commonly used strategies to produce biogenic nanoparticles of considerable size. Mokhtari et al. (2009) reported that the cell-free extract of *Klebsiella pneumonia* generated smaller-sized nanoparticles when premixed (300 rpm for 5 min) in the dark followed by irradiation with an effective dose of visible light (1000 mmol/m$^2$/s).

### 4.6.5 Incubation time

In most cases, the size of the nanoparticles increases with the length of reaction/incubation time as reported for *Rhodopseudomonas palustris* and *Rhodobacter sphaeroides* during the formation of ZnS and CdS nanoparticles (Bai et al., 2009). The increase in the size was associated with the nucleation effects where small particles clustered to form large multimers (Holmes et al., 1997). It was found that on elevating the reaction time, less spherical particles were formed in *B. megaterium* (Wen et al., 2009). A similar observation of large particles endowed by the extended incubation time was reported in the cell-free extract of *Shewanella* (Ogi et al., 2010).

## 4.7 Effect of reaction time

Earlier reports have revealed that microbial growth phases play a pivotal role in determining the synthesis of a maximum number of nanoparticles with higher stability. According to Sweeney et al. (2004), *E. coli* synthesized 20-fold more Cd nanoparticles during the stationary rather than the late exponential phase. Silver (Ag) nanoparticles synthesized by *Bacillus licheniformis* were more generous, safe, and secure during the stationary phase. A similar pattern of growth condition in nanoparticle synthesis was reported in *E. coli*, revealing the importance of growth phase in nanoparticle synthesis (Gurunathan et al., 2009).

## 4.8 Location of synthesis of nanoparticles

In addition to the above-mentioned factors, the locations of produced nanoparticles play a vital role in determining their size. Generally, it has been shown that nanoparticles synthesized at intracellular locations (cell membrane) are smaller in size as compared to those generated on extracellular locations (cell wall) and thus are more specific as far as their application requirements are concerned.

## 4.9 Application in sustainable agriculture

Nanotechnology has received considerable interest in the agricultural sector because it can enhance agricultural productivity without deteriorating the environment together with low input of cost and energy. Due to the size specificity of nanoparticles they have been used in vast applications in the form of nanofertilizers, nanopesticides, nanoherbicides, nanosensors, and smart delivery systems for controlled release of agrochemicals (Grillo et al., 2016; Oliveira et al., 2014). Furthermore, new agriculture research is focused on extending the application of nanotechnology in important crop improvement disciplines such as plant breeding and genetic engineering purposes by developing nanotechnology-based devices, etc. (Jiang et al., 2013). Evidence has shown the application of nanoparticles in the agriculture sector, particularly in the key areas of crop productivity and disease management (Mishra et al., 2014; Mishra and Singh, 2016). Other promising nanotechnology-based applications that could be beneficial in the agricultural sector are nanoencapsulation of seeds, pesticides, nanoparticle-mediated delivery of genetic material for crop improvement; carbon nanotube-assisted seed germination of rain-fed crops; nanofertilizer for enhanced crop nutrition and crop productivity; nanopesticides and nanoherbicides for weed elimination; and nanosensors for detection and forecasting of pathogens and soil monitoring (Campos et al., 2015a,b; Liu and Lal, 2015).

In plant pathology, nanoparticles, particularly the silver nanoparticles, have been widely noted for their antimicrobial potential against a myriad of plant diseases caused by phytopathogens. In management of plant diseases, it has been observed that silver nanoparticles are quite effective in controlling fungal and bacterial pathogens. In vitro assessment showed that silver nanoparticles suppressed the colony growth of tested pathogens, but *Magnaporthe grisea* growth was significantly affected by silver nanoparticle application. When tested in vivo with perennial ryegrass (*Lolium perenne*), silver ions, and nanoparticles brought significant reduction in disease severity when applied for 3 hours before the entry of pathogen (Jo and Kim, 2009). A similar effective application of biogenic nanoparticles synthesized from *Serratia* sp. BHU-S4 was shown to be effective against *Bipolaris sorokiniana* causing spot blotch disease in wheat (Mishra et al., 2014).

## 4.10 Applications in environmental management

Population, industrialization, and exposure of various toxic materials such as polyaromatic hydrocarbons, heavy metals, etc., either due to natural or anthropogenic activities, have deteriorated the environment. Hence, the demand for clean technology has become a global challenge. The expansion of nanosciences has been found to be an effective and safe technology for controlling the toxicity of xenobiotics or unwanted materials in the environment. For example, CNTs have been extensively used in water purification as they possess antimicrobial activity with no side effects; even chemical oxidation takes place (Gehrke et al., 2015). Silver nanoparticles implanted in zeolites are used in sanitary purposes as they inhibit the growth of harmful microbes multiplying within water (Dutta, 2011). Other nanoparticles like $TiO_2$ are broadly being used as decontaminating and disinfecting agents; they provide a lifetime coating and remain stable throughout the process of purification (Savage and Diallo, 2005).

Rapid industrialization and some anthropogenic activities have contaminated our natural resources with undesirable organic and inorganic materials. Large quantities of these substances in water have given rise to risks to human health that need to be addressed with ecofriendly technology. The organochlorine pesticide lindane and several heavy metals have been found in groundwater and other water bodies. The iron sulfide nanoparticles stabilized by fungus *Itajahia* sp.-derived biomolecules have been shown to facilitate the degradation of chlorinated compounds (Paknikar et al., 2005) in order to purify drinking water. The microbial cells isolated from lindane-enriched sites were capable of mineralizing the lindane (94%) within 8 h of incubation in a medium supplemented with iron and sulfur. Iron sulfide-based nanomaterials using the sulfate-reducing bacteria *Desulfovibrio vulgaris* can be successfully applied to mitigate the pollution problem caused by heavy metals (Watson et al., 1999). The promising potential is of industrial significance to treat waste water with exceptionally large concentrations of heavy metals/metalloids. Similarly, Das et al. (2009) has suggested the application of gold nanoparticles synthesized in vitro by *Rhizopus oryzae* to manage the contamination caused by a wide array of chlorinated pesticides. Interestingly, the good antimicrobial property associated with fabricated nanoparticles could also be applied to decontamination of pathogenic microbes from water. Recently Qu et al. (2018) reported on the efficacy of spherical shape gold nanoparticles synthesized by *Magnusiomyces ingens* LH-F1 for the reduction of hazardous nitrophenols. The catalytic reduction by culture-free supernatant in the presence of sodium tetra hydrobrate was noted to be dose dependent. Some metal oxidizers such as iron-oxidizing bacteria producing iron nanoparticles are also of immense importance (Singh et al., 2018). Biologically produced nanoparticles can be applied for remediation of arsenic, iron, manganese, copper, chromium, lead, zinc, and cadmium. Similarly, Matsushita et al. (2018) evaluated the application of manganese oxidizing bacteria in a reactor system to demonstrate the removal of cobalt, manganese, and nickel from contaminated systems. The optimal conditions can be utilized to treat the metal laden waste water. Bedi et al. (2018) recommended the application of *Aspergillus aculeatus* for the remediation of abandoned waste generated from mining activities. The fungus showed remarkable efficiency in extracellular synthesis of iron nanoparticles coated with protein under natural environmental conditions. In vitro studies with iron oxide nanomaterials resulted in enhanced growth of test plants.

Apart from the applications mentioned above, biologically fabricated nanoparticles can be employed to treat industrial dye effluents, to develop biosensors to detect the very low content of hazardous contaminants, and to degrade waste cooking oil. However, for field-scale applications more research is needed.

## 4.11 Green versus physicochemical synthesis of nanoparticles

Green syntheses involve the synthesis of nanoparticles through biological means (i.e., plants, bacteria, fungi, actinomycetes, and yeast). Although large quantities of nanoparticles with defined sizes and shapes can be produced by traditional methods (physical and chemical methods) in a very short time these are difficult, outdated, costly, and environmentally unsafe. However, biological tools-based processes are safe, free from hazardous substances, and ecologically sound for nanoparticle fabrication. This has given birth to the idea of green nanobiotechnology. Besides being ecofriendly and biocompatible, biological route is a better way to achieve monodispersity and well-defined dimensions of the particles. Microbial proteins or enzymes lead to the production of superior quality nanoparticles (Sintubin et al., 2009). These attractive features can help us create characteristic nanomaterials for specific applications.

## 4.12 Future perspectives

Different types of nanoparticles such as Au, Ag, Pd, and Cd, have emerged with rapid, stable techniques that use a wide range of nontoxic biomolecules low in cost for wide applications in the agriculture and environment sectors. While the major hurdle with this biological synthesis approach is the polydispersity of synthesized nanoparticles, by optimizing synthesizing conditions such as pH, temperature, and salt concentration, the shape, size, and dispersity of nanoparticles can be largely controlled.

Another major concern is the exact mechanism of synthesis and action of nanoparticles synthesized from fungal resources, which still needs to be explored. Using an extract of pathogenic fungal strain as a startup material for nanoparticle synthesis may risk the spread of disease. Therefore, risk analyses must be done before large-scale scale application of nanoparticles. In addition, the mycosynthesis of silver and gold nanoparticles needs to be studied in depth, since researchers still do not know how to control nanoparticle formation, shape, size, and size distribution.

Future research must focus on field applications rather than in laboratory conditions. Studying nanoparticle synthesis and granting few applications in laboratory conditions could not contribute to the complete acceptance of nanotechnology in agricultural sector as well as in human health. Therefore, scientific community should endeavor to make this approach more realistic and reliable.

## 4.13 Conclusions

Nanotechnology is an emerging field and nanosized particles have widely attracted researchers worldwide for industrial and agricultural applications. Therefore, much effort is being made toward exploiting natural resources and implementing biological synthesis methods that are ecofriendly, cost effective, and more efficient than physical and chemical methods. To the best of our knowledge, there is no specific and comparative studies between chemical and biological synthesis of magnetic and metallic nanoparticles. Therefore, these investigations are needed to enhance the reliability of results. There is also a need to develop methodology for biological synthesis of hybrid magnetic and metallic structure, which may enhance the diagnosis and therapeutic application of nanomaterials. However, the slow rate of biosynthesis and control of particle size and morphology of nanoparticles are still future challenges, in addition to research on particle size and monodispersity.

Another need is understanding the potential toxicities of these nanoparticles in biological systems. Since the use of nanoparticles is growing, concern about their toxicity is increasing for the environment as nanoparticles can spread in water bodies, air, and even in soil. Current nanotoxicological studies are insufficient to fully understand the extent of the hazardous effects caused by their exposure. Therefore, in order to make use of nanoparticles safely, there is a need to develop not only efficient systems to determine the hazardous events induced by nanoparticles both in vitro and in vivo but also strict regulatory mechanisms for their applications. Data sheets of all nanomaterials could inform consumers about the possible hazards associated with nanoparticle use.

## References

Ahmad, A., Satyajyoti, S., Khan, M.I., et al., 2003. Extracellular biosynthesis of monodisperse gold nanoparticles by a novel extremophilic actinomycete, *Thermomonospora* sp. Langmuir 19, 3550–3553.

Alghuthaymi, M.A., Almoammar, H., Rai, M., Said-Galiev, E., Abd-Elsalam, K.A., 2015. Myconanoparticles: synthesis and their role in phytopathogens management. Biotechnol. Biotechnol. Equip. 29, 221–236.

Baesman, S.M., Bullen, T.D., Dewald, J., et al., 2007. Formation of tellurium nanocrystals during anaerobic growth of bacteria that use Te oxyanions as respiratory electron acceptors. Appl. Environ. Microbiol. 73, 2135–2143.

Bai, H.J., Zhang, Z.M., Guo, Y., Jia, W., 2009. Biological synthesis of size-controlled cadmium sulfide nanoparticles using immobilized *Rhodobacter sphaeroides*. Nanosc. Res. Lett. 4, 717–723.

Bai, H.J., Yang, B.S., Chai, C.J., et al., 2011. Green synthesis of silver nanoparticles using *Rhodobacter sphaeroides*. World J. Microbiol. Biotechnol. 27, 2723–2728.

Banerjee, K., Rai, V.R., 2018. A review on mycosynthesis, mechanism, and characterization of silver and gold nanoparticles. BioNanoSci. 8, 17–31.

Bao, C., Jin, M., Lu, R., Zhang, T., Zhao, Y.Y., 2003. Preparation of Au nanoparticles in the presence of low generational poly (amidoamine) dendrimer with surface hydroxyl groups. Mater. Chem. Phys. 81, 160–165.

Bedi, A., Singh, B.R., Deshmukh, S.K., Adholeya, A., Barrow, C.J., 2018. An *Aspergillus aculateus* strain was capable of producing agriculturally useful nanoparticles via bioremediation of iron ore tailings. J. Environ. Manage. 215, 100–107.

Bulovic V., Mandell A., Perlman A. (2004). Molecular memory device. US 20050116256 A1.

Bunge, M., Sobjerg, L.S., Rotaru, A.E., et al., 2010. Formation of palladium(0) nanoparticles at microbial surfaces. Biotechnol. Bioeng. 107, 206–215.

Campos, E.V., Oliveira, J.L., Fraceto, L.F., Singh, B., 2015a. Polysaccharides as safer release systems for agrochemicals. Agron. Sustain. Dev. 35, 47–66. Available from: https://doi.org/10.1016/j.carbpol.2013.10.025.

Campos, E.V., Oliveira, J.L., Goncalves, C.M., Pascoli, M., Pasquoto, T., Lima, R., et al., 2015b. Polymeric and solid lipid nanoparticles for sustained release of carbendazim and tebuconazole in agricultural applications. Sci. Rep. 5, 13809.

Castro-Longoria, E., Vilchis-Nestor, A.R., Avalos-Borja, M., 2011. Biosynthesis of silver, gold and bimetallic nanoparticles using the filamentous fungus *Neurospora crassa*. Colloids Surf. B. Biointerf. 83, 42.

Castro-Longoria, E., Moreno-Velásquez, S.D., Vilchis-Nestor, A.R., Arenas-Berumen, E., Avalos-Borja, M., 2012. Production of platinum nanoparticles and nano aggregates using *Neurospora crassa*. J. Microbiol. Biotechnol. 22, 1000.

Chaturvedi, S., Kumar, A., 2018. Production of titanium dioxide nanoparticle using Baker's yeast. CSVTU Int. J. Biotechnol. Bioinf. Biomed. 3 (1), 01–05.

Das, S.K., Das, A.R., Guha, A.K., 2009. Gold nanoparticles: microbial synthesis and pplication in water hygiene management. Langmuir 25, 8192–8199.

Das, K.R., Kerkar, S., 2017. Biosynthesis of iron nanoparticles by sulphate reducing bacteria and its application in remediating chromium from water. J. Pharm. Biol. Sci. 8, 538–546.

Deplanche, K., Macaskie, L.E., 2008. Biorecovery of gold by *Escherichia coli* and *Desulfovibrio desulfuricans*. Biotechnol. Bioeng. 99, 1055–1964.

Dutta, P., 2011. Silver nanoparticles embedded in zeolite membranes: release of silver ions and mechanism of antibacterial action.. Int. J. Nanomed. 15, 1833.

Elahian, F., Reiisi, S., Shahidi, A., Mirzaei, S.A., 2017. High-throughput bioaccumulation, biotransformation, and production of silver and selenium nanoparticles using genetically engineered *Pichia pastoris*. Nanomed. Nanotechnol. 13, 853–861.

El Domany, E.B., Essam, T.M., Ahmed, A.E., Farghali, A.A., 2018. Biosynthesis physicochemical optimization of gold nanoparticles as anti-cancer and synergetic antimicrobial activity using *Pleurotus ostreatus* fungus. J. Appl. Pharm. Sci. 8, 119–128.

Fatemi, M., Mollania, N., Momeni-Moghaddam, M., Sadeghifar, F., 2018. Extracellular biosynthesis of magnetic iron oxide nanoparticles by *Bacillus cereus* strain HMH1: characterization and in vitro cytotoxicity analysis on MCF-7 and 3T3 cell lines. J. Biotechnol. 270, 1–11.

Gade, A.K., Bonde, P., Ingle, A.P., Marcato, P.D., Durán, N., Rai, M.K., 2008. Exploitation of *Aspergillus niger* for synthesis of silver nanoparticles. J. Biobased Mater. Bioenergy 2, 243.

Gan, L., Zhang, S., Zhang, Y., He, S., Tian, Y., 2018. Biosynthesis, characterization and antimicrobial activity of silver nanoparticles by a halotolerant *Bacillus endophyticus* SCU-L. Prep. Biochem. Biotechnol. 5, 1–7.

Gehrke, I., Geiser, A., Somborn-Schulz, A., 2015. Innovations in nanotechnology for water treatment. Nanotechnol. Sci. Appl. 14, 1.

Ghandehari, F., Fani, M., Rezaee, M., 2018. Biosynthesis of iron oxide nanoparticles by cytoplasmic extract of bacteria *Lactobacillus fermentum*. J. Med. Chem. Sci. 1, 28–30.

Ghorbani, H., 2016. ChemInform abstract: a review of methods for synthesis of Al nanoparticles.. ChemInform 47, 23–28.

Golinska, P., Wypij, M., Ingle, A.P., Gupta, I., Dahm, H., Rai, M., 2014. Biogenic synthesis of metal nanoparticles from actinomycetes: biomedical applications and cytotoxicity. Appl. Microbiol. Biotechnol. 98, 8083–8097.

Gowri, S., Gopinath, K., Arumugam, A., 2018. Experimental and computational assessment of mycosynthesized CdO nanoparticles towards biomedical applications. J. Photochem. Photobiol. B. Biol. 180, 166–174.

Grillo, R., Abhilash, P.C., Fraceto, L.F., 2016. Nanotechnology applied to bio-encapsulation of pesticides. J. Nanosci. Nanotechnol. 16, 1231–1234.

Gurunathan, S., et al., 2009. Biosynthesis, purification and characterization of silver nanoparticles using *Escherichia coli*. Colloids Surf. B. Biointerf. 74, 328–335.

Hasan, M., Ullah, I., Zulfiqar, H., Naeem, K., Iqbal, A., Gul, H., et al., 2018. Biological entities as chemical reactors for synthesis of nanomaterials: progress, challenges and future perspective. Mater. Today Chem. 8, 13–28.

He, S., Guo, Z., Zhang, Y., Zhang, S., Wang, J., Gu, N., 2007. Biosynthesis of gold nanoparticles using the bacteria *Rhodopseudomonas capsulata*. Mater. Lett. 61 (18), 3984–3987.

Hennebel, T., Van Nevel, S., Verschuere, S., et al., 2011. Palladium nanoparticles produced by fermentatively cultivated bacteria as catalyst for diatrizoate removal with biogenic hydrogen. Appl. Microbiol. Biotechnol. 91, 1435–1445.

Holmes, J.D., Richardson, D.J., Saed, S., et al., 1997. Cadmium-specific formation of metal sulfide "Q-particles" by *Klebsiella pneumoniae*. Microbiology 143, 2521–2530.

Hulkoti, N.I., Taranath, T.C., 2014. Biosynthesis of nanoparticles using microbes—a review. Colloids Surf. B. Biointerf. 121, 474–483.

Iravani, A., Akbari, M., Zohoori, M., 2017. Advantages and disadvantages of green technology; goals, challenges and strengths. Int. J. Sci. Eng. Appl. 6, 272–284.

Iravani, S., 2014. Bacteria in nanoparticle synthesis: current status and future prospects. Int. Sch. Res. Notices 1–18. Article ID 359316.

Jiang, S., Eltoukhy, A.A., Love, K.T., Langer, R., Anderson, D.G., 2013. Lipidoid-coated iron oxide nanoparticles for efficient DNA and siRNA delivery. Nano Lett. 13, 1059–1064.

Jo, Y.-K., Kim, B.H., 2009. Antifungal activity of silver ions and nanoparticles on phytopathogenic fungi. Plant Dis. 93, 1037–1043.

Kalpana, V.N., Kataru, B.A.S., Sravani, N., Vigneshwari, T., Panneerselvam, A., Rajeswari, V.D., 2018. Biosynthesis of zinc oxide nanoparticles using culture filtrates of *Aspergillus niger*: antimicrobial textiles and dye degradation studies. Open Nano 3, 48–55.

Kannan, N., Mukunthan, K.S., Balaji, S., 2011. A comparative study of morphology, reactivity and stability of synthesized silver nanoparticles using *Bacillus subtilis* and *Catharanthus roseus* (L.) G. Don. Colloids Surf. B. Biointerf. 86, 378–383.

Khalil, S., Zamir, R., Ahmad, N., 2014. Selection of suitable propagation method for consistent plantlets production in *Stevia rebaudiana* (Bertoni). Saudi J. Biol. Sci. 21, 566–573.

Khan, F., 2013. Chemical hazards of nanoparticles to human and environment (a review). Orient. J. Chem. 29, 1399–1408.

Kimber, R.L., Lewis, E.A., Parmeggiani, F., Smith, K., Bagshaw, H., Starborg, T., et al., 2018. Biosynthesis and characterization of copper nanoparticles using *Shewanella oneidensis*: application for click chemistry. Small 14, 1703145.

Kitching, M., Ramani, M., Marsili, E., 2015. Fungal biosynthesis of gold nanoparticles: mechanism and scale up. Microbial Biotechnol. 8, 904–917.

Kobashigawa, J.M., Robles, C.A., Ricci, M.L.M., Carmarán, C.C., 2018. Influence of strong bases on the synthesis of silver nanoparticles (AgNPs) using the ligninolytic fungi *Trametes trogii*. Saudi J. Biol Sci. Available from: https://doi.org/10.1016/j.sjbs.2018.09.006.

Konishi, Y., Ohno, K., Saitoh, N., et al., 2007. Bioreductive deposition of platinum nanoparticles on the bacterium Shewanella algae. J. Biotechnol. 128, 648–653.

Kulkarni, N., Muddapur, U., 2014. Biosynthesis of metal nanoparticles: a review. J. Nanotechnol. 1–8. Article ID 510246.

Li, Q., Gadd, G.M., 2017. Biosynthesis of copper carbonate nanoparticles by ureolytic fungi. Appl. Microbiol. Biotechnol. 101, 7397–7407.

Liu, R., Lal, R., 2015. Potentials of engineered nanoparticles as fertilizers for increasing agronomic productions. Sci. Total Environ. 514, 131–139.

Li, X., Xu, H., Chen, Z.S., Chen, G., 2011. Biosynthesis of nanoparticles by microorganisms and their applications. J. Nanomater. Available from: https://doi.org/10.1155/2011/270974.

Li, J., Tian, B., Li, T., Dai, S., Weng, Y., Lu, J., et al., 2018. Biosynthesis of au, ag and au–ag bimetallic nanoparticles using protein extracts of *Deinococcus radiodurans* and evaluation of their cytotoxicity. Int. J. Nanomed. 13, 411.

Lv, Q., Zhang, B., Xing, X., Zhao, Y., Cai, R., Wang, W., et al., 2018. Biosynthesis of copper nanoparticles using *Shewanella loihica* PV-4 with antibacterial activity: novel approach and mechanisms investigation. J. Hazard. Mater. 347, 141–149.

Malik, P., Shanker, R., Malik, V., Sharma, N., Mukherjee, T.K., 2014. Green chemistry based benign routes for nanoparticle synthesis. J. Nanopart 1–14. ID 302429.

Marshall, M.J., Plymale, A.E., Kennedy, D.W., et al., 2008. Hydrogenase-and outer membrane c-type cytochrome-facilitated reduction of technetium (VII) by *Shewanella oneidensis* MR-1. Environ. Microbiol. 10, 125–136.

Matsushita, S., Komizo, D., Cao, L.T.T., Aoi, Y., Kindaichi, T., Ozaki, N., et al., 2018. Production of biogenic manganese oxides coupled with methane oxidation in a bioreactor for removing metals from wastewater. Water Res. 130, 224–233.

Mazumdar, H., Haloi, N., 2017. A study on biosynthesis of iron nanoparticles by *Pleurotus* sp. J. Microbiol. Biotechnol. Res. 1, 39–49.

Mesbahi-Nowrouzi, M., Mollania, N., 2018. Purification of selenate reductase from *Alcaligenes* sp. CKCr-6A with the ability to biosynthesis of selenium nanoparticle: enzymatic behavior study in imidazolium based ionic liquids and organic solvent. J. Mol. Liq. 249, 1254–1262.

Mishra, S., Singh, H.B., 2016. Preparation of biomediated metal nanoparticles. Indian Patent Filed 201611003248.

Mishra, S., Singh, A., Keswani, C., Singh, H.B., 2014. Nanotechnology: exploring potential application in agriculture and its opportunities and constraints. Biotech. Today 4, 9–14.

Moghaddam, A.B., Moniri, M., Azizi, S., Rahim, R.A., Ariff, A.B., Saad, W.Z., et al., 2017. Biosynthesis of ZnO nanoparticles by a new *Pichia kudriavzevii* yeast strain and evaluation of their antimicrobial and antioxidant activities. Molecules 22, 872.

Mohanasrinivasan, V., Devi, C.S., Mehra, A., Prakash, S., Agarwal, A., Selvarajan, E., et al., 2018. Biosynthesis of Mgo nanoparticles using lactobacillus sp. and its activity against human leukemia cell lines hl-60. BioNanoScience 8 (1), 249–253.

Mokhtari, N., Daneshpajouh, S., Seyedbagheri, S., et al., 2009. Biological synthesis of very small silver nanoparticles by culture supernatant of *Klebsiella pneumonia*: the effects of visible-light irradiation and the liquid mixing process. Mater. Res. Bull. 44, 1415–1421.

Molnár, Z., Bódai, V., Szakacs, G., Erdélyi, B., Fogarassy, Z., Sáfrán, G., et al., 2018. Green synthesis of gold nanoparticles by thermophilic filamentous fungi. Sci. Rep. 8, 3943.

Moon, J.W., Roh, Y., Lauf, R.J., Vali, H., Yeary, L.W., Phelps, T.J., 2007. Microbial preparation of metal-substituted magnetite nanoparticles. J. Microbiol. Methods 70, 150–158.

Narayanan, K.B., Sakthivel, N., 2010. Biological synthesis of metal nanoparticles by microbes. Adv. Colloid Interf. Sci. 156, 1–13.

Nath, D., Banerjee, P., 2013. Green nanotechnology—a new hope for medical biology. Environ. Toxicol. Pharmacol. 36, 997–1014.

Ogi, T., Saitoh, N., Nomura, T., Konishi, Y., 2010. Room-temperature synthesis of gold nanoparticles and nanoplates using *Shewanella* algae cell extract. J. Nanopart. Res. 12, 2531–2539.

Oliveira, J.L., Campos, E.V., Bakshi, M., Abhilash, P.C., Fraceto, L.F., 2014. Application of nanotechnology for the encapsulation of botanical insecticides for sustainable agriculture: prospects and promises. Biotechnol. Adv. 32, 1550–1561.

Paknikar, K.M., Nagpal, V., Pethkar, A.V., Rajwade, J.M., 2005. Degradation of lindane from aqueous solutions using iron sulfide nanoparticles stabilized by biopolymers. Sci. Technol. Adv. Mater. 6, 370–374.

Perez-Gonzalez, T., Jimenez-Lopez, C., Neal, A.L., Rull-Perez, F., Rodriguez-Navarro, A., Fernandez-Vivas, A., et al., 2010. Magnetite biomineralization induced by *Shewanella oneidensis*. Geochim. Cosmochim. Acta 74, 967–979.

Phanjom, P., Ahmed, G., 2017. Effect of different physicochemical conditions on the synthesis of silver nanoparticles using fungal cell filtrate of *Aspergillus oryzae* (MTCC No. 1846) and their antibacterial effect. Adv. Nat. Sci. Nanosci. Nanotechnol. 8, 045016.

Prabhu, S., Poulose, E.K., 2012. Silver nanoparticles: mechanism of antimicrobial action, synthesis, medical applications, and toxicity effects. Int. Nano Lett. 2, 32.

Prakash, A., Sharma, S., Ahmad, N., et al., 2010. Bacteria mediated extracellular synthesis of metallic nanoparticles. Int. Res. J. Biotechnol. 1, 71–79.

Prasad, R., Bhattacharyya, A., Nguyen, Q.D., 2017. Nanotechnology in sustainable agriculture: recent developments, challenges, and perspectives. Front. Microbiol. 8, 1014.

Presentato, A., Piacenza, E., Anikovskiy, M., Cappelletti, M., Zannoni, D., Turner, R.J., 2018. Biosynthesis of selenium-nanoparticles and-nanorods as a product of selenite bioconversion by the aerobic bacterium *Rhodococcus aetherivorans* BCP1. New Biotechnol. 41, 1–8.

Priyanka, U., Akshay Gowda, K.M., Elisha, M.G., Nitish, N., 2017. Biologically synthesized PbS nanoparticles for the detection of arsenic in water. Int. Biodeter. Biodegrad. 119, 78–86.

Qu, Y., You, S., Zhang, X., Pei, X., Shen, W., Li, Z., et al., 2018. Biosynthesis of gold nanoparticles using cell-free extracts of *Magnusiomyces ingens* LH-F1 for nitrophenols reduction. Bioproc. Biosyst. Eng. 41 (3), 359–367.

Quester, K., Avalos-Borja, M., Castro-Longoria, E., 2013. Biosynthesis and microscopic study of metallic nanoparticles. Micron 54–55, 1–27.

Revati, K., Pandey, B.D., 2011. Microbial synthesis of iron-based nanomaterials—a review. Bull. Mater. Sci. 34, 191–198.

Sabri, M., Umer, A., Awan, G., Hassan, M., Hasnain, A., 2016. Selection of suitable biological method for the synthesis of silver nanoparticles.. Nanomater. Nanotechnol. 6, 29.

Sanguiñedo, P., Fratila, R.M., Estevez, M.B., de la Fuente, J.M., Grazú, V., Alborés, S., 2018. Extracellular biosynthesis of silver nanoparticles using fungi and their antibacterial activity. Nano Biomed. Eng. 10, 165–173.

Sarkar, J., Acharya, K., 2017. Alternaria alternata culture filtrate mediated bioreduction of chloroplatinate to platinum nanoparticles. Inorg. NanoMet. Chem. 47, 365–369.

Sastry, M., Ahmad, A., Khan, M.I., Kumar, R., 2003. Biosynthesis of metal nanoparticles using fungi and actinomycetes. Curr. Sci. 85, 162.

Savage, N., Diallo, M., 2005. Nanomaterials and water purification: opportunities and challenges. J. Nanopart. Res. 7, 331–342.

Shakibaie, M., Amiri-Moghadam, P., Ghazanfari, M., Adeli-Sardou, M., Jafari, M., Forootanfar, H., 2018. Cytotoxic and antioxidant activity of the biogenic bismuth nanoparticles produced by *Delftia* sp. SFG. Mater. Res Bull. 104, 155–163.

Shankar, P.D., Shobana, S., Karuppusamy, I., Pugazhendhi, A., Ramkumar, V.S., Arvindnarayan, S., et al., 2016. A review on the biosynthesis of metallic nanoparticles (gold and silver) using bio-components of microalgae: formation mechanism and applications. Enzyme Microb. Technol. 95, 28–44.

Sharma, K.D., 2017. Antibacterial activity of biogenic platinum nanoparticles: an invitro study. Int. J. Curr. Microbiol. Appl. Sci. 6, 801–808.

Shen, W., Qu, Y., Pei, X., Li, S., You, S., Wang, J., et al., 2017. Catalytic reduction of 4-nitrophenol using gold nanoparticles biosynthesized by cell-free extracts of *Aspergillus* sp. WL-Au. J. Hazard. Mater. 321, 299–306.

Silva, L.P.C., Oliveira, J.P., Keijok, W.J., da Silva, A.R., Aguiar, A.R., Guimarães, M.C.C., et al., 2017. Extracellular biosynthesis of silver nanoparticles using the cell-free filtrate of nematophagous fungus *Duddingtonia flagrans*. Int. J. Nanomed. 12, 6373.

Singh, O.V., 2015. Bio-Nanoparticles-Biosynthesis and Sustainable Biotechnological Implications.. Wiley Blackwell, New Jersey.

Singh, T., Jyoti, K., Patnaik, A., Singh, A., Chauhan, R., Chandel, S.S., 2017. Biosynthesis, characterization and antibacterial activity of silver nanoparticles using an endophytic fungal supernatant of *Raphanus sativus*. J. Gen. Eng. Biotechnol 15, 31–39.

Singh, V.K., Singh, A.L., Singh, R., Kumar, A., 2018. Iron oxidizing bacteria: insights on diversity, mechanism of iron oxidation and role in management of metal pollution. Environ. Sustain. 1–11.

Sinha, A., Khare, S.K., 2011. Mercury bioaccumulation and simultaneous nanoparticle synthesis by *Enterobacter* sp. cells. Bioresour. Technol. 102, 4281–4284.

Sintubin, L., De Windt, W., Dick, J., Mast, J., Van der Ha, D., Verstarete, W., et al., 2009. Lactic acid bacteria as reducing and capping agent for the fast and efficient production of silver nanoparticles. Appl. Microbiol. Biotechnol. 84, 741.

Stephen, J.R., Macnaughtont, S.J., 1999. Developments in terrestrial bacterial remediation of metals. Curr. Opin. Biotechnol. 10, 230–233.

Subramaniyan, S.A., Sheet, S., Vinothkannan, M., Yoo, D.J., Lee, Y.S., Belal, S.A., et al., 2018. One-pot facile synthesis of pt nanoparticles using cultural filtrate of microgravity simulated grown *P. chrysogenum* and their activity on bacteria and cancer cells. J. Nanosci. Nanotechnol. 18, 3110–3125.

Sweeney, R.Y., Mao, C., Gao, X., et al., 2004. Bacterial biosynthesis of cadmium sulfide nanocrystals. Chem. Biol. 11, 1553–1559.

Tanaka, M., Arakaki, A., Staniland, S.S., Matsunaga, T., 2010. Simultaneously discrete biomineralization of magnetite and tellurium nanocrystals in magnetotactic bacteria. Appl. Environ. Microbiol. 7616, 5526–5532.

Taran, M., Rad, M., Alavi, M., 2017. Antibacterial activity of copper oxide (cuo) nanoparticles biosynthesized by *Bacillus* sp. fu4: optimization of experiment design. Pharm. Sci. 23, 198–206.

Thakkar, K.N., Mhatre, S.S., Parikh, R.Y., 2010. Biological synthesis of metallic nanoparticles. Nanomed. Nanotechnol. Biol. Med. 6, 257.

Uddandarao, P., Balakrishnan, R.M., 2017. Thermal and optical characterization of biologically synthesized ZnS nanoparticles synthesized from an endophytic fungus *Aspergillus flavus*: a colorimetric probe in metal detection. Spectrochim. Acta A 175, 200–207.

Vetchinkina, E.P., Loshchinina, E.A., Vodolazov, I.R., Kursky, V.F., Dykman, L.A., Nikitina, V.E., 2017. Biosynthesis of nanoparticles of metals and metalloids by basidiomycetes. Preparation of gold nanoparticles by using purified fungal phenol oxidases. Appl. Microbiol. Biotechnol. 101, 1047–1062.

Vijayanandan, A.S., Balakrishnan, R.M., 2018. Biosynthesis of cobalt oxide nanoparticles using endophytic fungus *Aspergillus nidulans*. J. Environ. Manage. 218, 442–450.

Volesky, B., Holan, Z.R., 1999. Biosorption of heavy metals. Biotechnol. Progr. 11, 235.

Wadhwani, S.A., Shedbalkar, U.U., Singh, R., Chopade, B.A., 2018. Biosynthesis of gold and selenium nanoparticles by purified protein from *Acinetobacter* sp. SW 30. Enzyme Microbial. Technol. 111, 81–86.

Wang, W., Zhang, B., Liu, Q., Du, P., Liu, W., He, Z., 2018. Biosynthesis of palladium nanoparticles using *Shewanella loihica* PV-4 for excellent catalytic reduction of chromium (vi). Environ. Sci. Nano 5, 730–739.

Watson, J.H.P., Ellwood, D.C., Soper, A.K., Charnock, J., 1999. Nanosized strongly-magnetic bacterially-produced iron sulfide materials. J. Magnet. Magn. Mater. 203 (1-3), 69–72.

Wen, L., Lin, Z., Gu, P., et al., 2009. Extracellular biosynthesis of monodispersed gold nanoparticles by a SAM capping route. J. Nanopart. Res. 11, 279–288.

Xia, X., Wu, S., Li, N., Wang, D., Zheng, S., Wang, G., 2018. Novel bacterial selenite reductase CsrF responsible for Se (IV) and Cr (VI) reduction that produces nanoparticles in *Alishewanella* sp. WH16-1. J. Hazard. Mater. 342, 499–509.

Zhang, H., Hu, X., 2017. Rapid production of Pd nanoparticle by a marine electrochemically active bacterium *Shewanella* sp. CNZ-1 and its catalytic performance on 4-nitrophenol reduction. RSC Adv. 7, 41182–41189.

Zhang, H., Hu, X., 2018. Biosynthesis of Pd and Au as nanoparticles by a marine bacterium *Bacillus* sp. GP and their enhanced catalytic performance using metal oxides for 4-nitrophenol reduction. Enzyme Microb. Technol. 113, 59–66.

Zhu, Y., Ren, B., Li, H., Lin, Z., Bañuelos, G., Li, L., et al., 2018. Biosynthesis of selenium nanoparticles and effects of selenite, selenate, and selenomethionine on cell growth and morphology in *Rahnella aquatilis* HX2. Appl. Microbiol. Biotechnol. 102, 6191–6205.

# Sustainable agriculture and benefits of organic farming to special emphasis on PGPR

*Pooja Mishra*[1], *Prem Pratap Singh*[2], *Sandeep Kumar Singh*[2] *and Hariom Verma*[3]

[1]Microbial Technology Department, CSIR-Central Institute of Medicinal and Aromatic Plants (CSIR-CIMAP), Lucknow, India, [2]Center of Advanced Study in Botany, Institute of Science, Banaras Hindu University, Varanasi, India, [3]B.R.D. Government Degree College, Sonebhadra, India

## 5.1 Introduction

By the end of the century, about a third more people will inhabit the planet making about one out of every nine people unnourished around the globe (FAO, 2014). For sub-Saharan Africa, stats get tighter as this figure is about one out of four. About 98% of hungry people come from developing countries, with around 526, 227, and 37 million for Asia, Africa, and Latin America, respectively. In general, people of developing countries are hungry. Moreover, changing diets in the developing world let people draw on the natural resources for more dairy and meat products (Godfray et al., 2010; Seufert, 2012). The projected stats in the agriculture estimates that, by 2050, agro-product demand will grow by 1.1% annually due to the growing world's population (Alexandratos and Bruinsma, 2012). It is projected that in 2050 the world's population will be over 9 billion, which presents a major global agricultural challenge when considering sustainability (FAO, 2010). According to the FAO agriculture utilizes 11% of the land around the globe and uses 70% of total freshwater resources.

The Green Revolution, the historic movement that boosted the agricultural industry on a global level, has resulted in natural resource degradation (Altieri, 2009; Rundgren and Parrott, 2006; Bazuin et al., 2011). Hence, the recent sustainable development strategies were growing on the pillar of greater plant diversities within the agricultural systems (IAASTD, 2009; Davies et al., 2009). Agriculture has relied on technological advancements over the last few decades including various genetic improvement programs, etc., but also for preventing diversity loss as well as excessive fossil and agrochemical usage. They are avowed contributors of global warming, and cause various contamination, beneficial biodiversity loss, etc. (Kim and Dale, 2005). Rising concerns about environment sustainability have put attention on changing current cropping habits to improved, efficient, and more

**Role of Plant Growth Promoting Microorganisms in Sustainable Agriculture and Nanotechnology.**
**DOI: https://doi.org/10.1016/B978-0-12-817004-5.00005-1**

sustainable ways of agriculture (Cox and Atkins, 1979; Jackson and Piper, 1989; Vandermeer et al., 1998; Griffon, 2006).

Organic farming (OF) comprises a more advanced crop management system that aims at ecofriendly production of agriculture commodities and relies on strong on-farm nutrient cycling such as biological nitrogen fixation and crop rotations, which enrich the soil fertility with more organic matter. Green consumerism has also resulted in high demands for organic products. People are more knowledgeable of safe food and environmental issues. OF is the common ground between green consumerism and sustainable agriculture (Mäder et al., 2002).

On an international level, principles of drive the Federation of Organic Agricultural Movements (IFOAM), which is aimed at the goal of sustainable agriculture and its full diversity (IFOAM, 2005). The European Union (EU) passed the Regulation 2092/91/EEC two decades ago stating, "organic agro-production and indications referring thereto on agro-products and foodstuffs" to empower OF. The objective of this regulation can be stated as follows: (1) environmental protection through organic management practices and (2) consumer health protection through organic products. The "OF" decree of EU Member States holds the provision of financial support in accordance with the area of organic cultivation (Schwarz et al., 2010).

## 5.2 The genesis of the Green Revolution

Evolution in agricultural methods and techniques made civilization possible and also shaped worldwide socioeconomic modifications. In last few decades the growing population has demanded higher food production, leading to excessive use/consumption of high yielding cultivars, synthetic chemical fertilizers, and pesticides along with modern machinery. Increasing the use of our limited resources was the main concern of the Green Revolution. The Green Revolution is based on biochemical, mechanical, and social principles (Hazell, 2009). Biochemical elements comprise hybrid seed selection, use of chemical fertilizers, herbicides, and pesticides for higher yields and better weed and pest control. Mechanical elements include controlled water supply for better irrigation and accessibility of machines to manage arable land and minimize the labor cost and facilitate marketing. The social aspect of green revolution was land reformation and farm consolidation to encourage small farmers by providing better quality seeds and machinery (Pretty, 2002). Excessive use of susceptible varieties promotes the use extreme use of chemical fertilizers and pesticides, which increase the risk of soil and water as well as affect the soil fertility and water quality (Pretty and Hine, 2001; Pender and Mertz, 2006; Piment et al., 2000). The use of genetically modified seeds diminishes biodiversity of crops because they lack cross-pollination (Conner et al., 2003; Kruft, 2001). Land degradation is one more setback of the Green Revolution (Pretty and Hine, 2001).

## 5.3 Indian Agricultural System

Agriculture is the backbone of the Indian economy and a source of income for more than 70% of the Indian population. According to the FAO world 2010, in production of a number of fresh fruits and vegetables, fibrous crops, staple crops, and oil seeds India ranked first globally and second in wheat and rice production. India has also a remarkable contribution in production of other crops (fruits, roots and tuber crops, pulses, coconut, sugarcane) and cash crops (coffee and cotton) have been also seen (Murugasamy and Veerachamy, 2012). All these practices adversely affect soil fertility and health resulting in exploitation of natural resources that negatively affect human and plant health (Olson, 1972; Ramos and Martínez-Casasnovas, 2006; Savci, 2012). The Indian subcontinent has diverse topography and physiography allowing a variety of diversified agricultural systems such as those discussed in the following.

### 5.3.1 Subsistence and commercial farming

Subsistence farming is the most popular method used by small farmers with no modern equipment or chemical fertilizers. On the other hand, commercial farming is farming for economic profit and relies on heavy equipment, chemical fertilizers, pesticides, and irrigation facilities.

### 5.3.2 Intensive and extensive farming

Intensive farming involves farming in small lands when long stretches of open fields are not available. It is commonly practiced in India. Extensive farming involves vast stretches of land under single crop cultivation and resulting products are highly commercialized.

### 5.3.3 Plantation farming

Growing of a single cash crop for commercial purposes such as tea, coffee, rubber, banana, and spices in India is known as plantation farming.

### 5.3.4 Mixed farming

In this type of farming animals are reared along with crops. This type of farming is economically the most beneficial.

## 5.4 Why organic farming?

The high demand for high yields has led to nutrient loss in agricultural fields due to excessive erosion along with nutrient runoff and loss of organic matter. OF is a solution to these issues (Lossin, 1970; Bhattacharya and Chakraborty, 2005;

Mehta et al., 2012). OF is an ancient. About 10,000 years ago when agriculture system is totally dependent on natural resources, it uses organic inputs as the brief references were found in our ancient literature like Rigveda, Ramayana, Mahabharata, Kautilya Arthasashthra, etc. (Bhattacharya and Chakraborty, 2005).

According to the USDA National Organic Standards Board organic farming is ecologically important because it enhances biodiversity as well as soil-biological activity. OF is mainly based on negligible use of off-farm inputs and aimed at restoring, maintaining, and increasing ecological harmony. To maintain sustainability, health, and safety, natural fertilizers are used. A clean and pollution-free environment and increased soil fertility with chemical-free foods are the result of organic farming. As the system uses on-farm resources, the investment on off-farm inputs are minimized (Bhattacharya and Chakraborty, 2005; Mehta et al., 2012; de Bertoldi et al., 1983; Ramesh et al., 2010).

OF has gained more attention recently according to the Research Institute of Organic Agriculture, the International Federation of Organic Agricultural Movements, and Foundation Ecology and Agriculture (Ramesh et al., 2010). Organic agriculture is preferred in more than 130 countries, which is about 0.65% of the world's total land.

In India, the number of certified farms is increasing rapidly and is now about 0.3% of total farmland. Although plants require a number of macro- and micronutrients for their growth the nutrient level decreases at the time of harvesting. Moreover, the continuous use of chemicals pesticides adversly affect the texture and productivity of soil. Therefore, plant residue and its recycling add essential nutrients to the soil. Some of the ways in which the soil and plant health can be maintained and made available to plants include:

1. Use of plant growth-promoting microbes (PGPM): PGPM strains have the capacity to improve plant growth and yield and control plant pathogens (Kloepper et al., 2004; Leeman et al., 1995; McSpadden Gardener, 2004; Glick et al., 1995). Many mechanisms (production of growth regulators or other plant stimulants) have been proposed for growth promotion by PGPM strains (Olivares et al., 1997; Lucy et al., 2004; Ryu et al., 2003).
2. Limiting the populations of pathogenic or deleterious microorganisms through antibiosis (Handelsman and Stab, 1996; Jetiyanon and Kloepper et al., 2004), salicylic acid (Singh and Basu, 2004) and indole-3 acetic acid production (Mayak et al., 1999), which improve plant health.
3. Calliterpenone from the plant *Callicarpa macrophylla* Vahl (Verbenaceae) is a phyllocladane diterpenoid promoting plant growth (Haider et al., 2009). It works as a precursor of gibberellins (Hanson and White, 1969; Liu et al., 2003; Bottini et al., 2004; Singh et al., 2004).
4. Use of manure, compost, and vermicompost: The fertility of agricultural soil is enhanced by using compost and vermicompost as they stimulate plant growth by increasing the soil nutrient availability to plants.

## 5.5 Composting and vermicomposting

Composting is a microbial successional process which have progression in breakdown of substrates and the resulted product, work as a substrate for next successive population, whereas vermicomposting is a biotechnological process in which certain species of earthworms (*Lumbricuse terestris*, *Lumbricuse rebellious*, and *Eisenia foetida*) (Hortenstein et al., 1979; Collier, 1978) are used to enhance the process of waste degradation.

It is a mesophilic process that comprises bacterial and fungal degradation with the help of earthworms; all of them are active at 10°C–32°C. Initial degradation takes place in the gut of the earthworm so it is a fast process, but the exact procedures of transformation are still unknown. Earthworms are often called as friends of farmers because they convert waste into gold (Adhikary, 2012). The dominant phyla in compost are *actinobacteria* and *firmicutes* whereas in vermicompost uncultured *acidobacteria*, *chloroflexi*, *bacteroidetes*, and *gemmatimonadetes* are dominant. Highly active bacterial as well as fungal populations are found in compost and vermicompost (Ashraf et al., 2007). *Lactobacillus* and *acetobacter* have been reported in various studies (Partanen et al., 2010). At maturity composts, because of high temperature, are rich in Gram-negative bacteria such as *bacillus* and *actinobacteria* (Gopal et al., 2005). In composting, mixed culturing is easily carried out using bacteria and fungi (Partanen et al., 2010).

Use of compost and vermicompost increases the soil carbon ratio (soil organic matter; SOM) about 3%–5%, which maintain a good-quality soil structure and fertility. SOM promote the richness of beneficial microbes and accelerate microbial activities (Sinha et al., 2013). A comparative account by Suhane (2007) showed that availability of nutrients (kg/ha) like nitrogen, phosphorus, potash, and carbonic biomass (mg/kg of soil) in chemical fertilizer and vermicompost applied soil increased from 185 to 256, 28.5 to 50.5, 426.5 to 489.5, and 217.0 to 273.0 (mg/kg of soil), respectively (Hussaini, 2013). Earthworms contribute more than 20–40 kg nitrogen/ha/year and mineral nutrients with growth regulators. They also help in recycling nitrogen (20–200 kg N/ha/year) in short duration, which improves soil fertility and plant growth by 30%–200% (Darwin and van Wyhe, 2002).

Vermicompost amendment helps plant soil sodicity and salinity and encourages multiplication of microbial biocontrol agents in disease suppressive soils and suppress a variety of diseases; *Gaeumannomyces graminis* var. *tritici* on wheat (Elmer, 2009). The population of earthworms *Lumbricus terrestris* when amplified in soils contaminated with soil-borne pathogens greatly reduces disease incidence in susceptible cultivars of asparagus (*Asparagus officinalis*), eggplant (*Solanum melongena*), and tomato (*Solanum lycopersicum*) (Elmer, 2009).

## 5.6 Use of green manure/manure

Green manure, a crop used as a nutrient source for soil, grows on-site where inhibitive handling and transportation costs of other organic inputs are higher. Long-term cultivation increases SOM and microbial biomass as well (Goyal et al., 1992, 1999), in addition to improving the nutrient-holding property of soil and N-uptake efficiency. When used in uncultivated land, reduce erosion (Dapaah and Vyn, 1998), nutrient, or pesticide losses (Delgado et al., 2001; Gaston et al., 2003), and suppress weeds (Phatak et al., 1987; Dyck et al., 1995; Burgos and Talbert, 1996) and specific crop pests (Bugg et al.,1990; Caswell et al., 1991). Green manure can also offer a good habitat for beneficial organisms (Bugg et al., 1991; Nicholls and Altieri, 2001). Some green manure like *Sesbania* can fix nitrogen by both root and shoot nodules therefore they can fix atmospheric nitrogen even when there is sufficient amount of nitrogen present in soil. Whole plants used as fertilizers increase the organic matter of soil as well, which promotes the growth of microbial population.

## 5.7 Role of microbes in organic agriculture

Soil microbial populations play an influential role in the biological management of soil fertility and productivity. They are harnessed and processed in a way to hook the beneficial effects on the soil and structure the soil-biological relation in an ameliorating manner. Today's farmers are interested in using soil and plant microbial inoculants to maintain the microbial equilibrium to enhance soil fertility and promote agrocrop production. Microbes increase plant growth promotion and enhance resistance against phytopathogens, etc., which in turn influence the agro-production efficiency as all are closely linked. Until now, increasing agro-productivity was not correlated with sustainability or ecofriendly behavior, but recent agro-trends have pointed to the efficient use of soil microflora that yield enhanced growth, productivity, and agriculture quality.

The mechanisms involved in enhancing agro-productivity include nitrogen fixation, hormonal homeostasis, siderophore and phytohormone production, phytopathogen resistivity, nutrient availability, promotion of mycorrhizal functioning, and decreasing pollutant toxicity (Glick et al., 1999). The interactive nature of plants and microbes works in either direct or indirect stimulatory processes. The direct stimulatory processes include the phytohormones (auxin, gibberellin, and cytokinin), siderophores, and enzyme production along with elicitation of systemic resistance while indirect stimulation comprises antibiotic and extracellular enzyme production for further survival processes (Zahir et al., 2004). There were many reports which investigates the processes by which microbes can enhance the plant growth. Dey et al. (2004) reports the production of 1-aminocyclopropane-1-carboxylate (ACC) deaminase, which reduces the level of ethylene in the roots of developing plants. Narula et al. (2006), Saleem et al. (2007), Ortíz-Castro et al. (2008), and Mishra et al. (2010) reported the production of phytohormones such as gibberellic acid, ethylene, cytokinin, and indole-acetic acid, respectively. Pathma et al. (2011) reported pathogen resistivity by the production of siderophores, β-1,3-glucanase, chitinases, antibiotics, fluorescent pigment, and cyanide.

## 5.8 Challenges for developing countries and small-scale farmers

IFOAM defines OF on the basis of four basic principles: health, ecology, fairness, and care for humans as well as ecosystems (Jouzi et al., 2017; Rundgren and

Parrott, 2006). In some regions like in East Africa (UNEP-UNCTAD, 2008) evidence shows that OF plays a major role in food security (Azadi and Ho, 2010). On the other hand, in developing countries where small-scale farmers are the majority the expense of the extra labor required for organic agriculture is an issue (Reganold and Wachter, 2016), showing how poverty and food insecurity often go hand in hand (Mwaniki, 2006). Despite the advantages and opportunities, small-scale farmers are still experiencing serious challenges when trying to switch to organic systems. One issue is that the yields of organic farms are around 25% less than conventional farms (Seufert, 2012). Some studies also argue that OF is not a feasible option for small farm holders in many regions like Africa, who cannot produce sufficient amounts of compost and green manure. Since soil management practices are time consuming, soil fertility is depleted. On average, farmers need almost 5 years for the best return of investment (Lotter, 2015).

Finally, certification provides farmers the opportunity to benefit from the price premiums of their products. However, to help small farm holders access organic certification and marketing, IFOAM endorses strategies like Internal Control Systems (ICS via group certification) and Participatory Guarantee Systems (PGS), which are based on social trust and exchanging knowledge. Contract farming can also provide opportunities to participate on the market (Kirsten and Sartorius, 2002). The challenges and potential of are depicted in Fig. 1.

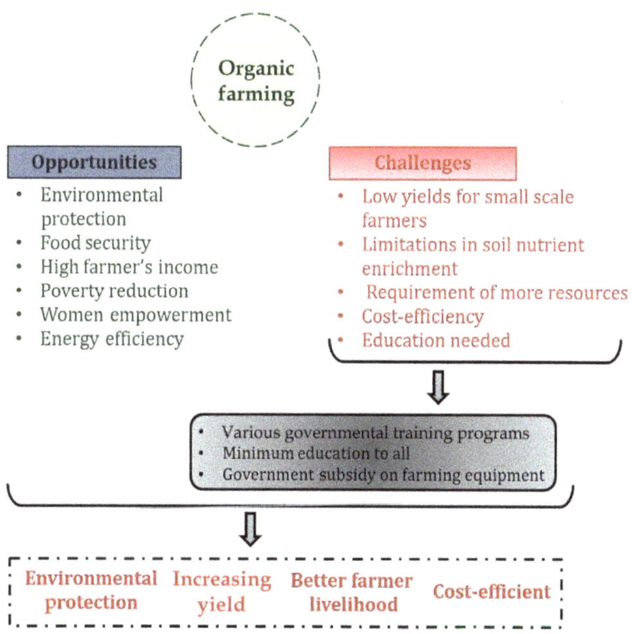


**Figure 1** Systemic framework for analyzing the potential of organic farming.

## 5.9 Conclusions

OF, the advanced crop management system that has gained increased attention in recent years, is in danger of being lost. The role of OF has been considered safety for the environment and the establishment of high-quality foodstuffs. But is it the agenda of OF be attained only by financing it? Where are the food-product standards for consumers? Would replacing conventional methods with organic methods help the chemical runoff from the environment? These questions needed to be answered for OF to move forward in a sustainable way. Policy on OF must also consider environmental practices, consumers' willingness to pay for products, and the social aspects of organic agriculture.

## References

Adhikary, S., 2012. Vermicompost, the story of organic gold: a review. Agric. Sci. 3 (7), 905–917.

Alexandratos, N., Bruinsma, J., 2012. World agriculture towards 2030/2050: the 2012 revision. ESA Working Paper 3.

Altieri, M.A., 2009. Agroecology, small farms, and food sovereignty. Mon. Rev. 61 (3), 102–113.

Ashraf, R., Shahid, F., Ali, T.A., 2007. Association of fungi, bacteria and actinomycetes with different composts. Pak. J. Bot. 39 (6), 2141–2151.

Azadi, H., Ho, P., 2010. Genetically modified and organic crops in developing countries: a review of options for food security. Biotechnol. Adv. 28 (1), 160–168.

Bazuin, S., Azadi, H., Witlox, F., 2011. Application of GM crops in sub-Saharan Africa: lessons learned from Green Revolution. Biotechnol. Adv. 29 (6), 908–912.

Bhattacharya, P., Chakraborty, G., 2005. Current status of organic farming in india and other countries. Indian J. Fertil. 1 (9), 111–123.

Bottini, R., Cassán, F., Piccoli, P., 2004. Gibberellin production by bacteria and its involvement in plant growth promotion and yield increase. Appl. Microbiol. Biotechnol. 65, 497–503.

Bugg, R.L., Wäckers, F.L., Brunson, K.E., Phatak, S.C., Dutcher, J.D., 1990. Tarnished plant bug (hemiptera: Miridae) on selected cool-season leguminous cover crops. J. Entomol. Sci. 25 (3), 463–474.

Bugg, R.L., Wackers, F.L., Brunson, K.E., Dutcher, J.D., Phatak, S.C., 1991. Cool-season cover crops relay intercropped with cantaloupe: Influence on a generalist predator, *Geocoris punctipes*. J. Econ. Entomol. 84, 408–416.

Burgos, N.R., Talbert, R.E., 1996. Weed control and sweet corn (*Zea mays var. rugosa*) response in a no-till system with cover crops. Weed Sci. 355–361.

Caswell, E.P., DeFrank, J., Apt, W.J., Tang, C.S., 1991. Influence of nonhost plants on population decline of *Rotylenchulus reniformis*. J. Nematol. 23 (1), 91.

Collier, J., 1978. Conversion of Municipal Wastewater Treatment Plant Residual Sludges into Earthworm Castings for Use as Topsoil, First Annual Report to National Science Foundation, NSF Grant ENV77-16832.

Conner, A.J., Glare, T.R., Nap, J.P., 2003. The release of genetically modified crops into the environment: part II. Overview of ecological risk assessment. Plant J. 33 (1), 19–46.

Cox, G.W., Atkins, M.D., 1979. Agricultural Ecology: An Analysis of World Food Production Systems. WH Freeman.

Darwin, C.R., van Wyhe, J., 2002. The Complete Work of Charles Darwin Online. University of Cambridge.

Dapaah, H.K., Vyn, Tony, 1998. Nitrogen fertilization and cover crop effects on soil structure stability and corn performance. Communications in Soil Science and Plant Analysis. Commun. Soil. Sci. Plant. Anal. 29, 2557–2569. Available from: https://doi.org/10.1080/00103629809370134.

Davies, B., Baulcombe, D., Crute I., Dunwell, J., Gale, M., Jones, J., et al. 2009. Reaping the benefits: science and the sustainable intensification of global agriculture. Royal Society, London.

de Bertoldi, M., Vallini, G., Pera, A., 1983. The biology of composting: a review. Waste Manage. Res. 1 (12), 157–176.

Delgado, J.A., Riggenbach, R.R., Sparks, R.T., Dillon, M.A., Kawanabe, L.M., Ristau, R.J., 2001. Evaluation of nitrate-nitrogen transport in a potato–barley rotation. Soil Sci. Soc. Am. J. 65 (3), 878–883.

Dey, R.K.K.P., Pal, K.K., Bhatt, D.M., Chauhan, S.M., 2004. Growth promotion and yield enhancement of peanut (*Arachis hypogaea* L.) by application of plant growth-promoting rhizobacteria. Microbiol. Res. 159 (4), 371–394.

Dyck, E., Liebman, M., Erich, M.S., 1995. Crop-weed interference as influenced by a leguminous or synthetic fertilizer nitrogen source: I. Doublecropping experiments with crimson clover, sweet corn, and lambsquarters. Agric. Ecosyst. Environ. 56 (2), 93–108.

Elmer, W.H., 2009. Influence of earthworm activity on soil microbes and soil-borne diseases of vegetables. Plant Dis. 93 (2), 175–179.

FAO, 2010. The state of the food insecurity in the world. Addressing food insecurity in protracted crises. Food and Agriculture Organization of the United Nations. <http://www.fao.org/docrep/013/i1683e/i1683e.pdf.> (accessed 11.07.14.).

FAO, 2014. The State of Food Insecurity in the World. FAO, IFAD and WFP, Rome.

Gaston, L.A., Boquet, D.J., Bosch, M.A., 2003. Fluometuron sorption and degradation in cores of silt loam soil from different tillage and cover crop systems. Soil Sci. Soc. Am. J. 67 (3), 747–755.

Glick, B.R., Karaturovic, D.M., Newell, P.C., 1995. A novel procedure for rapid isolation of plant growth promoting Pseudomonas. Can. J. Microbiol. 41 (6), 533–536.

Glick, B.R., Patten, C.L., Holguin, G., Penrose, D.M., 1999. Biochemical and Genetic Mechanisms Used by Plant Growth-Promoting Bacteria. Imperial College Press, London.

Godfray, H.C.J., Beddington, J.R., Crute, I.R., Haddad, L., Lawrence, D., Muir, J.F., et al., 2010. Food security: the challenge of feeding 9 billion people. Science 327 (5967), 812–818.

Gopal, M., Jha, S.K., Shukla, L., Rawat, R.V., 2005. In vitro detoxification of Pendimethalin by two *Actinomycetes* spp.. Bull. Environ. Contam. Toxicol. 75 (5), 1041–1045.

Goyal, S., Mishra, M.M., Hooda, I.S., Singh, R., 1992. Organic matter-microbial biomass relationships in field experiments under tropical conditions: effects of inorganic fertilization and organic amendments. Soil Biol. Biochem. 24 (11), 1081–1084.

Goyal, S., Chander, K., Mundra, M.C., Kapoor, K.K., 1999. Influence of inorganic fertilizers and organic amendments on soil organic matter and soil microbial properties under tropical conditions. Biol. Fertil. Soils 29 (2), 196–200.

Griffon, M., 2006. Feeding the Planet. Odile Jacob, Paris.

Haider, F., Bagchi, G.D., Singh, A.K., 2009. Effect of calliterpenone on growth, herb yield, and oil quality of *Mentha arvensis*. Int. J. Integr. Biol. 7 (1), 1–53.

Handelsman, J., Stab, E.V., 1996. Bio-control of soil borne plant pathogens. Plant Cell 8 (10), 1855–1869.

Hanson, J.R., White, A.F., 1969. Terpenoid biosynthesis IV. Biosynthesis of the kaurenolids and gibberellic acid. J. Chem. Soc. C 6, 981–985.

Hazell, P.B., 2009. The Asian Green Revolution. International Food Policy Research Institute, 911.

Hortenstein, R., Leaf, A.L., Neuhauser, E.F., Dickelhaupt, D.H., Hortenstein, F., 1979. Physiochemical Changes Accompanying the Conversion of Activated Sludge into Castings by the Earthworm. SUNY College of Environmental Science Forestry, Syracuse, New York.

Hussaini, A., 2013. Vermiculture bio-technology: an effective tool for economic and environmental sustainability. Afr. J. Environ. Sci. Technol. 7 (2), 56–60.

IAASTD (International Assessment of Agricultural Knowledge, Science and Technology for Development), 2009. Agriculture at a crossroads. International assessment of agricultural knowledge, science and technology for development. Global report. Island Press, Washington, DC.

IFOAM, 2005. Basic Standards for Organic Production and Processing. Germany <http://www.ifoam.org/about ifoam/standards/norms>.

Jackson, W., Piper, J., 1989. The necessary marriage between ecology and agriculture. Ecology 70 (6), 1591–1593.

Jouzi, Z., Azadi, H., Taheri, F., Zarafshani, K., Gebrehiwot, K., Van Passel, S., et al., 2017. Organic farming and small-scale farmers: main opportunities and challenges. Ecol. Econ. 132, 144–154.

Kim, S., Dale, B.E., 2005. Life cycle assessment of various cropping systems utilized for producing biofuels: bioethanol and biodiesel. Biomass Bioenergy 29 (6), 426–439.

Kirsten, J., Sartorius, K., 2002. Linking agribusiness and small-scale farmers in developing countries: is there a new role for contract farming?. Dev. South Afr. 19 (4), 503–529.

Kloepper, J.W., Ryu, C.M., Zhang, S., 2004. Induced systemic resistance and promotion of plant growth by *Bacillus* spp. Phytopathology 94 (11), 1259–1266.

Kruft, D., 2001. Impacts of Genetically-Modified Crops and Seeds on Farmers Prepared by Legal Research Assistant.

Leeman, M., Van Pelt, J.A., Den Ouden, F.M., Heinsbroek, M., Bakker, P.A.H.M., Schippers, B., 1995. Induction of systemic resistance against Fusarium wilt of radish by lipopolysaccharides of *Pseudomonas fluorescens*. Phytopathology 85 (9), 1021–1027.

Liu, G., Muller, R., Ruedi, P., 2003. Chemical transformation of phyllocladane (=13β-Kaurane) diterpenoids. Helv. Chim. Acta 86 (2), 420–438.

Lossin, R.D., 1970. Compost studies. Compost. Sci. 11, 16.

Lotter, D., 2015. Facing food insecurity in Africa: why, after 30 years of work in organic agriculture, I am promoting the use of synthetic fertilizers and herbicides in small scale staple crop production. Agric. Hum. Values 32 (1), 111–118.

Lucy, M., Reed, E., Glick, B.R., 2004. Applications of free-living plant growth-promoting rhizobacteria. Rev. Anton. Leeuw. 86, 1–25.

Mäder, P., Fliessbach, A., Dubois, D., Gunst, L., Fried, P., Niggli, U., 2002. Soil fertility and biodiversity in organic farming. Science 296 (5573), 1694–1697.

Mayak, S., Tirosh, T., Glick, B.R., 1999. Effect of wild type and mutant plant growth promoting rhizobacteria on the rooting of mung bean cuttings. J. Plant Growth Regul. 18 (2), 49–53.

McSpadden Gardener, B.B., 2004. Ecology of *Bacillus* and *Paenibacillus* spp. in agricultural systems. Phytopathology 94 (11), 1252–1258.

Mehta, C.M., Gupta, V., Singh, S., Srivastava, R., Sen, E., Romantschuk, M., et al., 2012. Role of microbiologically rich compost in reducing biotic and abiotic stresses. Microorganisms in Environmental Management. Springer, Dordrecht, pp. 113–134.

Mishra, R.K., Prakash, O., Alam, M., Dikshit, A., 2010. Influence of plant growth promoting rhizobacteria (PGPR) on the productivity of *Pelargonium Graveolens* l. Herit. Recent Res. Sci. Technol. 2 (5).

Murugasamy, M., Veerachamy, P., 2012. Resource Use Efficiency in Agriculture—A Critical Survey of the Literature. Language in India.

Mwaniki, A., 2006. Achieving Food Security in Africa: Challenges and Issues. Cornell University, New York.

Narula, N., Deubel, A., Gans, W., Behl, R.K., Merbach, W.P.S.E., 2006. Paranodules and colonization of wheat roots by phytohormone producing bacteria in soil. Plant Soil Environ. 52 (3), 119.

Nicholls, C.I., Altieri, M.A., 2001. Manipulating plant biodiversity to enhance biological control of insect pest. A case study of a northern California vineyard. In: Gliessman, S.R. (Ed.), Agroecosystem sustainability, developing practical strategies. CRC Press, Boca Raton, FL, pp. 29–50.

Olivares, F.L., James, E.K., Baldani, J.I., Dobereiner, J., 1997. Infection of molted stripe disease susceptible and resistant varieties of sugarcane by the endophytic diazotrophs *Herbaspirillum*. New Phytol. 135 (4), 723–737.

Olson, R.A., 1972. Effects of Fertilizer Use on Human health and Environment, Swedish Internation Development Authority, Food and Agricultural Organization of the United Nations.

Ortíz-Castro, R., Martínez-Trujillo, M., López-Bucio, J., 2008. N-acyl-L-homoserine lactones: a class of bacterial quorum-sensing signals alter post-embryonic root development in *Arabidopsis thaliana*. Plant Cell Environ. 31 (10), 1497–1509.

Partanen, P., Hultman, J., Paulin, L., Auvinen, P., Romantschuk, M., 2010. Bacterial diversity at different stages of the composting process. BMC Microbiol. 10 (1), 94.

Pathma, J., Kennedy, R.K., Sakthivel, N., 2011. Mechanisms of fluorescent pseudomonads that mediate biological control of phytopathogens and plant growth promotion of crop plants. Bacteria in Agrobiology: Plant Growth Responses. Springer, Berlin, pp. 77–105.

Pender, J. and Mertz, O., 2006. Soil fertility depletion in Sub-Saharan Africa: what is the role of organic agriculture. Global Development of Organic Agriculture: Challenges and Prospects, pp. 215–240.

Phatak, S.C., Callaway, M.B., Vavrina, C.S., 1987. Biological control and its integration in weed management systems for purple and yellow nutsedge (*Cyperus rotundus* and *C. esculentus*). Weed Technol. 1 (1), 84–91.

Piment, D., Westra, L., Noss, R.F., 2000. Ecological Integrity: Integrating Environment, Conservation and Health. Island Press, Washington DC, p. 428.

Pretty, J., Hine, R., 2001. Reducing food poverty with sustainable agriculture: a summary of new evidence. Final Report from the "SAFE World" Research Project. University of Essex.

Pretty, J.N., 2002. Agri-culture: Reconnecting People. Land and Nature. Earthscan, London.

Ramesh, P., Panwar, N.R., Singh, A.B., Ramana, S., Yadav, S.K., Shrivastava, R., et al., 2010. Status of organic farming in India. Curr. Sci. 98 (9), 1190–1194.

Ramos, M.C., Martínez-Casasnovas, J.A., 2006. Erosion rates and nutrient losses affected by composted cattle manure application in vineyard soils of NE Spain. Catena 68 (2–3), 177–185.

Reganold, J.P., Wachter, J.M., 2016. Organic agriculture in the twenty-first century. Nat. Plants 2 (2), 15221.

Rundgren, G., Parrott, N., 2006. Organic Agriculture and Food Security. IFOAM.

Ryu, C.M., Farag, M.A., Hu, C.H., Reddy, M.S., Wei, H.X., Paré, P.W., et al., 2003. Bacterial volatiles promote growth in Arabidopsis. Proc. Natl. Acad. Sci. 100 (8), 4927–4932.

Saleem, M., Arshad, M., Hussain, S., Bhatti, A.S., 2007. Perspective of plant growth promoting rhizobacteria (PGPR) containing ACC deaminase in stress agriculture. J. Ind. Microbiol. Biotechnol. 34 (10), 635–648.

Savci, S., 2012. An agricultural pollutant: chemical fertilizer. Int. J. Environ. Sci. Dev. 3 (1), 73.

Schwarz, G., Nieberg, H., Sanders, J., 2010. Organic Farming Support Payment in the EU. Agriculture and Forestry Research. Johan Heinrich von Thuennen-Institut. Programmes Support Organic Farming, Braunschweig, Germany.

Seufert, V., 2012. Organic agriculture as an opportunity for sustainable agricultural development. Research to Practice Policy Briefs, Policy Brief, p. 13.

Singh, N.B., Basu, M.S., 2004. Groundnut Research in India: An overview. Groundnut Research in India. National Research Centre for Groundnut, Junagadh, pp. 1–6.

Singh, A.K., Bagchi, G.D., Singh, S., Dwivedi, P.D., Gupta, A.K., Khanuja, S.P.S., 2004. Use of phyllocladane diterpenoids for plant growth promotion and alleviation of growth retardant allelochemicals and method therefore. US Patent 6, 673, 749.

Sinha, R.K., Soni, B.K., Agarwal, S., Shankar, B., Hahn, G., 2013. Vermiculture for organic horticulture: producing chemical-free, nutritivc & health protective foods by earthworms. Agric. Sci. 1 (1), 17–44.

Suhane, R.K., 2007. Vermicompost. Rajendra Agriculture University, Pusa, p. 88.

UNEP-UNCTAD. 2008. Organic Agriculture and Food Security in Africa (UNCTAD/DITC/TED/2007/15), available at http://www.unep-unctad.org/cbtf, Accessed 25 Feb 2009.

Vandermeer, J., van Noordwijk, M., Anderson, J., Ong, C., Perfecto, I., 1998. Global change and multi-species agroecosystems: concepts and issues. Agric. Ecosyst. Environ. 67 (1), 1–22.

Zahir, Z.A., Arshad, M., Frankenberger, W.T., 2004. Plant growth promoting rhizobacteria: applications and perspectives in agriculture. Adv. Agron. 81, 98–169.

## Further reading

Baulcombe, D., Crute, I., Davies, B., Dunwell, J., Gale, M., Jones, J., et al., 2009. Reaping the Benefits: Science and the Sustainable Intensification of Global Agriculture. The Royal Society.

Collier, J.E., Livingstone, D., 1981. Conversion of Municipal Wastewater Treatment Plant Residual Sludges Into Earthworm Castings for Use as Topsoil. U.S. National Technical Information Service, Springfield, p. 40, Report No. PB 81-206310, (08 G COL).

Connor, D.J., 2008. Organic agriculture cannot feed the world. Field Crop Res. 106 (2), 187–190.

German, R.N., Thompson, C.E., Benton, T.G., 2017. Relationships among multiple aspects of agriculture's environmental impact and productivity: a meta-analysis to guide sustainable agriculture. Biol. Rev. 92 (2), 716–738.

Hazell, P., Poulton, C., Wiggins, S., Dorward, A., 2010. The future of small farms: trajectories and policy priorities. World Dev. 38 (10), 1349–1361.

IFOAM, 2011. The role of small-holders in organic agriculture (positionpaper). <http://infohub.ifoam.bio/sites/default/files/page/files/position_paper_smallholders.pdf>.

Jetiyanon, K., Kloepper, J.W., 2002. Mixtures of plant growth-promoting rhizobacteria for induction of systemic resistance against multiple plant diseases. Biol. Contr. 24 (3), 285–291.

McIntyre, B.D., 2009. International assessment of agricultural knowledge, science and technology for development (IAASTD). Global Report.

Pretty, J.N., Morison, J.I., Hine, R.E., 2003. Reducing food poverty by increasing agricultural sustainability in developing countries. Agric. Ecosyst. Environ. 95 (1), 217–234.

# Plant growth-promoting microbes for abiotic stress tolerance in plants

*Rusi Lata and Surendra K. Gond*
Department of Botany, MMV, Banaras Hindu University, Varanasi, India

## 6.1 Introduction

Plant growth-promoting microbes (PGPM) may be either free living or symbiotic, and mostly composed of rhizospheric microbes, endophytes, or mycorrhizal fungi. Rhizosphere inhabits the highest density of microorganisms responsible for root activity and metabolism. This zone includes fungi, protozoa, and algae, with bacteria being the most abundant. Considering their great capability of root colonization, it is likely that bacteria have more control over plant physiology than other classes of microorganisms (Saharan and Nehra, 2011; Antoun and Kloepper, 2001; Barriuso et al., 2008). Bacteria colonizing plant roots have a positive effect ranging from direct influence mechanisms to an indirect effect. Thus bacteria inhabiting the rhizosphere that are beneficial to plants are called plant growth-promoting rhizobacteria (PGPR) (Kloepper et al., 1980). Endophytes are microorganisms that asymptomatically reside in plant tissues (roots, leaves, shoots, and seeds) without causing a disease. They can be bacteria, fungi, or actinomycetes. Endophytes fall into three categories: (1) nonpathogenic, (2) pathogenic in a host but nonpathogenic to another when in endophytic relation, and (3) pathogens that have been rendered nonpathogenic but still capable of colonization by selection methods or genetic alteration (Backman and Sikora 2008; Lata et al., 2018). Endophytes provide plants with resistance against stress conditions, alterations in physiological properties, and production of phytohormones and other compounds of biotechnological interest.

When fungi form a symbiotic association with the roots of higher plants, this is called a mycorrhizal association. They are broadly categorized as arbuscular mycorrhizae (AM) and ectomycorrhizae (ECM), in which AM have very high frequency in agricultural soils (Christie et al., 2004; Khan and Belik, 1995; Liu and Chen, 2007; Willis et al., 2013). Almost 8 out of 10 roots of terrestrial plants have mycorrhizal relationships (Giovannetti et al., 2006). Fungi get their nutritional needs met by penetrating the cortical cells of the root system and forming a special haustorium-like structure for extracting nourishment called the arbuscule, and in return, fungi provide minerals and water by proliferating the hyphal

**Role of Plant Growth Promoting Microorganisms in Sustainable Agriculture and Nanotechnology.**
**DOI: https://doi.org/10.1016/B978-0-12-817004-5.00006-3**

network into the soil (Oueslati, 2003; Bethlenfalvay and Linderman, 1992). Plant growth-promoting activity is also seen by the activity of the mycorrhizal relationship when plants face stress conditions. Beneficial soil microorganisms, endophytes such as bacteria, and fungi/or arbuscular mycorrhizal fungi (AMF) can adapt to specific environmental conditions and develop tolerance to stressful conditions.

## 6.2 Abiotic stresses affecting plant health

Abiotic and biotic factors affect plant growth in agricultural soil. Light, water, carbon, and mineral nutrients are the major factors that regulate plant growth, development, and reproduction (Lata et al., 2018). However, when these environmental conditions reach extremities plants undergo physiological and morphological modifications to adapt to these abrupt changes (Shukla et al., 2012). Major abiotic stresses that affect plant growth and agricultural productivity are drought, salinity, water logging, etc. In addition, heavy metal deposition, nutrient depletion, and fluctuation in optimum temperature also affect growth and yield of plants. Abiotic stress can lead to 20%−50% loss in crop yield (Shrivastava and Kumar, 2015), depending on the type of crop and place it is grown such as in arid and semiarid regions or where salinity is normally high, resulting in low growth rate of aboveground biomass, and also affecting the photosynthetic ability of crops due to lowered water uptake.

### *6.2.1 Drought*

Drought affects plants at various subcellular levels, in the cell organs and in the whole plant (Choluj et al., 2004; Rahdari and Hoseini, 2012). It has been estimated that by 2050 drought will cause serious plant growth problems for >50% of arable land (Vinocur and Altman, 2005). Furthermore, with global climate change frequent and for long-lasting droughts are expected (Overpeck and Cole, 2006). Naveed et al. (2014a,b) showed how drought adversely affects the physiological, biochemical, and growth parameters of wheat seedlings through decreased $CO_2$ assimilation, stomatal conductance, relative water content, transpiration rate, and chlorophyll content. Nutrient availability and transport are affected due to drought as nutrients are carried to the roots by water (Vurukonda et al., 2016) Water-soluble nutrients such as nitrate, sulfate, Ca, Mg, and Si become unavailable to due to obstruction of nutrient diffusion and their mass flow (Barber, 1995; Selvakumar et al., 2012). Due to lower uptake of nitrate from soil the activity of nitrate reductase declines (Caravaca et al., 2005). Generation of reactive oxygen species (ROS) and free radicals such as superoxide radicals, hydrogen peroxide, and hydroxyl radicals result in oxidative stress. ROS most likely cause lipid peroxidation and membrane deterioration and degrade proteins, lipids, and nucleic acids in plants at high concentrations (Hendry, 2005; Sgherri et al., 2000; Nair et al., 2008). Drought heightens biosynthesis of ethylene, which inhibits plant growth through several mechanisms affecting quantity and quality of yield in plants (Ali et al., 2009). High salt concentration and scarcity of water are the main causes of balance disturbance in osmotic maintenance in plant cells, which ultimately negatively affects growth. As can be seen, osmotic imbalance greatly affects the growth and development of plants (Zhu, 2002). A variety of responses (physiological, cellular, and molecular) are triggered due to the osmotic imbalance caused by biotic or abiotic stresses such as an increase in the development of the below-ground system, reduction in the growth of above-ground system, changes in transport of ions (uptake, extrusion, and compartmentalization) and changes in metabolic activities such as photosynthesis. These physiological, molecular, or cellular responses are the result of direct or indirect signals of stress. While osmotic imbalance acts as direct signal others such as the production of ROS, abscisic acid, ethylene, and phospholipid like secondary messengers, act as indirect signals. Root-derived abscisic acid (ABA) may not be confined to the primary stress sites and can ascend with transpiration flow to regulate stomatal aperture in leaves under drought. Maintaining optimum water content in rhizosphere is a prerequisite for good plant growth and development in a water-deficient stress environment.

### 6.2.2 Salinity

Among abiotic stresses, salinity is biggest stressor that suppresses agricultural productivity. Soil salinity is an enormous problem for 7% of the world's total land (Szabolcs, 1994). In salinity affected areas, most crops are grown under irrigation, and to exacerbate the problem, inadequate irrigation management leads to secondary salinization that affects 20% of irrigated land worldwide (Glick et al., 2007; Al-Maskri et al., 2010). Ion toxicity, nutrient deficiency (N, Ca, K, P, Fe, and Zn), and oxidative stress on plants are the major drawbacks of salinity, which eventually limits water uptake from soil (Shrivastava and Kumar, 2015). Soil salinity significantly reduces plant phosphorus (P) uptake because phosphate ions precipitate with Ca ions (Bano and Fatima, 2009). Some elements, such as sodium, chlorine, and boron, have specific toxic effects on plants. Excessive accumulation of sodium in cell walls can rapidly lead to osmotic stress and cell death (Munns, 2002). Salinity affects almost all aspects of plant development (germination, vegetative growth, and reproductive development) and photosynthetic efficiency (Netondo et al., 2004). Salinity adversely affects reproductive development by inhibiting microsporogenesis and stamen filament elongation, enhancing programed cell death in some tissue types, ovule abortion, and senescence of fertilized embryos. The saline growth medium causes many adverse effects on plant growth, due to a low osmotic potential of soil solution (osmotic stress), specific ion effects (salt stress), nutritional imbalances, or a combination of these factors. The negative effects of salinity on the growth of plants can also occur due to disturbance in the transport of phytohormones as well as photosynthetic products to the developing plant tissue (Ashraf, 2004).

### 6.2.3 *Heavy metal deposition in soil*

Out of 90 metals 53 are reported as heavy metals and not all are biologically important. Based on their solubility under physiological conditions only 17 heavy metals are available to living cells and are of importance to them. While Fe, Mo, and Mn are important as micronutrients, Zn, Ni, Cu, V, Co, W, and Cr have been reported to be toxic elements with high or low importance as trace elements and Hg, Ag, Sb, Cd, Pb, and U are toxic to plants and microorganisms (Godbold and Hüttermann, 1985; Breckle and Kahle, 1991; Nies, 1999). Naturally heavy metals occur in the underlying parent material and the atmosphere. Anthropogenic activities like mining, combustion of fossil fuels, metal-working industries, phosphate fertilizers, etc., add to the emission and accumulation of heavy metals in ecosystems (Angelone, 1992; Schutzendubel and Polle, 2002). Toxic heavy metals generally cause inhibition of cytoplasmic enzymes and damage cell membranes, thus reducing plant growth and development (Chibuike and Obiora, 2014).

### 6.2.4 *Fluctuations in temperature*

Both high- and low-temperature stresses can seriously impair of plant physiology. Membrane fluidity, nucleic acid, and protein structures, as well as metabolite and osmolyte concentrations are affected due to rapid change in temperature. High temperature causes elevation in ROS concentrations, thus leading to oxidative damage and cell death (Zinn et al., 2010). Low temperature impairs physiological and biochemical machineries in plants resulting in visible symptoms (wilting, necrosis, or chlorosis) (Ruelland and Zachowski, 2010), which may also lead to transformation in the structure of cell membrane and its lipid content (Uemura and Steponkus, 1999). These changes may cause loss of electrolytic content from the cytoplasm resulting in the activation of optional pathways to regulate electron flow (Seo et al., 2010). Knight et al. (1998) reported changes in the concentration of calcium ions inside the cell content is also caused by temperature fluctuations. Ruelland and Zachowski (2010) observed that variation in temperature may also affect the enzymatic activity of proteins and their overall content in cells, cause changes in the structure of cell organelles (thylakoid membrane and plastids), phosphorylation of the proteins found in mitochondria and thylakoid unit.

## 6.3 How do Plant growth growth-promoting microorganism help ameliorate abiotic stresses of plants?

The role of microorganisms in plant growth promotion, nutrient management, and disease control is well known and well established. They help in (1) supplying nutrients to crops; (2) stimulating plant growth (production of phytohormones); (3) biocontrol; (4) improving properties of soil; (5) and bioaccumulation or microbial leaching of inorganics. More recently, bacteria have also been used in soil for mineralization of organic pollutants (i.e., bioremediation of polluted soils) (Burd et al.,1998). (Burd et al., 1998; Zhuang et al., 2007) have highlighted the use of bacteria in bioremediation of organic pollutants. *Rhizobium*, *Bacillus*, *Pseudomonas*, *Pantoea*, *Paenibacillus*, *Burkholderia*, *Achromobacter*, *Azospirillum*, *Microbacterium*, *Methylobacterium*, *Variovorax*, *Enterobacter*, etc., are the most common genera reported to provide tolerance to host plants under different abiotic stress environments (Grover et al., 2011). Egamberdieva and Kucharova (2009) reported phytohormone production (GA and IAA) by microbes that enhance the root system by increasing their number, length, and surface area; this help plants to grow well due to increase in nutrient uptake. According to Štajner et al. (1997) some rhizobacterial species can produce antioxidants and cytokinin, which leads to higher concentrations of ABA and destruction of ROS. Evidence shows a link between higher activities of enzymes involved in antioxidant action and tolerance to oxidative stress in cells.

### 6.3.1 Drought tolerance

The role of PGPM in plant abiotic stress tolerance (such as drought stress) has been studied to understand the adaptation of living organisms to extreme environments (Marulanda et al., 2010) (Table 6.1). The plant growth-promoting endophytic bacteria *Burkholderia phytofirmans* PsJN was used to investigate the potential to ameliorate the effects of drought stress on the growth, physiology, and yield of wheat (*Triticum aestivum* L.) under natural field conditions. Inoculation of wheat with PsJN significantly diluted the adverse effects of drought on relative water content and $CO_2$ assimilation rate, thus improving the photosynthetic rate, water-use efficiency, and chlorophyll content over the uninoculated control. Grain yield was also decreased when plants were exposed to drought stress at the tillering and flowering stage, but inoculation resulted in better grain yield (up to 21% and 18% higher, respectively) than the respective uninoculated control (Naveed et al., 2014a). *Achromobacter piechaudii* ARV8, which produced 1-aminocyclopropane-1-carboxylate (ACC) deaminase, conferred IST (induced systemic tolerance) against drought and salt stress in pepper and tomato (Mayak et al., 2004). Waqas et al. (2012) reported an increase in plant biomass, assimilation of essential nutrients, and reduced the sodium toxicity in cucumber by endophytic *Phoma glomerata* and *Penicillium* sp. under sodium chloride and polyethylene glycol–induced salinity and drought stress when compared with control plants.

Nadeem et al. (2017) showed that application of *Pseudomonas fluorescens* alone and in combination with biochar and/or compost reduced the negative impact of water-deficit stress on cucumber growth, improved relative water content of leaves, and reduced the electrolyte leakage. The findings of Aslam et al. (2018) indicated that drought-tolerant carbonic anhydrase-producing bacteria improved chlorophyll content, photosynthetic rate, and relative water content and biomass of wheat seedlings under water-deficit conditions compared to uninoculated plants. Several

**Table 6.1** Some plant growth-promoting microorganisms involved in alleviating abiotic stress.

| Abiotic stresses | Plant growth-promoting microbe | Plant hosts | Physiological changes in plants | References |
|---|---|---|---|---|
| Drought | *Trichoderma hamatum* DIS 219b | *Theobroma cacao* | Delayed drought-induced changes in stomatal conductance and net photosynthesis | Bae et al. (2009) |
| | *Phyllobacterium brassicacearum* STM196 | *Arabidopsis thaliana* | Antagonizes high nitrate inhibition of lateral root development | Bresson et al. (2013) |
| | *Burkholderia phytofirmans* PsJN and *Enterobacter* sp. FD17 | *Zea mays* | Significant increase in shoot and root biomass, leaf area, chlorophyll content, photosynthesis, and photochemical efficiency of PSII | Naveed et al. (2014b) |
| | *B. phytofirmans* PsJN | Wheat | Improved the ionic balance, antioxidant levels, and also increased the nitrogen, phosphorus, potassium, and protein concentration in grains | Naveed et al.(2014a) |
| | *Proteus penneri* | *Z. mays* | Improved soil moisture contents, plant biomass, root-shoot length, and leaf area. inoculated plants showed increase in relative water content, protein, and sugar though the proline content and the activities of antioxidant enzymes were decreased | Naseem and Bano (2014) |
| Salinity | *Glomus mosseae* | *Z. mays* L. cv. Yedan 13 | Mycorrhizal plants maintained higher root and shoot dry weights. Concentrations of chlorophyll, P and soluble sugars were higher than in nonmycorrhizal plants | Feng et al. (2002) |
| | *Pseudomonas mendocina* Palleroni in combination with *Glomus intraradices* | *Lactuca sativa* L. cv. Tafalla | Salt stress decreased sugar accumulation and increased foliar proline concentration, particularly in plants inoculated with the PGPR | Kohler et al. (2009) |
| | *Bacillus megaterium* | *Z. mays* L. | Modify salt response in maize plants in terms of root growth, necrotic leaf area, leaf relative water | Marulanda et al. (2010) |
| | *Brachybacterium saurashtrense* (JG-06), *Brevibacterium casei* (JG-08), and *Haererohalobacter* (JG-11) | *Arachis hypogaea* | Plant length, shoot length, root length, shoot dry weight, root dry weight, and total biomass were significantly higher in inoculated plants | Shukla et al. (2012) |
| | *Bacillus subtilis* SU47 and *Arthrobacter* sp. SU18 | | Antioxidant enzymes in wheat leaves decreased under salinity stress after PGPR coinoculation | Upadhyay et al. (2012) |
| | *Bacillus licheniformis* B2r | *Lycopersicon esculentum* | Significant increase in the germination percentage, germination index, root length and dry weight of seedling at saline condition | Chookietwattana and Maneewan (2012) |

| | | | | |
|---|---|---|---|---|
| **Temperature** | | | | |
| High temperature | *Pseudomonas* AKM-P6 | Pigeon pea | Reduced membrane injury, and improved the levels of cellular metabolites like proline, chlorophyll, sugars, amino acids, and proteins | Ali et al. (2009) |
| | *Glomus etunicatum* | *Z. mays* L. genotype Zhengdan 958 | Enhanced the net photosynthetic rate, stomatal conductance and transpiration rate in the maize leaves at 25°C, 35°C, and 40°C | Zhu et al. (2011) |
| Low temperature | *G. mosseae* | *Citrus tangerina* | Mycorrhizal inoculation significantly increased the root length and the Ca content of the seedlings grown at 15°C | Wu and Zou (2010) |
| | *Pseudomonas vancouverensis* OB155-gfp *P. fredriksbergensis* OS261-gfp | *Solanum lycopersicum* Mill. | Codes for proteins that protect cells against cold/chilling stress. Reduced membrane damage and ROS levelTomato lipoxygenase | Subramanian et al. (2015) |
| **Heavy metal** | | | | |
| Cd | *Brassica napus* | *Kluyvera ascorbata* SUD165 | Plant demonstrated normal growth under high levels of $Ni^{2+}$, $Pb^{2+}$, $Zn^{2+}$, and $CrO_4{}^{2-}$ | Burd et al. (1998) |
| | *G. intraradices* | *Pisum sativum* L. (*cv. Frisson*, VIR4788, VIR7128) | Decreased Cd accumulation in roots and pods | Rivera-Becerril et al. (2002) |
| Cr | *Mesorhizobium* RC3 | *C. arietinum* | Nitrogen in roots and shoots were increased, decreased the uptake of chromium by plant | Wani et al. (2008) |
| Cu | *Pseudomonas asplenii* | *B. napus* | Plant showed normal growth under high level of $Cu^{2+}$ | Reed et al. (2005) |
| Zn, Cd, As, and Pb | *Pseudomonas koreensis* AGB-1 | *Miscanthus sinensis* | High tolerance to Zn, Cd, As, and Pb through extracellular sequestration, increased catalase and SOD activities in plants by ~42% and 33% | Babu et al. (2015) |

(*Continued*)

**Table 6.1** (Continued)

| Abiotic stresses | Plant growth-promoting microbe | Plant hosts | Physiological changes in plants | References |
|---|---|---|---|---|
| **Nutrient deficiency** | | | | |
| Phosphate | *Mesorhizobium mediterraneum* | Chickpea and barley | Mobilize phosphorous efficiently in both plants when tricalcium phosphate was added to the soil. The dry matter, nitrogen, potassium, calcium and magnesium content in both plants was significantly increased | Peix et al. (2001) |
| | *Prosopis strombulifera* (halophyte) | *Oryza sativaTriticum aestivum* | Gibberelic acid used by bacteria to solubilize phosphate | Choi et al. (2008) |
| P and Fe acquisition | *G. mosseae* | *Arachis hypogea* L. and *Sorghum bicolor* L. | Increased uptake of P and Fe from the soil | Caris et al. (1998) |

ecophysiological studies have demonstrated that AM symbiosis often results in altered rates of water movement into, through, and out of host plants, with consequent effects on tissue hydration and plant physiology. Zuccarini and Save (2016) showed that inoculation of *Spinacia oleracea* L. with three species of AM *Glomus* sp. enhanced tolerance to water stress.

### 6.3.2 Sequestration of heavy metals

The plant growth promoting microbes growing on trace metal contaminated soils play an important role in phytoremediation (Khan, 2005; Saharan and Nehra, 2011). They secrete acids, proteins, phytoantibiotics, and other chemicals to mitigate toxic effects of heavy metals (Denton, 2007). Dell'Amico et al. (2008) in their work showed improved growth of *Brassica napus* by a cadmium-resistant rhizobacteria probably through accumulation of ACC in roots. The PGPB *Kluyvera ascorbata* SUCD165 was reported to decrease nickel toxicity in canola seedlings (Burd et al., 1998). Another bacterium *Streptomyces acidiscabies* E13 promotes cowpea growth under nickel contamination by producing hydroxamate type siderophores (Dimkpa et al., 2008). A greenhouse study on *Brassica juncea* by Wu et al. (2006) evaluated the effects of bacterial inoculation on the uptake of heavy metals from Pb–Zn mine tailings by plants. The presence of these beneficial bacteria had little influence on the metal concentrations in plant tissues, but produced a much larger above-ground biomass and altered metal bioavailability in the soil. Many plant species, especially trees, are associated with mycorrhiza under natural conditions. Ouziad et al. (2005) conducted an experiment growing tomato in either "Breinigerberg" soil, which has a high content of Zn and of other heavy metals, or in nonpolluted soil enriched with up to 1 mM $CdCl_2$. Plants colonized with the AMF *Glomus intraradices* grew distinctly better than nonmycorrhizal controls. AMF colonization results in downregulation of plant gene coding for products potentially involved in heavy metal tolerance.

### 6.3.3 Salinity

Plants have inbuilt mechanisms to tolerate salinity such as synthesis of osmolytes and polyamines, reducing ROS, the antioxidant defense mechanism, ion transport (Zhu, 2002), and compartmentalization (Gaxiola et al., 1999). Tank and Saraf (2010) isolated a few PGPR strains from tomato fields and adapted their cultures to 6% NaCl concentration after which they showed phosphate solubilization potential and produced phytohormones, siderophores, and ACC deaminase enzyme in pot-cultured tomato plants. The three PGPR isolates, *Pseudomonas alcaligenes* PsA15, *Bacillus polymyxa* BcP26, and *Mycobacterium phlei* MbP18, conferred tolerance against high temperature and salt stress to maize growing in nutrient-deficient calcisol soil (Egamberdiyeva, 2007). On constant exposure to 500 mM NaCl solution (seawater levels to mimic exposure of plants in their native beach habitat), nonsymbiotic plants *Leymus mollis* (dunegrass) became severely wilted and desiccated within 7 days and were dead after 14 days. In contrast, symbiotic plants infected with endophytic *Fusarium culmorum* did not show wilting symptoms until they were exposed to 500 mM NaCl solution for 14 days (Rodriguez et al., 2008; Lata et al., 2018). Al-Karaki (2006) demonstrated that inoculation of tomato plants with *Glomus mosseae* (AMF) improved plant growth under salt stress. PGPM produce microbial polysaccharides that bind soil particles to form microaggregates and macroaggregates (Feng et al., 2002). Exopolysaccharides produced by bacteria impart increased resistance to water and salinity stress by improving soil structure (Sandhya et al., 2009).

### 6.3.4 Tolerating changes in temperature

As noted earlier, fluctuating temperature causes damage to the physiochemical properties of plant cell membrane. To avoid oxidative damage due to high temperature, plants adapt to de novo synthesis of organic compatible solutes or osmolytes and accumulate several secondary metabolites like polyphenols to ameliorate stress responses (Cheruiyot et al., 2007). Antioxidants like reduced glutathione prevent damages to important cellular components upon exposure to ROS invasion. When cucumber plants inoculated with bacteria *Paecilomyces formosus* LHL10 were exposed to high temperature (38°C), the plants showed higher growth attributes. The endophytic fungi increased heat tolerance in wheat in terms of height, weight of grain, as well as germination of second-generation seeds (Hubbard et al., 2014). The ability of thermotolerance (38°C and 65°C) in grass *Dichanthelium lanuginosum* growing in Yellowstone National Park is due to the tripartite symbiosis that it shares with the fungi *Curvularia protuberata* and its mycovirus *Curvularia* thermal tolerance virus (Redman et al., 2002).

### 6.3.5 Combating nutrient deficiency

The key to increasing agricultural production is improvement of soil fertility. Drought and salinity cause water deficiency and eventually lead to unavailability of nutrients affecting crop yield. Iron, inorganic phosphate, fixed nitrogen, and phytohormones are essential for growth of plants. Under iron-limiting conditions PGPB produce siderophores to sequester ferric ions from soil and provide them to the plants (Whipps, 2001; Compant et al., 2005). Nitrogen fixation is very important in enhancing soil fertility. Plant-root exudates are metabolized by nitrogen-fixing bacteria and in turn provide nitrogen to the plant for amino acid synthesis. Phosphate solubilization is just as important as biological nitrogen fixation. The phosphate solubilizing microorganisms, composed largely of bacteria and fungi, serve as alternatives to meet the phosphate needs of plants. Among bacteria *Bacillus*, *Rhizobium*, and *Pseudomonas* while fungi belonging to genera *Aspergillus* and *Penicillium* are most efficient phosphate solubilizers. Rivas et al. (2006) analyzed biodiversity of phosphate solubilizing microbes from *Cicer arietinum* and identified two rhizobia species as *Mesorhizobium ciceri* and *Mesorhizobium mediterraneum*. *Pseudomonas indica* and *Azotobacter chroococcum* improved uptake of mineral nutrients (especially Zn) in wheat (Abadi and Sepehri, 2016). Mycorrhiza have the maximum share in the phosphorous uptake by plants root (Marschner and Dell, 1994). Besides providing P, mycorrhiza also take part in nutrient cycling in soil and provide other essential nutrients (N, K, Cu, Mg, and Zn) to plants by transforming them into soluble forms that can be easily taken up by plants (Smith and Read, 2010). Nitrogen fixation is also affected by availability of other micronutrients; so by providing other essential nutrients, mycorrhiza can enhance nitrogen fixation (Guo et al., 2010; Requena et al., 2001; Tavasolee et al., 2011). According to Gadkar and Rillig (2006), mycorrhiza produce a glycoprotein called glomalin that plays a big role in stabilizing soil particles by aggregating them; thus, mycorrhizal association also affects soil characteristics. Plants of *Miscanthus sacchariflorus* inoculated with AM fungi *Gigaspora margarita* were shown to have efficient nutrient acquisition (N, P, K, Mg, Fe, Cu, and Zn) (Lehmann and Rillig, 2015).

## 6.4 Conclusion

Many tools of modern science have been extensively applied for crop improvement under stress. This extensive research has clearly helped us understand the importance of PGPM in agriculture as well as in natural ecosystems in helping plants survive in harsh climates. Work is continuously being done for in-depth analysis of the mechanisms by which these PGPM benefit the plants and once an insight is achieved it would become easier to handle and manipulate these microbes for developing efficient microbial formulation for boosting plant performance under abiotic stresses.

## References

Abadi, V.A.J.M., Sepehri, M., 2016. Effect of *Piriformospora indica* and *Azotobacter chroococcum* on mitigation of zinc deficiency stress in wheat (*Triticum aestivum* L.). Symbiosis 69 (1), 9–19.

Ali, S.Z., Sandhya, V., Grover, M., Kishore, N., Rao, L.V., Venkateswarlu, B., 2009. *Pseudomonas* sp. strain AKM-P6 enhances tolerance of sorghum seedlings to elevated temperatures. Biol. Fertil. Soils 46 (1), 45–55.

Al-Karaki, G.N., 2006. Nursery inoculation of tomato with arbuscular mycorrhizal fungi and subsequent performance under irrigation with saline water. Sci. Horticult. 109 (1), 1–7.

Al-Maskri, A.H.M.E.D., Al-Kharusi, L., Al-Miqbali, H., Khan, M.M., 2010. Effects of salinity stress on growth of lettuce (*Lactuca sativa*) under closed-recycle nutrient film technique. Int. J. Agric. Biol 12 (3), 377–380.

Angelone, M., 1992. Trace elements concentrations in soils and plants of Western Europe. Biogeochem. Trace Met. 19–60.

Antoun, H., Kloepper, J., 2001. Plant growth promoting rhizobacteria (PGPR). In: Brenner, S., Miller, J.H. (Eds.), Encyclopedia of Genetics., 2001. Academic, New York, pp. 1477–1480.

Ashraf, M., 2004. Some important physiological selection criteria for salt tolerance in plants. Flora—Morphol. Distrib. Funct. Ecol. Plants 199 (5), 361–376.

Aslam, A., Ahmad Zahir, Z., Asghar, H.N., Shahid, M., 2018. Effect of carbonic anhydrase-containing endophytic bacteria on growth and physiological attributes of wheat under water-deficit conditions. Plant Prod. Sci. 21 (3), 244–255.

Backman, P.A., Sikora, R.A., 2008. Endophytes: an emerging tool for biological control. Biol. Contr. 46 (1), 1–3.

Babu, A.G., Shea, P.J., Sudhakar, D., Jung, I.B., Oh, B.T., 2015. Potential use of *Pseudomonas koreensis* AGB-1 in association with *Miscanthus sinensis* to remediate heavy metal (loid)-contaminated mining site soil. J. Environ. Manage. 151, 160–166.

Bae, H., Sicher, R.C., Kim, M.S., Kim, S.H., Strem, M.D., Melnick, R.L., et al., 2009. The beneficial endophyte *Trichoderma hamatum* isolate DIS 219b promotes growth and delays the onset of the drought response in *Theobroma cacao*. J. Exp. Bot. 60 (11), 3279–3295.

Bano, A., Fatima, M., 2009. Salt tolerance in *Zea mays* (L). following inoculation with *Rhizobium* and *Pseudomonas*. Biol. Fertil. Soils 45 (4), 405–413.

Barber, S.A., 1995. Soil Nutrient Bioavailability: A Mechanistic Approach. John Wiley & Sons.

Barriuso, J., Solano, B.R., Lucas, J.A., Lobo, A.P., García-Villaraco, A., Mañero, F.J.G., 2008. Ecology, genetic diversity and screening strategies of plant growth promoting rhizobacteria (PGPR). J. Plant Nutr. 1–17.

Bethlenfalvay, G.J., Linderman, R.G., 1992. Mycorrhizae in Sustainable Agriculture. ASA Special Publication No. 54, Madison: WI, pp. 8–13.

Breckle, S.W., Kahle, H., 1991. Ecological geobotany/autecology and ecotoxicology. Progress in Botany. Springer, Berlin, pp. 391–406.

Bresson, J., Varoquaux, F., Bontpart, T., Touraine, B., Vile, D., 2013. The PGPR strain *Phyllobacterium brassicacearum* STM196 induces a reproductive delay and physiological changes that result in improved drought tolerance in Arabidopsis. New Phytol. 200 (2), 558–569.

Burd, G.I., Dixon, D.G., Glick, B.R., 1998. A plant growth-promoting bacterium that decreases nickel toxicity in seedlings. Appl. Environ. Microbiol. 64 (10), 3663–3668.

Caravaca, F., Alguacil, M.M., Hernández, J.A., Roldán, A., 2005. Involvement of antioxidant enzyme and nitrate reductase activities during water stress and recovery of mycorrhizal *Myrtus communis* and *Phillyrea angustifolia* plants. Plant Sci. 169 (1), 191–197.

Caris, C., Hördt, W., Hawkins, H.J., Römheld, V., George, E., 1998. Studies of iron transport by arbuscular mycorrhizal hyphae from soil to peanut and sorghum plants. Mycorrhiza 8 (1), 35–39.

Cheruiyot, E.K., Mumera, L.M., Ng'etich, W.K., Hassanali, A., Wachira, F., 2007. Polyphenols as potential indicators for drought tolerance in tea (*Camellia sinensis* L.). Biosci. Biotechnol. Biochem. 71 (9), 2190–2197.

Chibuike, G.U., Obiora, S.C., 2014. Heavy metal polluted soils: effect on plants and bioremediation methods. Appl. Environ. Soil Sci. 2014.

Choi, O., Kim, J., Kim, J.G., Jeong, Y., Moon, J.S., Park, C.S., et al., 2008. Pyrroloquinoline quinone is a plant growth promotion factor produced by *Pseudomonas fluorescens* B16. Plant Physiol. 146 (2), 657–668.

Choluj, D., Karwowska, R., Jasinska, M., Haber, G., 2004. Growth and dry matter partitioning in sugar beet plants (*Beta vulgaris* L.) under moderate drought. Plant Soil Environ. 50 (6), 265–272.

Chookietwattana, K., Maneewan, K., 2012. Selection of efficient salt-tolerant bacteria containing ACC deaminase for promotion of tomato growth under salinity stress. Soil Environ. 31 (1).

Christie, P., Li, X., Chen, B., 2004. Arbuscular mycorrhiza can depress translocation of zinc to shoots of host plants in soils moderately polluted with zinc. Plant Soil 261 (1–2), 209–217.

Compant, S., Duffy, B., Nowak, J., Clément, C., Barka, E.A., 2005. Use of plant growth-promoting bacteria for biocontrol of plant diseases: principles, mechanisms of action, and future prospects. Appl. Environ. Microbiol. 71 (9), 4951–4959.

Denton, B., 2007. Advances in phytoremediation of heavy metals using plant growth promoting bacteria and fungi. MMG 445 Basic Biotechnol. 3, 1–5.

Dell'Amico, E., Cavalca, L., Andreoni, V., 2008. Improvement of Brassica napus growth under cadmium stress by cadmium-resistant rhizobacteria. Soil Biol. Biochem. 40 (1), 74–84.

Dimkpa, C., Svatoš, A., Merten, D., Büchel, G., Kothe, E., 2008. Hydroxamate siderophores produced by *Streptomyces acidiscabies* E13 bind nickel and promote growth in cowpea (*Vigna unguiculata* L.) under nickel stress. Can. J. Microbiol. 54 (3), 163–172.

Egamberdieva, D., Kucharova, Z., 2009. Selection for root colonising bacteria stimulating wheat growth in saline soils. Biol. Fertil. Soils 45 (6), 563–571.

Egamberdiyeva, D., 2007. The effect of plant growth promoting bacteria on growth and nutrient uptake of maize in two different soils. Appl. Soil Ecol. 36 (2–3), 184–189.

Feng, G., Zhang, F., Li, X., Tian, C., Tang, C., Rengel, Z., 2002. Improved tolerance of maize plants to salt stress by arbuscular mycorrhiza is related to higher accumulation of soluble sugars in roots. Mycorrhiza 12 (4), 185–190.

Gadkar, V., Rillig, M.C., 2006. The arbuscular mycorrhizal fungal protein glomalin is a putative homolog of heat shock protein 60. FEMS Microbiol. Lett. 263 (1), 93–101.

Gaxiola, R.A., Rao, R., Sherman, A., Grisafi, P., Alper, S.L., Fink, G.R., 1999. The *Arabidopsis thaliana* proton transporters, AtNhx1 and Avp1, can function in cation detoxification in yeast. Proc. Natl. Acad. Sci. 96 (4), 1480–1485.

Giovannetti, M., Avio, L., Fortuna, P., Pellegrino, E., Sbrana, C., Strani, P., 2006. At the root of the wood wide web: self-recognition and nonself-incompatibility in mycorrhizal networks. Plant Signal. Behav. 1 (1), 1–5.

Glick, B.R., Cheng, Z., Czarny, J., Duan, J., 2007. Promotion of plant growth by ACC deaminase-producing soil bacteria. New Perspectives and Approaches in Plant Growth-Promoting Rhizobacteria Research. Springer, Dordrecht, pp. 329–339.

Godbold, D.L., Hüttermann, A., 1985. Effect of zinc, cadmium and mercury on root elongation of *Picea abies* (Karst.) seedlings, and the significance of these metals to forest dieback. Environ. Pollut. 38 (4), 375–381.

Grover, M., Ali, S.Z., Sandhya, V., Rasul, A., Venkateswarlu, B., 2011. Role of microorganisms in adaptation of agriculture crops to abiotic stresses. World J. Microbiol. Biotechnol. 27 (5), 1231–1240.

Guo, Y.J., Ni, Y., Huang, J., 2010. Effects of rhizobium, arbuscular mycorrhiza and lime on nodulation, growth and nutrient uptake of lucerne in acid purplish soil in China. Trop. Grassl. 44, 109–114.

Hendry, G.A., 2005. Oxygen free radical process and seed longevity. Seed Sci. J. 3, 141–147.

Hubbard, M., Germida, J.J., Vujanovic, V., 2014. Fungal endophytes enhance wheat heat and drought tolerance in terms of grain yield and second-generation seed viability. J. Appl. Microbiol. 116 (1), 109–122.

Khan, A.G., 2005. Role of soil microbes in the rhizospheres of plants growing on trace metal contaminated soils in phytoremediation. J. Trace Elem. Med. Biol. 18 (4), 355–364.

Khan, A.G., Belik, M., 1995. Occurrence and ecological significance of mycorrhizal symbiosis in aquatic plants. Mycorrhiza. Springer, Berlin, pp. 627–666.
Kloepper, J.W., Leong, J., Teintze, M., Schroth, M.N., 1980. Enhanced plant growth by siderophores produced by plant growth-promoting rhizobacteria. Nature 286 (5776), 885.
Kohler, J., Caravaca, F., Roldán, A., 2009. Effect of drought on the stability of rhizosphere soil aggregates of *Lactuca sativa* grown in a degraded soil inoculated with PGPR and AM fungi. Appl. Soil Ecol. 42 (2), 160–165.
Knight, H., Brandt, S., Knight, M.R., 1998. A history of stress alters drought calcium signalling pathways in Arabidopsis. Plant J. 16 (6), 681–687.
Lata, R., Chowdhury, S., Gond, S.K., White Jr, J.F., 2018. Induction of abiotic stress tolerance in plants by endophytic microbes. Lett. Appl. Microbiol. 66 (4), 268–276.
Lehmann, A., Rillig, M.C., 2015. Arbuscular mycorrhizal contribution to copper, manganese and iron nutrient concentrations in crops—a meta-analysis. Soil Biol. Biochem. 81, 147–158.
Liu, R.J., Chen, Y.L., 2007. Micorrhizology (China). Science Press, Beijing.
Marschner, H., Dell, B., 1994. Nutrient uptake in mycorrhizal symbiosis. Plant Soil 159 (1), 89–102.
Marulanda, A., Azcón, R., Chaumont, F., Ruiz-Lozano, J.M., Aroca, R., 2010. Regulation of plasma membrane aquaporins by inoculation with a *Bacillus megaterium* strain in maize (*Zea mays* L.) plants under unstressed and salt-stressed conditions. Planta 232 (2), 533–543.
Mayak, S., Tirosh, T., Glick, B.R., 2004. Plant growth-promoting bacteria that confer resistance to water stress in tomatoes and peppers. Plant Sci. 166 (2), 525–530.
Munns, R., 2002. Comparative physiology of salt and water stress. Plant Cell Environ. 25 (2), 239–250.
Nadeem, S.M., Imran, M., Naveed, M., Khan, M.Y., Ahmad, M., Zahir, Z.A., et al., 2017. Synergistic use of biochar, compost and plant growth-promoting rhizobacteria for enhancing cucumber growth under water deficit conditions. J. Sci. Food Agric. 97 (15), 5139–5145.
Naseem, H., Bano, A., 2014. Role of plant growth-promoting rhizobacteria and their exopolysaccharide in drought tolerance of maize. J. Plant Interac. 9 (1), 689–701.
Nair, A.S., Abraham, T.K., Jaya, D.S., 2008. Studies on the changes in lipid peroxidation and antioxidants in drought stress induced cowpea (*Vigna unguiculata* L.) varieties. J. Environ. Biol. 29, 689–691.
Naveed, M., Hussain, M.B., Zahir, Z.A., Mitter, B., Sessitsch, A., 2014a. Drought stress amelioration in wheat through inoculation with *Burkholderia phytofirmans* strain PsJN. Plant Growth Regul. 73 (2), 121–131.
Naveed, M., Mitter, B., Reichenauer, T.G., Wieczorek, K., Sessitsch, A., 2014b. Increased drought stress resilience of maize through endophytic colonization by *Burkholderia phytofirmans* PsJN and Enterobacter sp. FD17. Environ. Exp. Bot. 97, 30–39.
Netondo, G.W., Onyango, J.C., Beck, E., 2004. Sorghum and salinity. Crop Sci. 44 (3), 797–805.
Nies, D.H., 1999. Microbial heavy-metal resistance. Appl. Microbiol. Biotechnol. 51 (6), 730–750.
Oueslati, O., 2003. Allelopathy in two durum wheat (*Triticum durum* L.) varieties. Agric. Ecosyst. Environ. 96 (1–3), 161–163.
Ouziad, F., Hildebrandt, U., Schmelzer, E., Bothe, H., 2005. Differential gene expressions in arbuscular mycorrhizal-colonized tomato grown under heavy metal stress. J. Plant Physiol. 162 (6), 634–649.

Overpeck, J.T., Cole, J.E., 2006. Abrupt change in Earth's climate system. Annu. Rev. Environ. Resour. 31, 1–31.

Peix, A., Rivas-Boyero, A.A., Mateos, P.F., Rodriguez-Barrueco, C., Martınez-Molina, E., Velazquez, E., 2001. Growth promotion of chickpea and barley by a phosphate solubilizing strain of *Mesorhizobium mediterraneum* under growth chamber conditions. Soil Biol. Biochem. 33 (1), 103–110.

Rahdari, P., Hoseini, S.M., 2012. Drought stress: a review. Int. J. Agron. Plant Prod. 3 (10), 443–446.

Redman, R.S., Sheehan, K.B., Stout, R.G., Rodriguez, R.J., Henson, J.M., 2002. Thermotolerance generated by plant/fungal symbiosis. Science 298 (5598), 1581-1581.

Reed, M.L.E., Warner, B.G., Glick, B.R., 2005. Plant growth–promoting bacteria facilitate the growth of the common reed *Phragmites australisin* the presence of copper or polycyclic aromatic hydrocarbons. Curr. Microbiol. 51 (6), 425–429.

Requena, N., Perez-Solis, E., Azcón-Aguilar, C., Jeffries, P., Barea, J.M., 2001. Management of indigenous plant-microbe symbioses aids restoration of desertified ecosystems. Appl. Environ. Microbiol. 67 (2), 495–498.

Rivas, R., Peix, A., Mateos, P.F., Trujillo, M.E., Martínez-Molina, E., Velázquez, E., et al., 2006. Biodiversity of populations of phosphate solubilizing rhizobia that nodulates chickpea in different Spanish soils. Plant Soil 287 (1–2), 23–33.

Rivera-Becerril, F., Calantzis, C., Turnau, K., Caussanel, J.P., Belimov, A.A., Gianinazzi, S., et al., 2002. Cadmium accumulation and buffering of cadmium–induced stress by arbuscular mycorrhiza in three *Pisum sativum* L. genotypes. J. Exp. Bot. 53, 1177–1185.

Rodriguez, R.J., Henson, J., Van Volkenburgh, E., Hoy, M., Wright, L., Beckwith, F., et al., 2008. Stress tolerance in plants via habitat-adapted symbiosis. ISME J. 2 (4), 404.

Ruelland, E., Zachowski, A., 2010. How plants sense temperature. Environ. Exp. Bot. 69 (3), 225–232.

Saharan, B.S., Nehra, V., 2011. Plant growth promoting rhizobacteria: a critical review. Life Sci. Med. Res. 21 (1), 30.

Sandhya, V.Z.A.S., Grover, M., Reddy, G., Venkateswarlu, B., 2009. Alleviation of drought stress effects in sunflower seedlings by the exopolysaccharides producing *Pseudomonas putida* strain GAP-P45. Biol. Fertil. Soils 46 (1), 17–26.

Schutzendubel, A., Polle, A., 2002. Plant responses to abiotic stresses: heavy metal-induced oxidative stress and protection by mycorrhization. J. Exp. Bot. 53 (372), 1351–1365.

Selvakumar, G., Panneerselvam, P., Ganeshamurthy, A.N., 2012. Bacterial mediated alleviation of abiotic stress in crops. Bacteria in Agrobiology: Stress Management. Springer, Berlin, pp. 205–224.

Seo, P.J., Kim, M.J., Park, J.Y., Kim, S.Y., Jeon, J., Lee, Y.H., et al., 2010. Cold activation of a plasma membrane-tethered NAC transcription factor induces a pathogen resistance response in *Arabidopsis*. Plant J. 61 (4), 661–671.

Sgherri, C.L.M., Maffei, M., Navari-Izzo, F., 2000. Antioxidative enzymes in wheat subjected to increasing water deficit and rewatering. J. Plant Physiol. 157 (3), 273–279.

Shrivastava, P., Kumar, R., 2015. Soil salinity: a serious environmental issue and plant growth promoting bacteria as one of the tools for its alleviation. Saudi J. Biol. Sci. 22 (2), 123–131.

Shukla, P.S., Agarwal, P.K., Jha, B., 2012. Improved salinity tolerance of *Arachis hypogaea* (L.) by the interaction of halotolerant plant-growth-promoting rhizobacteria. J. Plant Growth Regul. 31 (2), 195–206.

Smith, S.E., Read, D.J., 2010. Mycorrhizal Symbiosis. Academic press.

Štajner, D., Kevrešan, S., Gašić, O., Mimica-Dukić, N., Zongli, H., 1997. Nitrogen and Azotobacter chroococcum enhance oxidative stress tolerance in sugar beet. Biologia plantarum 39 (3), 441.

Subramanian, P., Mageswari, A., Kim, K., Lee, Y., Sa, T., 2015. Psychrotolerant endophytic *Pseudomonas* sp. strains OB155 and OS261 induced chilling resistance in tomato plants (*Solanum lycopersicum* Mill.) by activation of their antioxidant capacity. Mol. Plant Microb. Interact. 28 (10), 1073–1081.

Szabolcs, I., 1994. Salt affected soils as the ecosystem for halophytes. Halophytes as a Resource for Livestock and for Rehabilitation of Degraded Lands. Springer, Dordrecht, pp. 19–24.

Tank, N., Saraf, M., 2010. Salinity-resistant plant growth promoting rhizobacteria ameliorates sodium chloride stress on tomato plants. J. Plant Interact. 5 (1), 51–58.

Tavasolee, A., Aliasgharzad, N., SalehiJouzani, G., Mardi, M., Asgharzadeh, A., 2011. Interactive effects of arbuscular mycorrhizal fungi and rhizobial strains on chickpea growth and nutrient content in plant. Afr. J. Biotechnol. 10 (39), 7585–7591.

Uemura, M., Steponkus, P.L., 1999. Cold acclimation in plants: relationship between the lipid composition and the cryostability of the plasma membrane. J. Plant Res. 112 (2), 245–254.

Upadhyay, S.K., Singh, J.S., Saxena, A.K., Singh, D.P., 2012. Impact of PGPR inoculation on growth and antioxidant status of wheat under saline conditions. Plant Biol. 14 (4), 605–611.

Vinocur, B., Altman, A., 2005. Recent advances in engineering plant tolerance to abiotic stress: achievements and limitations. Curr. Opin. Biotechnol. 16 (2), 123–132.

Vurukonda, S.S.K.P., Vardharajula, S., Shrivastava, M., SkZ, A., 2016. Enhancement of drought stress tolerance in crops by plant growth promoting rhizobacteria. Microbiol. Res. 184, 13–24.

Wani, P.A., Khan, M.S., Zaidi, A., 2008. Chromium-reducing and plant growth-promoting *Mesorhizobium* improves chickpea growth in chromium-amended soil. Biotechnol. Lett. 30 (1), 159–163.

Waqas, M., Khan, A.L., Kamran, M., Hamayun, M., Kang, S.M., Kim, Y.H., et al., 2012. Endophytic fungi produce gibberellins and indoleacetic acid and promotes host-plant growth during stress. Molecules 17 (9), 10754–10773.

Whipps, J.M., 2001. Microbial interactions and biocontrol in the rhizosphere. J. Exp. Bot. 52 (Suppl_1), 487–511.

Willis, A., Rodrigues, B.F., Harris, P.J.C., 2013. The ecology of arbuscular mycorrhizal fungi. Crit. Rev. Plant Sci. 32 (1), 1–20.

Wu, S.C., Cheung, K.C., Luo, Y.M., Wong, M.H., 2006. Effects of inoculation of plant growth-promoting rhizobacteria on metal uptake by *Brassica juncea*. Environ. Pollut. 140 (1), 124–135.

Wu, Q.S., Zou, Y.N., 2010. Beneficial roles of arbuscular mycorrhizas in citrus seedlings at temperature stress. Sci. Horticult. 125 (3), 289–293.

Zhu, J.K., 2002. Salt and drought stress signal transduction in plants. Annu. Rev. Plant Biol. 53 (1), 247–273.

Zhu, X.C., Song, F.B., Liu, S.Q., Liu, T.D., 2011. Effects of arbuscular mycorrhizal fungus on photosynthesis and water status of maize under high temperature stress. Plant Soil 346 (1–2), 189–199.

Zhuang, X., Chen, J., Shim, H., Bai, Z., 2007. New advances in plant growth-promoting rhizobacteria for bioremediation. Environ. Int. 33 (3), 406–413.

Zinn, K.E., Tunc-Ozdemir, M., Harper, J.F., 2010. Temperature stress and plant sexual reproduction: uncovering the weakest links. J. Exp. Bot. 61 (7), 1959–1968.

Zuccarini, P., Save, R., 2016. Three species of arbuscular mycorrhizal fungi confer different levels of resistance to water stress in *Spinacia oleracea* L. Plant Biosyst.—Int. J. Deal. Aspect. Plant Biol. 150 (5), 851–854.

## Further reading

Belimov, A.A., Hontzeas, N., Safronova, V.I., Demchinskaya, S.V., Piluzza, G., Bullitta, S., et al., 2005. Cadmium-tolerant plant growth-promoting bacteria associated with the roots of Indian mustard (*Brassica juncea* L. Czern.). Soil Biol. Biochem. 37 (2), 241–250.

Chacko, S., Ramteke, P.W., John, S.A., 2009. Amidase from plant growth promoting rhizobacterium. Afr. J. Bacteriol. Res. 1 (4), 046–050.

da Silva Araújo, A.E., Baldani, V.L.D., de Souza Galisa, P., Pereira, J.A., Baldani, J.I., 2013. Response of traditional upland rice varieties to inoculation with selected diazotrophic bacteria isolated from rice cropped at the Northeast region of Brazil. Appl. Soil Ecol. 64, 49–55.

Hayat, R., Ali, S., Amara, U., Khalid, R., Ahmed, I., 2010. Soil beneficial bacteria and their role in plant growth promotion: a review. Ann. Microbiol. 60 (4), 579–598.

Hildebrandt, U., Regvar, M., Bothe, H., 2007. Arbuscular mycorrhiza and heavy metal tolerance. Phytochemistry 68 (1), 139–146.

Lantzy, R.J., Mackenzie, F.T., 1979. Atmospheric trace metals: global cycles and assessment of man's impact. Geochim. Cosmochim. Acta 43 (4), 511–525.

Sarkar, A., Asaeda, T., Wang, Q., Rashid, M.H., 2015. Arbuscular mycorrhizal influences on growth, nutrient uptake, and use efficiency of *Miscanthus sacchariflorus* growing on nutrient-deficient river bank soil. Flora—Morphol. Distrib. Funct. Ecol. Plants 212, 46–54.

Shoebitz, M., Ribaudo, C.M., Pardo, M.A., Cantore, M.L., Ciampi, L., Curá, J.A., 2009. Plant growth promoting properties of a strain of *Enterobacter ludwigii* isolated from *Lolium perenne* rhizosphere. Soil Biol. Biochem. 41 (9), 1768–1774.

Terekhova, V.A., Semenova, T.A., 2005. The structure of micromycete communities and their synecologic interactions with basidiomycetes during plant debris decomposition. Microbiology 74 (1), 91–96.

# Legal issues in nanotechnology

7

*Kshitij Kumar Singh*
Campus Law Centre, Faculty of Law, University of Delhi, Delhi, India

## 7.1 Introduction

Modern advancements in nanotechnology pose new challenges to the existing intellectual property (IP) laws and these challenges are heightened as advancements evolve. This brings uncertainty and confusion to the eligibility of such advancements for IP protection. The huge investment made in modern fields of technology such as biotechnology and nanotechnology requires strong IP protection in order to recoup the investment and encourage innovation in the field. However, there has always been a dispute between IP protection and open, free sharing of nanotechnology; the openness is more demanded in basic research to ensure the accessibility of the technology to downstream research. To strike a proper balance between the IP protection and the open and free accessibility of technology, it is important to examine the nature and scope of nanotechnology in light of the various IP protections available.

IP rights are gaining importance in the knowledge economy with the growing value of intangible assets. These rights are becoming increasingly important, especially with regard to how the information goods produced by universities and affiliate academic industries are changing over time. The basic or fundamental research that was confined to universities for academic/noncommercial purpose has started taking a commercial route. Modern technologies, biotechnology and nanotechnology included, are based on elemental research usually carried on in universities; however, due to the increasing commercial value placed on these technologies a new correlation between universities and industries has become pervasive. The commercial possibilities created by nanotechnology have brought it closer to intellectual property rights (IPR). However, since it is ever evolving, all forms of IPR have not yet become relevant, as compared to patents and, to some extent, trade secrets. The other rights are gearing up to offer respective forms of protection on different aspects of nanotechnology. Nevertheless, a general discussion on traditional IPR such as copyrights and trademarks along with patents and trade secrets emerges. Before delving into the discussion of nanotechnology—IPR interface, it is necessary to inquire into the nature and scope of nanotechnology and its convergence with IPR.

**Role of Plant Growth Promoting Microorganisms in Sustainable Agriculture and Nanotechnology.**
**DOI: https://doi.org/10.1016/B978-0-12-817004-5.00007-5**

## 7.2 Nature and scope of nanotechnology

Materials often used in nanotechnological developments express distinct behavior at the nanoscale level, and create the potential for the cheap construction of rare molecules for a range of uses, such as the production of light, exceptionally strong microfibers, and ultrasensitive detectors (Zekos, 2006). Nanodevices can also trigger changes in the medical/biomedical field. Distributed through the brain, these devices may allow for the copying of thought patterns, and ultimately the copying of a person's personality, leading to the creation of artificial intelligence. Nanodevices have the potential to produce new products, ranging from novel drugs and related devices to "nanorobots that travels through the body finding and diagnosing illness" (Zekos, 2006). Drug companies may indeed start focusing on nanotechnology research with hopes of developing nano-sized technologies that can detect, diagnose, and treat diseases. Preparing for implementation, the resulting nanomedicine has a function defined as "the monitoring, repair, construction and control of human biological systems at the molecular level, using engineered nanodevices and nanostructures" (Zekos, 2006). Nanotechnology applications are not only expanding the horizon of science but also stretching the boundaries of IPR (Nano werk, 2006). It is considered as "the main driver of economic growth in the foreseeable future" (Lemley, 2005).

## 7.3 Nanotechnology and intellectual property rights

Due to the interdisciplinary nature of nanotechnology, there is often enormous difficulty in filing an IP application. Nanotechnology creation is a process that involves a team of scientists across various scientific disciplines collaborating on technology comprising multiple components from each discipline, each potentially requiring multiple IP licenses. Nanotechnology is typically developed through multidisciplinary expertise from different fields such as biology, chemistry, engineering, and material science (Nano werk, 2006). The emergence of nanotechnology created commercial possibilities for universities, stemming from university research, just like biotechnology. However, it differs from the former in the inclusion of many of the building block patents. These building block patents, in an emerging technology such as nanotechnology have high economic worth and are prerequisites for downstream innovations. The previous trend was that the building block technologies remained outside of the reach of patents, and were not heavily concentrated in universities (Nano werk, 2006). However, following the current trend, new technology that has been brought to the market has immense commercial value and, just as has been seen with the biotech industry where there was a financial reward for innovation, this can lead to high-profile legal battles over patents. Given the relatively younger and evolutionary nature of nanotechnology, such high-profile battles have not yet been seen, but the introduction of this technology to various markets has the advantage of creating great interest among established industries to come up with improved versions of current products rather than attempting to create new products (Nano werk, 2006). Companies are also seeking new collaborators. For example, "big pharma" pharmaceutical companies are showing interest in either collaborating or acquiring nanotechnology startups (e.g., Elan acquired Nanosystems) for the aforementioned reason. Alliance and acquisitions among academic institutions, private companies, and the government will shape the future of nanotechnology (Nano werk, 2006).

Given the evolutionary character of nanotechnology, it is difficult to select an appropriate approach to IP. The IPR relating to nanotechnology have yet to be substantiated and the traditional doctrines of IP law lag behind the pace of technological development (Dickson, 2010). Despite an unclear situation surrounding the protection of nanotechnology IP, numerous attempts have been made under the existing mechanism; many have been successful. Expert drafters in closely associated fields (biotechnology being one such field) are engaged to come up with an appropriate claim to be made under the most appropriate IP protection. Companies cannot afford to let IP to be settled at some point in the future, as the rapid pace of technology could negate the benefit. To maximize the value and protection of IPR relating to nanotechnology, proper IP portfolio development is required that focuses on various components including processes, products, brands, and other potentially worthy subject matter of IPR (Dickson, 2010). Developing a proper strategy for IP protection is a complex phenomenon given the definitional conundrum and uncertain scope of nanotechnology; it is, thus, pertinent to explore the appropriate form of IP protection to cover these various aspects.

### 7.3.1 Copyrights

Copyrights protect original expressions of artistic, musical, literary, and dramatic works. The scope of copyrights has widened with new technological developments such as databases, software, and rights for cinematographs and broadcasting constitute copyrightable subject matter. Originality and authorship are two essentials for copyright protection with the condition that the expression must have fixation in tangible form. Though it is patents and not copyrights that primarily cover scientific and technological breakthroughs, the latter may have important implications given the protection extended to software code and chip design. Copyright protection could be extended to nanotechnology designs that incorporate such elements or any element that is analogous to software code or chip design (Dickson, 2010). Companies have started exploiting copyrights in these areas and have found it potentially instrumental in combating or deterring exact copying (Dickson, 2010).

### 7.3.2 Trademarks

Trademarks offer protection to a distinct mark of a trademark holder to establish and sustain his goodwill in the market and enable consumers to go by their choice. Distinctiveness and indication of source are two essentials of trademarking. In the contemporary world, trademarks are not only sought for goods or services currently existing but for the future goods of trade as well. Goods or services relating to the emerging technologies need to be protected by the entities making inroads in the new business. An emerging technology such as nanotechnology, branding needs to be established at the earliest point possible. The United States Patent and Trademark Office (USPTO) is flooded with trademark applications seeking to protect all things nano (Dickson, 2010). Digitization also exerts great influence on trademark regimes: it opens the gates for online trade by providing the Internet as a platform where traders may get trademark protection for their websites and domain names to launch their products in the market. In an emerging field like nanotechnology, establishing a brand value of the products or services through trademarks is critical; it may also be helpful to develop this value in well-known brands. This may create incredible marketing opportunity to companies beginning to develop nanotechnology. It also could provide an effective protection against counterfeiting of products. Although "distinctiveness" is the hallmark of trademarks, descriptive words such as "nanosilver" and "nanoparticle" lack distinctiveness and are difficult to protect (Dickson, 2010). On the other hand, fanciful or arbitrary marks offer high levels of protection due to its distinctive character (e.g., "Apple" for computers, "Camel" for stationary, etc.). Therefore, in the emerging field of nanotechnology, the trademark holders must be vigilant to protection themselves from being hijacked by competitors or imposters (Dickson, 2010).

### 7.3.3 Patents

Patents are the most suitable IPR for nanotechnology, as is true for other types of technology. Patents offer an effective protection to subject matter that has a functional value. Patents offer protection to inventions conforming to essential criteria such as novelty, nonobviousness (inventive step), and utility (industrial applicability). It is based on policymakers striking a balance between the interest of the individual inventor and the public good. Patents have an ultimate objective to promote scientific and technological growth. In order to strike such a balance, there is an attached statutory requirement of public disclosure of the invention in question. The requirement of public disclosure works on the principle that invention spurs invention and disclosure may lead to innovation. Through public disclosure many scientists and potential inventors may learn the technological information needed to improve upon the current patented invention, come up with a new and improved product/process or to invent an altogether new invention. Not only is the requirement of public disclosure through written description instrumental in promoting the balancing act but also essential are the various exemptions provided within the same legal framework in the form of research exemptions, compulsory licensing, and governmental use.

In the patentability debate, the patentable subject matter criterion is the first filter to be passed. The issue of patentable subject matter can be determined through an objective and subjective approach. The first approach involves the distinction between a discovery and an invention, as it is only the invention that forms the subject matter of patents and not the discovery. Many countries follow this distinction.

The other approach involves the subjective assessment of the invention to determine whether it fits in the legal framework of patents for the given country. In other words, if it does not fall under the list of nonpatentable subject matter included in the patent law of that country, it can be patentable. Even if it qualifies all the criteria of patents but has been excluded from the purview of patenting, it could not be made patentable in a particular patent system. This is based on the territoriality principle of patents stating that each country may develop its own patent law while considering its own socioeconomic condition. Trade related aspects of intellectual property rights (TRIPS) agreement sets minimum standards to be complied with by the member countries of world trade organization (WTO). It contains mandatory and optional provisions, providing reference for countries to shape their law by using the optional clauses for flexibility. It depends on the caliber of a particular country how one can make use of these flexibilities in alignment with its socioeconomic conditions. With regards to patentable subject matter, nanotechnology usually contains the building block technologies and deals with very basic science, which makes this filter difficult to pass. However, how a claim is drafted will determine how to distinguish it from product of nature.

Along with subject matter criterion, the other criteria to be met are novelty, nonobviousness (inventive step) and utility (industrial applicability). Novelty of an invention is adjudged against the prior art available at the date of the patent application. The invention in question must not be present in the prior art through documentation, public knowledge or public use. Countries vary regarding the degree of novelty required, as few prefer absolute novelty while the others relative novelty. After establishing the novelty the other test to be passed is nonobviousness or inventive step. Nonobviousness is established when the invention provides a new technical solution to a problem or brings advancement in technology that could not be foreseen by a person skilled in the art. The solution or advancement should not be obvious to the person skilled in the art. Here, the person skilled in the art entails a person of ordinary prudence working in a particular field of technology. The degree of nonobviousness criterion also varies from one country to another. The next filter that an invention has to pass is utility (industrial applicability). Under the utility test, the invention must have real industrial use—otherwise there is no cause in granting patent to it. This is based on the ultimate policy of patents to spur scientific and technological progress and economic growth. The USPTO had to issue guidelines for utility to be substantial, specific and credible after being flooded with patent applications with minimal utility. Therefore, the degree of utility also varies among countries and may change with time. The last statutory requirement for an invention to be patentable is its full and complete disclosure that enables other inventors to learn from it and infer the state of art involved in it. In case of biological materials, if the material cannot be properly disclosed through writings, drawings or model, a biological sample is to be deposited in the International Depository Authority.

The distinction between discovery and invention becomes blurred in the case of nanotechnology. Nanotechnology patents claims are overly broad and, due to absence of prior art, allows patent holders to lock up huge areas of technology (building blocks of the technology), affecting the downstream research and innovation in the field. Additional issues related to patenting is whether simply reducing the size of a known product to an atomic level would suffice to meet the criteria of novelty and inventive step [WIPO (a)], and whether the rights of a patent holder who was granted a patent without mentioning the size of the invention could be infringed by a corresponding nanotechnology version of the invention—can this become a basis for claiming royalties from the inventor of nanotechnology invention [WIPO (a)]?

The definitional conundrum of nanotechnology has far-reaching implications for patent search and classification, making it extremely difficult to track patent trends in this area. This makes the prior art uncertain for a person skilled in the field, and carries the risk of overlapping and conflicting patents. To overcome this difficulty the European Patent Office (EPO), Japan Patent Office (JPO), and USPTO restrict invention to a length scale of less than 100 nm (WIPO Magazine, 2011). Despite this restriction numerous terminologies used for compositions still create confusion for patent examiners, such as "nanoagglomerates" (WIPO Magazine, 2011). Due to its multidisciplinary nature, a patent application relating to a nanotechnology invention may involve more than one field of science and engineering. This makes the task of the patent examiner even more difficult as he may not be an expert of all the fields. In such a situation, a patent examiner may overlook the prior art, which results in the grant of a substandard patent. To rectify this situation, EPO, JPO, and USPTO came up with new nanotechnology tags in patent classification systems —"Y01N" (EPO), "ZNM" (Japan), and "977" (USPTO). This ensures quality patent searches to a deeper extent in the nanotechnology field (WIPO Magazine, 2011).

On the novelty count, simply reducing the size of an already patented product without any change in the properties of that product would negate novelty, however, the difference in the properties are sufficient to import novelty (BASF, 2002). In the field of nanotechnology, patents for inventions fall within overlapping ranges and create a fragmented patent proprietorship landscape with multiple blocking patents on the same invention. This leads to a problem of patent thickets that creates a "dense web of overlapping patent rights that creates uncertainty and makes it difficult for other inventors in designing around existing patents," with serious implications for innovation (WIPO Magazine, 2011).

Nonobviousness (inventive step) requires a technology to produce a new and unexpected result or serves previously unrecognized functions that overcomes a technical problem relating to prior art. In Smithkline Beecham Biologicals v. Wyeth Holdings Corporation, Boards of Appeal of the EPO reflected upon this situation where "the vaccine adjuvant was held to be inventive because of its unexpectedly improved effect and the fact that nothing in the prior art had suggested that a skilled person might consider reducing the particle size to achieve that advantage" (Smithkline 2003; WIPO Magazine, 2011). The novelty and nonobviousness tests heavily depend on how the claims in patent applications are construed, as it should specifically demarcate the nature and scope of the invention. Given the uncertain and unpredictable state of the art, nanotechnology creates complexity in establishing utility for its inventions. Enforcement of nanotechnology patents is also a complex phenomenon as it is difficult to detect its abuse, given its multidisciplinary nature. Sophisticated and expensive microscopy techniques and equipment are needed to determine the infringement of a nanotechnology patent. There is also a need to enable researchers and inventors freedom to operate so that they can come up with novel nanotechnology innovations. A balanced patent approach is required for sustained growth of nanotechnology (WIPO Magazine, 2011).

### 7.3.4 Trade secrets

Since it currently remains difficult to conform to the standards of patents in the nanotechnology field, many companies are trying to protect their products and processes through trade secrets until such a time that a fair amount of certainty is established for patents. Trade secrets are most suited to companies that are in the grassroots phase, as it is cost-effective and less time consuming than other means of protection. Numerous factors determine whether a company should pursue a patent or trade secrets protection. In cases where there is a probability of reverse engineering of the product or process by the competitors and consumers, trade secret protection would not be a very good idea. On the other hand, if there is less probability of getting a patent, or patents prove to be costly, it may be the right option (Dickson, 2010). Furthermore, it also depends on whether the goods could be effectively protected against the leak of trade secrets as well as the time after which the technology remains valuable (i.e., the lifecycle of the technology) (Dickson, 2010).

The term "trade secret" refers to confidential or undisclosed information that has trade value. The value of the information resides in its secrecy rather than disclosure. Such secret business information provides companies a cutting edge over its competitors. A range of information is included in trade secrets, including technological and nontechnological information such as business patterns, manufacturing formulas, and consumer profiles. Legal systems of countries vary as to the form of protection of trade secrets. Few legal systems protect trade secrets through special legislations but the US legal system did just that with the enactment of the Uniform Trade Secrets Act of 1986. However, there are other countries like India that regulate trade secrets through existing laws, depending on the nature of activity (through contract law, law of crimes, information and technology law, etc., or through common law remedy). By and large the main objective of trade secrets law is to regulate unfair trade practices including industrial or commercial espionage, breach of contract, and breach of confidence [WIPO (b)].

Trade secret protection has significant advantages and disadvantages over patent protection. In comparison to trade secrets, patents require a formal procedure of registration and compliance with technical requirements, which usually takes 3–4 years to receive a granted patent. Patents also come with an annual renewal fee. On the other hand, with trade secrets no procedural formality is required and the trade secret can go into effect immediately. The period of protection that a subject matter could be protected is subject to the disclosure of trade secret: this is a strategic choice made after assessing the situation at hand. Article 39 of the TRIPS Agreement describes the general requirements for trade secret protection:

- The information must be secret (i.e., it is not generally known among, or readily accessible to, circles that normally deal with the kind of information in question).
- It must have commercial value because it is a secret.
- It must have been subject to reasonable steps by the rightful holder of the information to keep it secret (e.g., through confidentiality agreements) [WIPO (c)].

However, there are certain risks which are associated with trade secrets. In many cases, if a trade secret is embodied in an innovative product, others may find the secret by reverse engineering after which they become entitled to use it. Once the secret becomes public the protection is lost and it will be open for anyone to use it. In comparison to patent, it is considered as a weak protection [WIPO (d)]. Though nanotechnology usually takes place in universities, which cannot derive benefits by keeping the inventions secret, there are still numerous benefits for choosing this alternative. Where the cost of maintaining a patent is high in comparison to making it secret, where the chances of reverse engineering and independent discovery is low and where the technology is not likely to generate significant licensing revenues, universities as well as many corporations prefer trade secrets as an appropriate strategy. Numerous cases reveal the significance of trade secrets in the United States relating to nanotechnology (Ouellette, 2015).

## 7.4 Nanotechnology and intellectual property rights: Indian perspective

In India, the huge growth in nanotechnology research and the increasing trend of commercialization is evident. Wherever IP protection of nanotechnology is concerned, patents play a predominant role. India contains an exhaustive list of patent exclusions under Section 3 and Section 4 of the Patents Act 1970. In order to be patentable, the patentable subject matter must not fall within the aforementioned list of exclusions. One of the exclusions to patents that is widely recognized by most of the countries except United States and Australia, is the method of treatment (Rostrum's Law Review, 2016).

Article 27(3)(b) of the TRIPS agreement maintains that a Member nation can make a subject matter patentable if the invention fulfill the criteria of novelty, inventive step and industrial applicability. However, it allows that Member countries may exclude the "diagnostic, therapeutic and surgical methods for the treatment of humans or animals" from the scope of patentability. Sticking to this provision, India shaped its law. Section 3(i) excludes from patentability "any process for the medicinal, surgical, curative, prophylactic, diagnostic, therapeutic, or other treatment of human beings or any process for a similar treatment of animals to render them free of disease or to increase their economic value or that of their products." It clearly establishes that a diagnostic or surgical tool could be patented but a diagnostic and surgical method is not patentable. With the introduction of nanomedicines, the scope of patent protection for methods becomes an important discussion. In developing both a medicinal product and treatment, a huge amount of investment in research and development is required, which can only be ensured and protected through an incentive-based model. Recognizing the patentability of the methods used for human and animal treatment may promote nanotechnology development in India (Rostrum's Law Review, 2016).

Patents receiving medical treatment require safety assessments through practical experiments and tests. Nanotechnology has made great impacts not only on products such as medical equipment, apparatuses, etc., but also on the processes or methods themselves. The interdependence of product and process in the nanotechnology field is well established; a nanotechnology-based product may be designed with a particular method of treatment in mind whereas without the product that method of treatment could not be done. For example, laser treatment is the best way for treating eye diseases and disorders but in India the laser device could be patentable but the method of treatment could not (Rostrum's Law Review, 2016).

Indian patent law recognize the absolute novelty requirement, which necessitates that in order to establish the novelty of an invention it shall not be present in the prior art of India or any part of the world. The interdisciplinary nature of nanotechnology enlarges the scope of prior art, which raises objections against novelty. In nanotechnology, novelty is difficult to establish as "the technology mainly concentrates on reduction of size in existing inventions in a compatible and enhanced manner" (Rostrum's Law Review, 2016). Strict construction of novelty criterion may preclude a huge number of nanotech inventions derived from a preexisting technology anticipated by the prior art.

The next criterion to be met is inventive step, which is described under Section 2(1)(ja) of the Patents Act 1970. It defines inventive step as "a feature of an invention that involves technical advance compared to the existing knowledge, or having economic significance, or both, that makes the invention not obvious to a person skilled in the art." The technical advancement or economic significance is subjected to an important condition that it makes the invention nonobvious to a person skilled in the art, or in other words the person skilled in the art cannot anticipate it. Even if technical advancement criterion is satisfied, the obviousness criterion makes things complex. In nanotechnology inventions, most inventions are with respect to optimization and advancement of features of product and curtailing the size; therefore the criteria of inventive step becomes tricky (Rostrum's Law Review, 2016). Philip Morris Co.'s patent reflects this complexity where a patent was related to the conversion of carbon monoxide to carbon dioxide while smoking cigarette. This invention found it difficult to pass the test of inventive step, as it may be obvious to a person skilled in the art given the fact that the conversion of carbon monoxide to carbon dioxide through catalyst reaction of metal oxide is already available in the prior art. Therefore, despite contributing in the technical advancement in the field, the invention falls short of inventive step (Rostrum's Law Review, 2016). There is a need to redefine the inventive step criterion by prescribing the mode of technological advancements and nonobviousness in the light of nanotechnology inventions; otherwise many useful inventions may be precluded from patenting.

Industrial applicability has been defined under Section 2(ac) of the Patents Act 1970 as "the invention is capable of being made or used in an industry." India adopts substantial utility or a real-world use. Indian patent law seeks to prevent inventions that fall short of industrial application that cannot be put into use. However, if "capable of being used in an industry" is given a very strict interpretation it will create problems for nanotechnology patents. Since nanotechnology is an evolving technology the precision of its industrial use cannot be easily established as certain discrepancies may arise from lab use to actual use in the market and in the real world (Rostrum's Law Review, 2016). As compared to the lab results, which happen in a controlled environment in which it is relatively easy to determine the external factors, this is not the case in the real world, where external influences may affect the test results, keeping them from conforming to the intended results (Rostrum's Law Review, 2016). For example, consider a hypothetical situation in the case of a lithium battery. The chemical composition and specific conductivity make the battery last longer, however, the result varies from the lab to the real world as in the real world, the external influence in the form of heat, humidity, etc., affects the performance of the battery and may be a hurdle in proving the claimed industrial use (Rostrum's Law Review, 2016). As compared to India, the United States and China maintain a lower threshold of utility. While the United States USA its utility criterion relatively stricter when the USPTO issued guidelines in 2001, which state that the utility required for an invention to be patentable must be credible, specific, and substantial, unlike India, it does not mandate for a real-world utility. Credible utility indicates that the person skilled in the art should believe its existence on the basis of the reasoning provided. Specific utility maintains that the utility in question must not be general in nature but must be specific to a particular subject matter. China also maintains a lower threshold for utility and requires mere laboratory results to overcome this criterion (Rostrum's Law Review, 2016). Attaching the real-world utility with the nanotechnology patents in India may prove to be difficult and it may restrict the scope of nanotechnology patents in India.

Section 3(d) is another filter to be passed in order to make a subject matter patentable. It was introduced in the Patents Act 1970 to check the ever-greening of patents, specifically for pharmaceutical substance through the Patents (Amendment) Act 2005. It states that a new form of a known substance is not patentable unless it has an enhanced efficacy as compared to its previous version.

Section 3(d) enunciates:

[T]he mere discovery of a new form of a known substance which does not result in the enhancement of the known efficacy of that substance or the mere discovery of any new property or new use for a known substance or of the mere use of a known process, machine or apparatus unless such known process results in a new product or employs at least one new reactant.

Explanation—for the purposes of this clause, salts, esters, ethers, polymorphs, metabolites, pure form, particle size, isomers, mixtures of isomers, complexes, combinations, and other derivatives of known substance will be considered to be the same substance, unless they differ significantly in properties with regard to efficacy.

The term "efficacy" has not been defined and in the famous case of Novartis (Novartis, 2013), the Supreme Court of India interpreted Section 3(d) in the context of pharmaceutical patents by invoking the efficacy as "therapeutic efficacy" and established that unless and until the product in question is a significant improvement upon a known substance that enhances the treatment value (efficiency) of such substance it could not pass the filter of Section 3(d). Given the uncertainty of the definition of efficacy, Section 3(d) may prove difficult to pass. Furthermore, since Section 3(d) creates a presumption against novelty of combination patents, it would be a difficult criterion for nanotechnology patents as in most cases the inventions under nanotechnology are "combinations of many particles, technologies or a nanoparticle of an existing material" and patents for the newly found substance will not be possible without substantial difference in character, efficacy, and industrial application (Rostrum's Law Review, 2016). Change in size is of no value unless and until the product in question demonstrates substantial differences in properties of the components increasing its efficacy. The term "efficacy" is difficult to define in an evolving field of nanotechnology and may create barriers in getting the patents. It would also make an adverse impact on nanotechnology research and may delay innovations in this field (Rostrum's Law Review, 2016).

## 7.5 Problems and prospects

There is a dearth of skilled or qualified examiners for nanotech patent examination in India, and the existing ones are overloaded. Likewise, there is no inclusive prior art database. The overly broad patent claims and multidisciplinary applications are prevalent in the case of nanotech patent applications. To lessen the burden from its shoulders, the Indian Patent Office is moving to outsourcing its normal patent searches to council of scientific and industrial research (CSIR) (Rostrum's Law Review, 2016). India adopted International Patent Classification and recently incorporated "the special nanotechnology patent classification B82Y, inspired from the EPO's Y01N classification" (Rostrum's Law Review, 2016). However, it lacks proficient automation tools for nanotech prior art searches as well as an ample database, causing it to fall well behind the world technology race. The USPTO maintains an inclusive list of nanotechnology terms in the 977 classifications to avoid any sort of confusion relating to conflicting definition. There are no such comprehensive guidelines in India that may define the terminology relating to nanotech inventions. (Rostrum's Law Review, 2016). It is not only the administrative processes that cause delay, but applicants are also equally responsible as the clerical errors found in applications and the subsequent delay in responses from applicants is a large factor. (Rostrum's Law Review, 2016).

There is a need for a balanced mechanism through which the bottlenecks of patents may be smoothened and a progressive path for nanotechnology patents may be opened. There is a need to develop comprehensive guidelines explaining the terminology relating to nanotechnology patents, related parameters, and inclusive prior art and nanotech classification in India. Special provisions for the nanotechnology patents are needed to be included in Indian patent law (Rostrum's Law Review, 2016). The growth of nanotechnology patents is required to encourage innovation.

## 7.6 Progressive trends

India has adopted a progressive trend in terms of R&D, as it ranks third in the number of publications in nanotechnology after China and the United States. The government has also been very supportive while granting substantial funding for the Nano Mission. These are reasons for India not to "miss the bus" as was the case of the electronic revolution of the 1970s, where other countries including China, Taiwan, and South Korea reaped the benefit of entering the trend of electronic development and became aligned with the wave of development. Nano Mission is not only focusing on research but also on training for the advancement of nanotechnology (Sibal, 2016). The intent of India is reflected in a research report prepared by ASSOCHAM and TechSci 2014: "From 2015 onwards, global nanotechnology industry would require about two million professionals and India is expected to contribute about 25% professionals in the coming years." India needs to prepare these professionals through proper training to position them to take the lead in nanotechnology development (Sibal, 2016).

Along with potential benefits, nanotechnology also poses enormous risks to human, animals, and the environment if not properly assessed and regulated. The majority of Indian practitioners recognize the ethical issues/constraints in nanotech research. These concerns include using the technology as a form of undetectable weaponry in warfare and the incorporation of nanodevices as performance enhancers in human beings (Sibal, 2016). Despite the fact that the Indian government itself has acknowledged that nanoparticles can be deposited in lungs and "may cause damage by acting directly at the site of deposition by translocating to other organs or by being absorbed through the blood," various products such as "nanosilver washing machines or insecticides with nanoparticles may continue to be sold in the Indian market without any proper risk assessment of their use" (Sibal, 2016). The risk relating to nanotechnology is well established as one of the studies made by the Massachusetts Institute of Technology found that carbon nanoparticles inhaled by rats "reached the olfactory bulb and also the cerebrum and cerebellum, suggesting that translocation to the brain occurred through the nasal mucosa along the olfactory nerve to the brain" (Sibal, 2016). India's lack of adequate investment in risk studies must be rectified through policy interventions (Sibal, 2016). Department of Science and Technology has released the "Guidelines and best practices for safe handling of nanomaterials in research laboratories and industries." This document describes precautionary measures for handling and disposal of nanoparticles by researchers and the industry (Sibal, 2016). The aforementioned efforts are in line with India's focus on taking the lead in the field of nanotechnology; however, it has to stand firm to meet the growing challenges.

## 7.7 Conclusion

There is no doubt that nanotechnology will be the next wave of technological revolution. Given the multidisciplinary nature of nanotechnology, it has wide implications in all walks of life and across a wide range of industries. No country can afford to avoid nanotechnology, as it is pervasive and seen as the technology of the future. Legal issues, particularly IP issues, raised by nanotechnologists are new and complex, challenging existing IP laws. Given the evolutionary and multidisciplinary nature of technology, there persists enormous confusion as to the application of IP laws and its doctrines. Dealing with nanotechnology requires a balanced legal approach: strict regulation or overregulation may dismantle the shape of nanotechnology and preclude the economic advantage that may be gained through this technology, while underregulation of these factors may pose enormous risks. The existing IP laws should be revised and interpreted based on the new technological developments of nanotechnology to produce constructive results.

## References

BASF v. Orica Australia Boards of Appeal of the EPO, T-0547/99 (8 Jan 2002).

Dickson, C.R., Creating and Protecting Intellectual Property Rights for Nanotechnology. Available from: <https://www.jonesday.com/practiceperspectives/nanotechnology/protecting_rights.html>.

Lemley, M.A., 2005. Patenting nanotechnology. Stanford Law Review 601–630.

Nano werk. 2006. Nanotechnology and intellectual property issues. Available from: <https://www.nanowerk.com/news/newsid = 1187.php>.

*Novartis AG v. Union of India (UOI) and Ors*, Civil Appeal No. 2706-2716 of 2013.

Ouellette, L.L., 2015. Nanotechnology and innovation policy 29:1 Harvard. Journal of Law & Technology 33–75.

Rostrum's Law Review, 2016. Nanotechpatentability issues in India. Available from: <https://rostrumlegal.com/journal/nanotech-patentability-issues-in-india/>.

Sibal, P. 2016. Why India needs nanotechnology regulation before it is too late. The wire. Available from: <https://thewire.in/law/why-india-needs-nanotechnology-regulation-before-it-is-too-late>.

Smithkline *Beecham Biologicals v Wyeth Holdings Corporation* Boards of Appeal of the EPO, T-0552/00 (30 October 2003).

WIPO (a), Patent expert issues: nanotechnology. Available from: <http://www.wipo.int/patents/en/topics/nanotechnology.html>.

WIPO (b), What is a trade secret? Available from: <http://www.wipo.int/sme/en/ip_business/trade_secrets/trade_secrets.htm>.

WIPO (c), How are trade secrets protected? Available from: <http://www.wipo.int/sme/en/ip_business/trade_secrets/protection.htm>.

WIPO (d), Patents or trade secrets? Available from: <http://www.wipo.int/sme/en/ip_business/trade_secrets/patent_trade.htm>.

WIPO Magazine, Patenting nanotechnology: exploring the challenges. Available from: <http://www.wipo.int/wipo_magazine/en/2011/02/article_0009.html>.

Zekos, G.I., 2006. Nanotechnology and Biotechnology Patents. International Journal of Law and Information Technology 310–369.

# Applying nanotechnology to bacteria: an emerging technology for sustainable agriculture

*Baby Kumari*[1], *M.A. Mallick*[2], *Manoj K. Solanki*[3], *Amandeep Hora*[4] *and Mahendra Mani*[4]

[1]Department of Biotechnology, Vinoba Bhave University, Hazaribag, India, [2]Faculty of Science, Vinoba Bhave University, Hazaribag, India, [3]Department of Food Quality and Safety, Institute for Post-Harvest and Food Sciences, The Volcani Center, Agricultural Research Organization, Rishon LeZion, Israel, [4]Department of Biotechnology, Guru Nanak College, Chennai, India

## 8.1 Introduction

The rapidly growing world population would cross about 9.6 billion by the year 2050 (as per UN report 2013), a 30% increase with respect to that in the 2010s. In this scenario, the increased demand for the high yield of crop production with lesser adverse effect on soil system is a big challenge for the modern agricultural system. The changing climate and overpopulation have led to the crisis of nutrient availability and food security for humans, especially in developing countries (Gouda et al., 2018; Prasad et al., 2017; Ram et al., 2014).

Soil is a robust living matrix and it is an important source of in agricultural practice and food security; it is also equally important for maintenance of all life processes (Dinesh et al., 2015; Reeves, 2017). Intensive farming practices that accomplish high yields need chemical fertilizers, mainly nitrogenous and phosphorus ones (Gupta and Dikshit, 2010; Rubin, van Groenigen, and Hungate, 2017), which are not only costly but also lead to soil, air, and water pollution. The intensive use of pesticides globally is about 2 million tons per year; out of which 45% Europe alone uses 45%, the United States consumes 25%, and 25% is used by the rest of the world (Pretty and Pervez Bharucha, 2015; Wani and Kothari, 2018). Indiscriminate use of these chemicals had led to pathogen and pest resistance, soil security and major disadvantage is the biomagnification of pesticides, pollinator decline and destroys natural habitat of farmer's field (Tilman et al., 2002). Disproportionate use of these chemicals has lethal effects on soil microorganisms, disturbs the fertility status of soil, and also pollutes the environment (Gupta et al., 2015; Ma et al., 2011; Tilman et al., 2002). The use of these fertilizers on a

**Role of Plant Growth Promoting Microorganisms in Sustainable Agriculture and Nanotechnology.**
**DOI: https://doi.org/10.1016/B978-0-12-817004-5.00008-7**

long-term basis often leads to decrease in pH and exchangeable bases, thus making them inaccessible to crops and the productivity of crop drops. Because of current public concern about the side effects of agrochemicals, there is increasing interest in understanding cooperative activities among plants and rhizosphere microbial populations. Therefore, there is an urgent need for biological agents (Duhan et al., 2017; Huang et al., 2015).

### 8.1.1 Bacteria are reservoirs of bioactive compounds

Plant growth peimoting rhizobacteria (PGPR) and rhizosphere has interaction that is, rhizoengineering and other techniques are the recent advances in this sector to meet global food and ecofriendly strategies for green earth/global warming (Duhan et al., 2017; Gupta et al., 2015; Huang et al., 2015; Ma et al., 2011; Vacheron et al., 2013). Advanced scientific research involves multidisciplinary approaches to understand the adaptation of PGPR, the effects on plant physiology and growth, induced systemic resistance, biocontrol of plant pathogens and biofertilization (Mma and Mfm, 2014; Singh et al., 2014). But there is an urgent need to develop technologies for formulation and mass production of bacteria at a commercial scale for field application (Gouda et al., 2018; Prasad et al., 2017).

### 8.1.2 Nanotechnology and agriculture

Nanotechnology is the branch of science that involves nanoparticles (NPs) in the order of 100 nm or less. Nanotechnology has been widely used in different fields of agriculture monitoring, food preservation, and other branches of physical, chemical, and medicinal sciences (Iavicoli et al., 2017; Wani and Kothari, 2018). A special field known as precision agriculture is a farming management process of measuring and responding to inter- and intrafield variations in crops to build a system of whole farm management and to make the most of the available resources (Duhan et al., 2017; Servin and White, 2016).

### 8.1.3 Role of bacteria in nanotechnology-based agricultural systems

Among the latest line of scientific innovations, nanotechnology holds an important place in transforming agriculture and food production (Cheng et al., 2016; Mukhopadhyay, 2014). The development of nanodevices and nanomaterials could open up potential applications in plant biotechnology and agriculture (Thakur et al., 2018; Timmusk et al., 2018). Nanotechnology permits broad advances in agricultural research, such as reproductive science and technology, conversion of agricultural and food wastes to energy and other useful byproducts through enzymatic nano-bioprocessing, disease prevention, and treatment in plants using various nanocides (Cheng et al., 2016; Duhan et al., 2017; Roco, 2003).

Potential aim and interest of nanotechnology is enormous (Prasad et al., 2017). These include improvement of plant growth through bacterial bioformulations and manifold increase in agricultural productivity using NP-encapsulated fertilizers for slow and sustained release of nutrients and water (Duhan et al., 2017; Prasad et al., 2017; Wani and Kothari, 2018). Nanobioformulation has emerged as a potential tool to enhance plant growth conditions by making the soil environment more suitable for plant rhozosphere; it can also control the contamination level of soil (Duhan et al., 2017; Iavicoli et al., 2017; Kashyap et al., 2013; Prasad et al., 2017; Vejan et al., 2016; Wani and Kothari, 2018).

Precision farming involves reaching farmers directly to educate them about research and development techniques, such as nanotechnology-based agricultural devices (Duhan et al., 2017). Nanoscale devices with novel properties make agricultural systems "smart." Smart systems deliver chemicals in a controlled and targeted manner similar to the proposed use of nanodrug delivery in humans. "Smart delivery systems" in agriculture have combinations of time-controlled, specifically targeted, highly controlled, remotely regulated/preprogrammed/self-regulated, and multifunctional characteristics to avoid biological barriers for successful targeting (Gouda et al., 2018; Ram et al., 2014).

In recent years, agricultural research has focused on nanotechnological aspects. The literature shows that there is much focus on developing bacterial-based NPs and delivery systems for application in nanoagriculture (Fig. 8.1A–C). This review summarizes current research efforts, new opportunities, and challenges for PGPR-based NP synthesis and potential applications in agriculture and related sectors (Fig. 8.2).

## 8.2 Biogenesis of nanoparticles

### *8.2.1 As an alternative to available inorganic nanoparticles*

In the last decade nanotechnology-based agricultural transformations have influenced the scientific community. Nanotechnology-based synthesis of biomolecules and modified microorganisms are widely applicable in almost all areas of agricultural systems and environment cleanup programs. Traditional synthesis of inorganic NPs using physical and chemical methods (Pantidos, 2014; Pooja Bansal and Gahlawat, 2014) have been replaced by biological methods.

Replacement of biogenetic NPs with inorganic NPs is advantageous as there is no need for complex and hazardous chemicals, sol gels, and laser applications, which lead to toxicity in the environment (Iavicoli et al., 2017; Pooja Bansal and Gahlawat, 2014; Siddiqi et al., 2018) and are costly (Pooja Bansal and Gahlawat, 2014). Biosynthesis of NPs overcome these disadvantages and are safe, cost-effective, and ecofriendly (Mukhopadhyay, 2014; Wilson et al., 2008). Microorganisms provide a suitable environment for the synthesis of NPs through biological methods. Physical conditions like pH, temperature, and substrate concentration impact the orientation, size, and shape of the synthesized NPs (Oliveira et al., 2018).

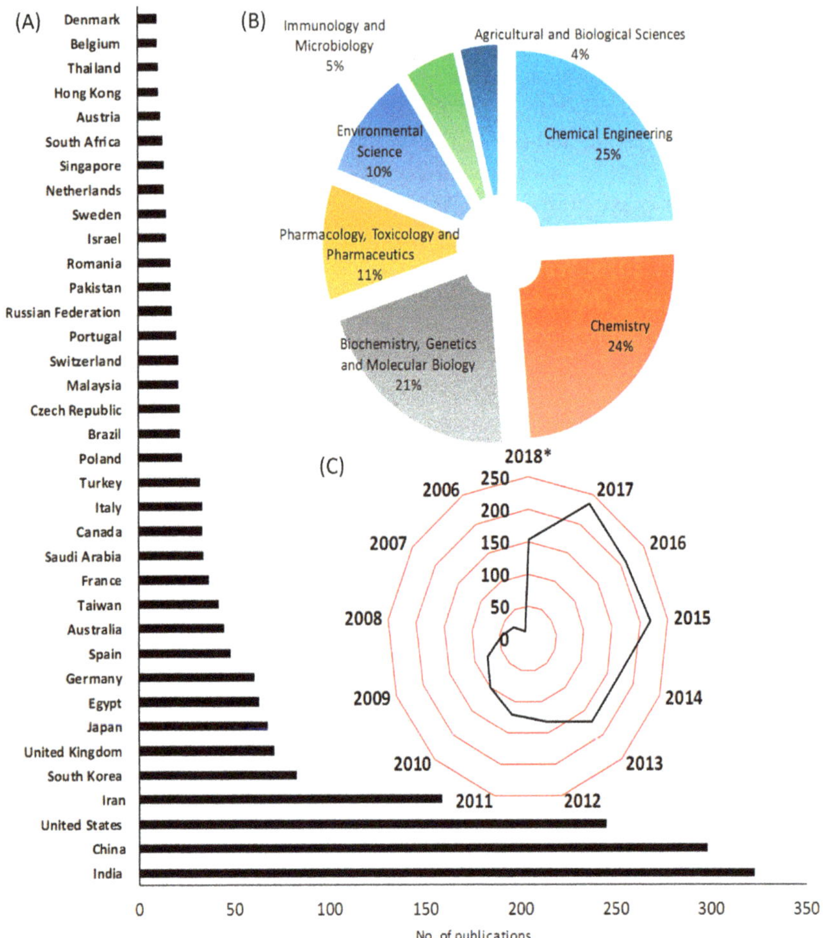


**Figure 8.1** Current status of bacteria-mediated nanoparticle research (country wise) (A), subject wise (B), and year wise. (C) Publications on NPs.
*Source*: Data: Scopus, Access date 13.07.18 by searching (bacterial and nano particles).

## 8.2.2 Bacteria as the source of biogenesis of nanoparticles

As bacteria are omnipotent and adaptable to extreme conditions, they are a good choice for study. The fast growth of bacteria is advantageous for nanoparticles fabrication and can be maipulate easily accordingly. Physical growth conditions like temperature, oxygenation, and incubation time can be easily controlled. Various studies have proven that altering the physical conditions of the growth medium during incubation results in NPs with differing size and shape (Pantidos, 2014; Pooja Bansal and Gahlawat, 2014).

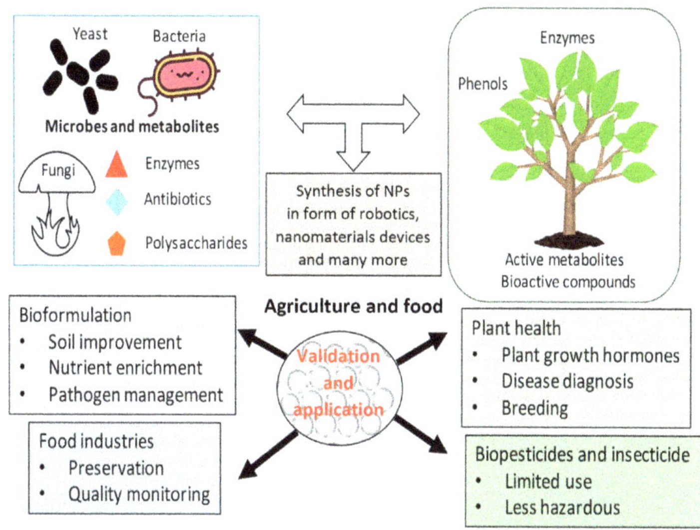


**Figure 8.2** Applications of NPs in agricultural, environmental, and food sectors.

As is known, bacteria can actively uptake and reduce metal ions. This can be achieved by exploiting the ability of bacteria to oxidize, reduce, and absorb metals, intracellularly or extracellularly (Siddiqi et al., 2018). Researchers around the globe are focused on this alternative green technology to fabricate NPs as it is an ecofriendly and cost-effective method. In a recent study, Yadav et al. (2013) reported the synthesis of silver NPs using culture supernatant of *Pseudomonas* sp. ARS22 (a well-known PGPR). This synthesis was achieved by extracellular reduction of Ag particles by *Pseudomonas* bacteria. This was validated by fourier transform infrared spectroscopy (FTIR) and atomic force microscopy (Yadav et al., 2013). In the past two decades, a large number of bacteria (*Pseudomonas*, *Bacillus* sps., *Cornybacterium*, *Shewanella oneidensis*, etc.) have been used as sources of NP synthesis using different inorganic metals like Ag, Au, Al, CdS, MnO, ZnO, $TiO_2$, MiO, etc. (Bhattacharyya et al., 2010; Bucur et al., 2018; Gui et al., 2017; Timmusk et al., 2018; Yadav et al., 2013). These bacteria-mediated NPs have shown positive effects on agricultural production, minimizing losses during cultivation of crops, transportation, and storage (Table 8.1).

### 8.2.3 Validation of nanoparticles

The demand of NPs products in consumers receiving significant attention from the past two decades, therefore validation of these products needed to avoid the uncertain side effects (Vance et al., 2015). Few metals are ruinously used to fabricate nanoformulation and devices, like silver and gold. Despite the increasing use, the risks of NPs are still not completely known (Wilson et al., 2008). NPs may lead to

**Table 8.1** Nano-based commercial products used in agriculture and food industry.

| Commercial Nano-products | Year of publication/ registration | Institute | Applications |
|---|---|---|---|
| Nano-sized nutrients (ZnO and $TiO_2$ NPs) | 2015 | Washington University in St. Louis | Boost in growth and antioxidants in tomatoes |
| Biodegradable thermoplastic starch (TPS) | 2002 | Pusan National University, Korea | Good tensile strength and lowered water permeability |
| Hydrolyzed collagen/sodium alginate nanocomposite | 2008 | Sichuan University, Sichuan, China | Preservation of loquat and cherry |
| Macronutrient fertilizers coated with zinc oxide NPs | 2012 | University of Adelaide, AU, CSIRO Land and Water, AU, Kansas State University, US | Enhancement of nutrients absorption by plants and the delivery of nutrients to specific sites |
| Primo MAXX | 2011 | Syngenta, Greensboro, NC, USA | Grass growth regulatory |
| Nanoemulsion | 2012 | VIT University, India | Neem oil (*Azadirachta indica*) nanoemulsion as larvicidal agent |
| Zeolites and Nano-clays | – | Geo humus-Frankfurt DE | Water retention and slow release of agrochemicals for proper absorption by the plants |
| Nanosensors | 2007 | University of Crete, GR | Pesticide detection with a liposome-based nano-biosensor |
| Acetamprid loaded alginate-chitosan nanocapsules | 2015 | GJUS & T, Hisar, India | Improved delivery of agrochemicals in the field, better efficacy, and better control of application/dose |
| Nano-Gro™ | 2016 | Agro Nanotechnology Crop, USA | Plant immunity improvement, growth regulator |
| The Nano-Ag Answer | 2016 | Urth agriculture, USA | Algal resistance, nutrient uptake enhancer |

(*Continued*)

**Table 8.1** (Continued)

| Commercial Nano-products | Year of publication/ registration | Institute | Applications |
|---|---|---|---|
| Nano Green | 2016 | Nano Green Sciences, Inc., India | Increases nutrient uptake |
| Rich Herba Green | 2016 | Richfield Fertilizer Pvt. Ltd., India | Plant resistance enhancement, growth enhancement, stress tolerance enhancement |
| Rich Vitaflora | 2016 | Richfield Fertilizer Pvt. Ltd., India | Plant resistance enhancement, growth enhancement, stress tolerance enhancement |
| Nano fertilizer | 2016 | Lazuriton Nano Biotechnology Co., Ltd., Taiwan | Growth regulator, pesticide usage reduction |

toxic effects on consumer health, and thus appropriate validation methods are needed to ascertain the safety of NPs in agriculture (Linsinger et al., 2012).

#### *8.2.3.1 Validation approaches*

##### Use of spiked samples

To overcome the problem of quantifying the amount and size of NPs in fabricated agricultural products, spiked samples are used in validation studies.

- Spiking should be optimized as per required size and concentration of NPs in nanoformulation.
- Spiking should not break up in agglomerates afterward.
- The sample preparation should incorporate spiking for proper homogenization.
- The spike should undergo as much as possible all sample preparation steps that might significantly change the materials.
- The method of spiking selection must be quick.

##### Determination of selectivity

Selectivity is defined as "the extent to which other substances interfere with the determination of a substance according to a given procedure" (Thompson et al., 2006). "Other substances" may refer to other intentionally added NPs as well as to materials already present in the sample (Kruve et al., 2015).

### Calibration curve, linearity, and working range

Mass fraction should be ranged from 50%-150% (lowest-highest) and the same should be applied for food matrix also.

### Precision

Testing precision should be done analogous to classical approaches in analytical chemistry, comprising determination of both repeatability as well as intermediate precision (within laboratory reproducibility). Ideally, two studies should be run in parallel; one using simple suspensions of the material in a solvent or water, and one using spiked food samples (https://sisu.ut.ee/lcms_method_validation/node/10721).

# 8.3 Application of bacterial nanoparticles

## 8.3.1 *Bacteria as nanobiopesticide/biocontrol agent*

The most recent advancement of nanobiopesticides (nanocides) in agriculture is the controlled release of pesticides, which results in fewer negative impacts on the soil's biophysical properties while still providing pest control in crops. Integrated uses of NPs with pesticides envisaged to reduce the application volume or quantity, and kinetics of development of resistance in pests. It has been reported that nanoencapsulated antimicrobial polypeptides can be released in the stomachs of insects, and may play a vital role in the protection of the environment by reducing leaching and evaporation of harmful substances. Botanical repellents encapsulated in nonmaterial have also been reported to minimize the toxicity level of synthetic pesticides. These designed bipropellants have controlled release and antipremature degradation qualities (Oliveira et al., 2018). *Escherichia coli* ATCC 8739, *Bacillus subtilis* ATCC 6633, *Streptococcus thermophiles* ESh1, and *Pseudomonas aerogenosa*) treated with gold, aluminum, ZnO, silica, and silver NPs have been reported as effective biocontrol agents against well-known biopests including *Fusarium* sps, (Iravani, 2014; Linsinger et al., 2012; Oliveira et al., 2018; Policy, 2017; Timmusk et al., 2018; Yadav et al., 2013). Recently, bacterial mediated Ag nanoparticle was reported (Siddiqi et al., 2018). These were synthesized by different methods, extracellularly or intracellularly, and were mediated NPs synthesized from a wide well-known novel bacterial species like *Brevibacterium*, *Bacillus*, *Salmonella*, *Gluconacetobacter*, *Pseudomonas*, *Lactoballicus*, etc. (Contado, 2015; Oliveira et al., 2018; Ram et al., 2014). These biosynthetic NPs have shown tremendous bactericidal effects. The antibacterial activity of SS-capped Ag NPs against gram-positive and gram-negative bacteria has been reported (Siddiqi et al., 2018). The antibacterial activity was studied in terms of minimum inhibitory concentration (MIC) that falls between 0.001 and 0.008 mM for all microorganisms namely, *Staphylococcus aureus*, *B. subtilis*, *Pseudomonas aeruginosa*, *Acinetobacter baumannii and E. coli* (Siddiqi et al., 2018).

### 8.3.2 Bacterial nanoparticles as nanobiofertilizers

The use of nanotechnology-based agricultural devices have revolutionized modern agricultural systems, leading to increased plant productivity and soil security. Nano-bioformulations are today considered as the most promising tool in environmental cleanup strategies (Huang et al., 2015). Reducing the bioavailability of metal contaminants in the rhizosphere (phytostabilization) as well as improving plant establishment, growth, and health could significantly affect plant growth and productivity (Ma et al., 2011).

PGPR (*Pseudomonas fluorescens*, *B. subtilis*, *Paenibacillus elgii*, and *Pseudomonas putida*) treated with gold, aluminum, and silver NPs have been reported recently to support plant growth yield and increase pathogen resistance. Nanoencapsulated biofertizers release NPs into target cells in a controlled manner, without any harmful effects. These NP-based formulations increase the adhesion of beneficial bacteria in the root rhizosphere (Mishra and Kumar, 2009). The rate of seed germination in different monocots and dicots has also been shown to be improved by pretreatment with ZnO NPs (Mishra and Kumar 2009). As can be seen, nanobiofertilizers are ecofriendly compounds that could be used in place of chemical pesticides (Caraglia et al., 2011).

Nanoencapsulation of fertilizers using with biodegradable materials also makes the concentrated active ingredients safe and easy to handle by the growers. The agronomic application of nanotechnology in plants (phyto-nanotechnology) has the potential to alter conventional plant production systems, allowing for the controlled release of agrochemicals (e.g., fertilizers, pesticides, herbicides) and target-specific delivery of biomolecules (e.g., nucleotides, proteins, activators). Improved understanding of the interactions between NPs and plant responses, including their uptake, localization, and activity, could revolutionize crop production through increased disease resistance, nutrient utilization, and crop yield (Wang et al., 2016).

### 8.3.3 Bacterial nanoparticles as biosensors

Biosensors have revolutionized agricultural systems by pathogen detection, high throughput analysis, and high quality monitoring of agricultural crops (Gui et al., 2017; Justino et al., 2017). Nanobiosensors are special sensors fabricated using an assembly of different transducers (physical, chemical, biological, electrochemical, etc.). They are classified on the basis of their transduction principle such as optical, piezoelectric, or electrochemical. Biosensors have also been classified based on their most sensitive carrier/recognition elements, and could be immunosensors, aptasensors, genosensors, and enzymatic biosensors (De et al., 2014; Huang et al., 2015; Mocan et al., 2017; Oliveira et al., 2018; Pretty and Pervez Bharucha, 2015). These sensors are rapid, specific, and selective and used for the detection of various toxic substances in agricultural systems and to detect plants diseases as well. Biologically originated (especially bacterial) biosensors are now developed by incorporating different NPs (Ag, Au, Cu, Zn, etc.) in microbes (bacteria, virus, and fungus). These NPs based devices (nanowires, nanoformulated particles, nanoencapsulated beds) are specialized in characterization of diseases in plants and plays and important role in cleanup strategies related to pesticide and insecticides accumulation in agricultural system.

Peiyan et al. (2018) described the role of NPs in the diagnostic of many pathogenic bacteria. There is a need for more flexible, reliable, and sensitive targeting of pathogens that could be used in biosensor systems. NPs with specific optical, electrochemical, or magnetic properties could increase the speed, sensitivity, and detection capacity of diagnostic methods (Contado, 2015; Nair et al., 2010; Subramanian and Tarafdar, 2011; Vance et al., 2015; Yadav et al., 2013). Furthermore, it is required to explore variety of configurations that allows us to envisage their implementation as point-of-care systems or multiplexed devices. Recently, Bucur et al. (2018) reviewed the role of biosensors based on enzymatic inhibition (microbial origin). These biosensors could be helpful in the quantitative toxicity detection of numerous insecticides currently being used such as organophosphorus compounds, carbamate compounds, etc.; these insecticides are almost banned or reported as having highest risk as reported by European Food Safety Authority (EFSA) for foods (Linsinger et al., 2012; Peiyan et al., 2018; Roco, 2003; Singh et al., 2014; Subramanian and Tarafdar, 2011). Gui et al. (2017) revealed that *Pseudomonas putida* (BMM-PL) could be applied as a whole cell-based biosensors. They concluded the role of these bacteria in detecting phenantherene in contaminated soil.

A variety of bacterial species converted to biogenetic nanobiosensors have been field tested and implemented in disease diagnosis and toxicity assessment (Table 8.2). Quality maintenance, one of the major constraint in Indian agriculture can be resolved through identity preservation (IP) and tracking. Proper monitoring of production systems through nanotechnology could ensure quality of organic products. IP is a system that creates increased value by providing customers with information about practices and activities used to produce a particular crop or other agricultural product. Certifying inspectors can take advantage of IP as a way of recording, verifying, and certifying agricultural practices. Through IP, it is possible to provide stakeholders and consumers with access to information, records, and supplier protocols. Nanoscale IP could provide continuous tracking and recording of the history of a particular agricultural product. The nanoscale monitors interconnected to recording and tracking devices to improve IP of food and agricultural products. The IP system is highly useful to discriminate between organic and conventional agricultural products.

## 8.4 Limitations and future of nanotechnology-based agricultural applications

Sustainable agriculture is the most suitable method in which abiotic and biotic live in coordinated way without harming the nature. Bacterial-based nonagricultural systems hold great promise for the agriculture and food sectors. The most important

**Table 8.2** List of bacterial resources for the synthesis of biological NPs and their applications in agricultural sector.

| Bacteria | Isolation source | Nanoparticle (s) | Size (nm) | Morphology mode of synthesis | Properties | References |
|---|---|---|---|---|---|---|
| *Corynebacterium* SH09 | Silver mine soil | Ag | 10–15 | – | – | Zhang et al. (2005) |
| *Bacillus licheniformis* KK2 | Sewage | Ag[a] | 50 | Spherical | – | Kalimuthu et al. (2008) |
| *Bacillus cereus* PGN1 | Effluent | Ag[a] | 4 and 5 | Spherical | – | Ganesh Babu and Gunasekaran (2009) |
| *Stenotrophomonas malophilia* AuRed02 | Gold mines soil sample | Au[a] | ~40 | – | – | Nangia et al. (2009) |
| *Brevibacterium casei* | Dairy industrial waste | Ag[a] | 10–50 | Spherical | – | Kalishwaralal et al. (2010) |
| *Rhodobacter sphaeroides* | – | Ag | 3–15 | Spherical | – | Bai et al. (2011) |
| *Escherichia coli* ATCC 8739, *B. subtilis* ATCC 6633, and *Streptococcus thermophiles* ESh1 | – | Ag | 5–25 | Spherical | Antimicrobial activity (*E. coli*, *Bacillus subtilis*, *Salmonella typhimurium*, *Klebsiella pneumoniae*, *Staphylococcus aureus*, *Pseudomonas aeruginosa*, and *Candida albicans*) | El-Shanshoury et al. (2011) |
| *P. aeruginosa* strain BS-161R | Petroleum-contaminated sludge | Ag | 13 | Mono-dispersed | Antimicrobial activity (Gram-positive, Gram- | Kumar and Mamidyala (2011) |

(*Continued*)

**Table 8.2** (Continued)

| Bacteria | Isolation source | Nanoparticle (s) | Size (nm) | Morphology mode of synthesis | Properties | References |
|---|---|---|---|---|---|---|
| | | | | and spherical | negative, and different *Candida* species) | |
| *Bacillus megaterium* NCIM 2326 | – | Ag | 80–98.56 | Irregular shapes | Antimicrobial activity (*Streptococcus pneumonia* and *Salmonella typhi*) | Saravanan et al. (2011) |
| *B. subtilis* | soil | Ag | – | Triangular, hexagonal | – | Kannan et al. (2011) |
| *Geobacillus stearothermophilus* | – | Ag | 5–35 | Spherical | – | Mohammed Fayaz et al. (2011) |
| *G. stearothermophilus* | – | Au | 5–8 | Spherical | – | Mohammed Fayaz et al. (2011) |
| *B. cereus* | *Garcinia xanthochymus* | Ag | 20–40 | Spherical | Antibacterial activity (*E. coli*, *P. aeruginosa*, *S. aureus*, *Salmonella typhi*, and *K. pneumonia*) | Sunkar and Nachiyar (2012) |
| *Lactobacillus mindensis* | Fixer solution | Ag | 2–20 | Spherical | – | Dhoondia and Chakraborty (2012) |
| *P. aeruginosa* | – | Ag | 20–50 | Spherical | Antimicrobial activity (*P. aeruginosa*, *S. aureus*, *and E. coli*) | Oza et al. (2012) |
| *Bacillus flexus* | Waste dump sites of silver mining | Ag | 12 and 65 | Spherical and triangular | Antimicrobial activity (*E. coli*, *B. subtilis*, *Streptococcus pyogenes*, and *P. aeruginosa*) | Priyadarshini et al. (2013) |

| | | | | | | |
|---|---|---|---|---|---|---|
| *Serratia nematodiphila* | | Ag | 10–31 | – | *B. subtilis*, *Klebsiella planticola*, and *P. aeruginosa* | Malarkodi et al. (2013) |
| *Stenotrophomonas* strain GSG-2 | Coral | Ag and Au | 40–60 (Ag), 10–50 (Au) | Circular, triangular, pentagonal, and hexagonal (Ag), spherical and irregular shapes (Au) | – | Malhotra et al. (2013) |
| *P. aeruginosa* | Saw mill | Ag | 70–80 | spherical | *E. coli* and *S. aureus* | Jeyaraj et al. (2013) |
| *Bacillus* sp. GP-23 | Marine soils | Ag | 7–21 | Spherical | Antifungal activity against *Fusarium oxysporum* | Gopinath and Velusamy (2013) |
| *Thermoactinomyces* sp. | Mangrove soil | Ag | 20–40 | Spherical | *S. aureus* and *B. subtilis* | Deepa et al. (2013) |
| *Stenotrophomonas maltophilia* OS4 | Rhizosphere of sweet pea | Ag | ~93 | Cuboidal | *S. aureus*, *E. coli*, and *Serratia marcescens* | Oves et al. (2013) |
| *Nocardiopsis* sp. MBRC-1 | Marine sediment | Ag | 45 ± 0.15 | Spherical | *E. coli ATCC 10536, B. subtilis ATCC 6633, Enterococcus hirae ATCC 10541, P. aeruginosa ATCC 27853, Shigella flexneri ATCC 12022, S. aureus ATCC 6538, Aspergillus niger ATCC 1015, A. brasiliensis ATCC 16404, A. fumigates ATCC 1022, C. albicans ATCC 10231* | Manivasagan et al. (2013) |

*(Continued)*

**Table 8.2** (Continued)

| Bacteria | Isolation source | Nanoparticle (s) | Size (nm) | Morphology mode of synthesis | Properties | References |
|---|---|---|---|---|---|---|
| *Gluconobacter roseus* NCIM No. 2049 | – | Ag | 10 | – | – | Krishnaraj and Berchmans (2013) |
| *Shewanella oneidensis* MR-1 | – | Ag | 2–16 | Spherical | – | Debabov et al. (2013) |
| *B. subtilis* MTCC 3053 | – | AgCl | | Polydispersed | *C. albicans* and *A. niger* | Paulkumar et al. (2013) |
| *Lactobacillus fermentum* ATCC 9338 | – | Ag | 13 | Spherical | – | Omidi et al. (2014) |
| *Bacillus thuringiensis* | Rhizosphere soil of cotton | Ag | 43.52–142.97 | Irregular | Larvicidal activity against dengue vector, *Aedes aegypti* | Najitha Banu et al. (2014) |
| *P. aeruginosa* KUPSB12 | Jute mill effluent contaminated site | Ag | 50–85 | Spherical | Antimicrobial activity (*E. coli*, *Vibrio cholerae*, *S. flexneri*, *B. subtilis*, *S. aureus*, and *Micrococcus luteus*) | Paul and Sinha (2014) |
| *Bacillus* strain CS 11 | Metal contaminated soil | Ag | 42–92 | Spherical | – | Das et al. (2014) |
| *Pseudomonas mandelii* SR1 | River bank | Ag | 1.9–10 | Spherical | Larvicidal activity, against *Anopheles subpictus* and *Culex tritaeniorhynchus* larvae | Mageswari et al. (2015) |

| | | | | | | |
|---|---|---|---|---|---|---|
| *Pseudomonas veronii* AS41G | Endophyte of *Annona squamosal* L. | Au | 5–25 | Irregular | *E. coli* MTCC 7410 and *S. aureus* MTCC 7443 | Baker and Satish (2015) |
| *Lactobacillus acidophilus* 58p | – | Ag[a] | 30.65 ± 5.81 | – | *Staphylococcus epidermidis*, *E. coli*, *K. pneumonia*, *S. flexneri*, and *S. sonnei* | Garmasheva et al. (2016) |
| *Pseudomonas fluorescens* CA 417 | Endophyte of *A. squamosal* L. | Ag | 20.66 | Spherical | *K. pneumonia* and *Xanthomonas campestris* | Baker et al. (2016) |
| *Lactobacillus plantarum* 92T | | Ag[a] | 19.92 ± 3.4 | – | *S. epidermidis*, *E. coli*, *K. pneumonia*, *S. flexneri*, and *S. sonnei* | Garmasheva et al. (2016) |
| *B. licheniformis* PTCC1320 | – | CdS | 2–10 | cubic | – | Bakhshi and Hosseini (2016) |
| *B. cereus* PTCC | – | Au | 20–50 | Spherical, hexagonal, and octagonal with irregular contours | Toxicity against cell lines | Pourali et al. (2017) |
| *P. aeruginosa* strain SN5 | Mangroves water | Ag | 35–60 | Spherical | *S. aureus* ATCC 6538, *B. subtilis* ATCC 6633, *E. coli* ATCC 8739, and *Staphylococcus epidermis* ATCC 12228 | Naik et al. (2017) |

(*Continued*)

**Table 8.2** (Continued)

| Bacteria | Isolation source | Nanoparticle (s) | Size (nm) | Morphology mode of synthesis | Properties | References |
|---|---|---|---|---|---|---|
| *P. aeruginosa* ATCC 27853 | – | Ag | 33–300 | Spherical | Antimicrobial activity (*E. coli*, *P. aeruginosa*, *S. typhimurium*, *S. aureus*, methicillin-resistant *S. aureus*, *Acinetobacter baumannii*, and *C. albicans*) | Peiris et al. (2017) |
| *Lactobacillus johnsonii* | Human gut | $TiO_2$ | 4–9 | Irregular | – | Al-Zahrani et al. (2018) |
| *L. johnsonii* | Human gut | ZnO | 5–9 | Spherical | – | Al-Zahrani et al. (2018) |
| *B. cereus* strain HMH1 | Chromite mines | MIO | 29.3 | Spherical | Cytotoxicity of NPs on the viability of MCF7 and 3T3 cell lines | Fatemi et al. (2018) |

others are extracellular; –, not available.
[a]Intracellular

thing about this modern technology is its application from "field to plate" (in biofertilizers, biopesticides, biosensors, and food packaging and preservation, etc.).

The biggest limitation of nanobiotechnology-based agricultural advancement is ethical acceptance. In India, the farmers are not well educated and they have less knowledge about the nanotechnological agricultural facilities. Another major issue is toxicity, the accumulation of NPs in biological systems like the food chain. Recently interest in NP-based agricultural research has focused on delayed exposure of NPs in the natural system. Scientist and researchers are working to find out the adaptive ways for the acceptability of NPs and nonaccumulation in a cell or system which leads it magnification. Although bacteria provide an excellent system for NPs to be applied in defferent sectors, yet there is need to investigate the method of NPs entry into plant cell system and its accumulation in serial part. In addition the transistion methods of biosynthetic NP delivery systems from the laboratory to the field should also be studied in the coming years (Bucur et al., 2018; Fatemi et al., 2018; Pourali et al., 2017; Servin and White, 2016; Thakur et al., 2018).

## References


Al-Zahrani, H., El-Waseif, A., El-Ghwas, D., 2018. Biosynthesis and evaluation of TiO2 and ZnO nanoparticles from in vitro stimulation of *Lactobacillus johnsonii*. J. Innov. Pharm. Biol. Sci. 5, 16–20.

Bai, H.-J., Yang, B.-S., Chai, C.-J., Yang, G.-E., Jia, W.-L., Yi, Z.-B., 2011. Green synthesis of silver nanoparticles using *Rhodobacter sphaeroides*. World J. Microbiol. Biotechnol. 27, 2723–2728. Available from: https://doi.org/10.1007/s11274-011-0747-x.

Baker, S., Satish, S., 2015. Biosynthesis of gold nanoparticles by *Pseudomonas veronii* AS41G inhabiting *Annona squamosa* L. Spectrochim. Acta A. Mol. Biomol. Spectrosc. 150, 691–695. Available from: https://doi.org/10.1016/J.SAA.2015.05.080.

Baker, S., Nagendra Prasad, M.N., Dhananjaya, B.L., Mohan Kumar, K., Yallappa, S., Satish, S., 2016. Synthesis of silver nanoparticles by endosymbiont *Pseudomonas fluorescens* CA 417 and their bactericidal activity. Enzyme Microb. Technol. 95, 128–136. Available from: https://doi.org/10.1016/J.ENZMICTEC.2016.10.004.

Bakhshi, M., Hosseini, M.R., 2016. Synthesis of CdS nanoparticles from cadmium sulfate solutions using the extracellular polymeric substances of *B. licheniformis* as stabilizing agent. Enzyme Microb. Technol. 95, 209–216. Available from: https://doi.org/10.1016/J.ENZMICTEC.2016.08.011.

Bhattacharyya, A., Bhaumik, A., Rani, P.U., Mandal, S., Epidi, T.T., 2010. Nano-particles—a recent approach to insect pest control. Afr. J. Biotechnol. 9, 3489–3493. Available from: https://doi.org/10.5897/AJBx09.021.

Bucur, B., Munteanu, F.D., Marty, J.L., Vasilescu, A., 2018. Advances in enzyme-based biosensors for pesticide detection. Biosensors 8, 1–28. Available from: https://doi.org/10.3390/bios8020027.

Caraglia, M., Rosa, G.D., Abbruzzese, A., Leonetti, C., 2011. Nanotechnologies: new opportunities for old drugs the case of amino-bisphosphonates. Nanomedic. Biotherapeu Dis 1, 103e.

Cheng, H.N., Klasson, K.T., Asakura, T., Wu, Q., 2016. Nanotechnology in Agriculture. ACS Symposium Series 1224, 233–242. Available from: https://doi.org/10.1021/bk-2016-1224.ch012.

Contado, C., 2015. Nanomaterials in consumer products: a challenging analytical problem. Front. Chem. 3, 1–20. Available from: https://doi.org/10.3389/fchem.2015.00048.

Das, V.L., Thomas, R., Varghese, R.T., Soniya, E.V., Mathew, J., Radhakrishnan, E.K., 2014. Extracellular synthesis of silver nanoparticles by the Bacillus strain CS 11 isolated from industrialized area. 3 Biotech 4, 121–126. Available from: https://doi.org/10.1007/s13205-013-0130-8.

De, A., Bose, R., Kumar, A., Mozumdar, S., 2014. Targeted delivery of pesticides using biodegradable polymeric nanoparticles. https://doi.org/10.1007/978-81-322-1689-6.

Debabov, V.G., Voeikova, T.A., Shebanova, A.S., Shaitan, K.V., Emel'yanova, L.K., Novikova, L.M., et al., 2013. Bacterial synthesis of silver sulfide nanoparticles. Nanotechnol. Russ. 8, 269–276. Available from: https://doi.org/10.1134/S1995078013020043.

Deepa, S., Kanimozhi, K., Panneerselvam, A., 2013. Antimicrobial activity of extracellularly synthesized silver nanoparticles from marine derived actinomycetes. Int. J. Curr. Microbiol. Appl. Sci 2, 223–230.

Dhoondia, Z.H., Chakraborty, H., 2012. Lactobacillus mediated synthesis of silver oxide nanoparticles. Nanomater. Nanotechnol. 2, 15. Available from: https://doi.org/10.5772/55741.

Dinesh, R., Anandaraj, M., Kumar, A., Bini, Y.K., Subila, K.P., Aravind, R., 2015. Isolation, characterization, and evaluation of multi-trait plant growth rhizobacteria for their growth promoting and disease suppressing effects on ginger. Microbiol. Res. 173, 34–43. Available from: https://doi.org/10.1016/j.micres.2015.01.014.

Duhan, J.S., Kumar, R., Kumar, N., Kaur, P., Nehra, K., Duhan, S., 2017. Nanotechnology: the new perspective in precision agriculture. Biotechnol. Rep. 15, 11–23. Available from: https://doi.org/10.1016/j.btre.2017.03.002.

El-Shanshoury, A.E.-R.R., ElSilk, S.E., Ebeid, M.E., 2011. Extracellular biosynthesis of silver nanoparticles using *Escherichia coli* ATCC 8739, *Bacillus subtilis* ATCC 6633, and *Streptococcus thermophilus* ESh1 and their antimicrobial activities. ISRN Nanotechnol. 2011, 1–7. Available from: https://doi.org/10.5402/2011/385480.

Fatemi, M., Mollania, N., Momeni-Moghaddam, M., Sadeghifar, F., 2018. Extracellular biosynthesis of magnetic iron oxide nanoparticles by *Bacillus cereus* strain HMH1: characterization and in vitro cytotoxicity analysis on MCF-7 and 3T3 cell lines. J. Biotechnol. 270, 1–11. Available from: https://doi.org/10.1016/J.JBIOTEC.2018.01.021.

Ganesh Babu, M.M., Gunasekaran, P., 2009. Production and structural characterization of crystalline silver nanoparticles from *Bacillus cereus* isolate. Colloids Surf. B. Biointerf. 74, 191–195. Available from: https://doi.org/10.1016/J.COLSURFB.2009.07.016.

Garmasheva, I., Kovalenko, N., Voychuk, S., Ostapchuk, A., Livins'ka, O., Oleschenko, L., 2016. Lactobacillus species mediated synthesis of silver nanoparticles and their antibacterial activity against opportunistic pathogens in vitro. Bioimpacts 6, 219–223. Available from: https://doi.org/10.15171/bi.2016.29.

Gopinath, V., Velusamy, P., 2013. Extracellular biosynthesis of silver nanoparticles using *Bacillus* sp. GP-23 and evaluation of their antifungal activity towards *Fusarium oxysporum*. Spectrochim. Acta A. Mol. Biomol. Spectrosc. 106, 170–174. Available from: https://doi.org/10.1016/J.SAA.2012.12.087.

Gouda, S., Kerry, R.G., Das, G., Paramithiotis, S., Shin, H.S., Patra, J.K., 2018. Revitalization of plant growth promoting rhizobacteria for sustainable development in agriculture. Microbiol. Res. 206, 131–140. Available from: https://doi.org/10.1016/j.micres.2017.08.016.

Gui, Q., Lawson, T., Shan, S., Yan, L., Liu, Y., 2017. The application of whole cell-based biosensors for use in environmental analysis and in medical diagnostics. Sensors 17, 1623. Available from: https://doi.org/10.3390/s17071623.

Gupta, S., Dikshit, A.K., 2010. Biopesticides: an ecofriendly approach for pest control. J. Biopestic. 3, 186–188. Available from: https://doi.org/10.4172/2168-9881.S1.012.

Gupta, G., Parihar, S.S., Ahirwar, N.K., Snehi, S.K., Singh, V., 2015. Plant growth promoting Rhizobacteria (PGPR): current and future prospects for development of sustainable agriculture. . Microb. Biochem. Technol. 7, 96–102. Available from: https://doi.org/10.4172/1948-5948.1000188.

Huang, S., Wang, L., Liu, L., Hou, Y., Li, L., 2015. Nanotechnology in agriculture, livestock, and aquaculture in China. A review.. Agron. Sustain. Dev. . Available from: https://doi.org/10.1007/s13593-014-0274-x.

Iavicoli, I., Leso, V., Beezhold, D.H., Shvedova, A.A., 2017. Nanotechnology in agriculture: opportunities, toxicological implications, and occupational risks. Toxicol. Appl. Pharmacol. 329, 96–111. Available from: https://doi.org/10.1016/j.taap.2017.05.025.

Iravani, S., 2014. Bacteria in nanoparticle synthesis: current status and future prospects. Int. Sch. Res. Not. 2014, 1–18. Available from: https://doi.org/10.1155/2014/359316.

Jeyaraj, M., Varadan, S., Anthony, K.J.P., Murugan, M., Raja, A., Gurunathan, S., 2013. Antimicrobial and anticoagulation activity of silver nanoparticles synthesized from the culture supernatant of *Pseudomonas aeruginosa*. J. Ind. Eng. Chem. 19, 1299–1303. Available from: https://doi.org/10.1016/J.JIEC.2012.12.031.

Justino, C., Duarte, A., Rocha-Santos, T., 2017. Recent progress in biosensors for environmental monitoring: a review. Sensors 17, 2918. Available from: https://doi.org/10.3390/s17122918.

Kalimuthu, K., Suresh Babu, R., Venkataraman, D., Bilal, M., Gurunathan, S., 2008. Biosynthesis of silver nanocrystals by *Bacillus licheniformis*. Colloids Surf. B. Biointerf. 65, 150–153. Available from: https://doi.org/10.1016/J.COLSURFB.2008.02.018.

Kalishwaralal, K., Deepak, V., Ram Kumar Pandian, S., Kottaisamy, M., BarathManiKanth, S., Kartikeyan, B., et al., 2010. Biosynthesis of silver and gold nanoparticles using *Brevibacterium casei*. Colloids Surf. B. Biointerf. 77, 257–262. Available from: https://doi.org/10.1016/J.COLSURFB.2010.02.007.

Kannan, N., Mukunthan, K.S., Balaji, S., 2011. A comparative study of morphology, reactivity and stability of synthesized silver nanoparticles using *Bacillus subtilis* and *Catharanthus roseus* (L.) G. Don. Colloids Surf. B. Biointerf. 86, 378–383. Available from: https://doi.org/10.1016/J.COLSURFB.2011.04.024.

Kashyap, P.L., Kumar, S., Srivastava, A.K., Sharma, A.K., 2013. Myconanotechnology in agriculture: a perspective. World J. Microbiol. Biotechnol. 29, 191–207. Available from: https://doi.org/10.1007/s11274-012-1171-6.

Krishnaraj, R.N., Berchmans, S., 2013. In vitro antiplatelet activity of silver nanoparticles synthesized using the microorganism *Gluconobacter roseus*: an AFM-based study. RSC Adv. 3, 8953. Available from: https://doi.org/10.1039/c3ra41246f.

Kruve, A., Rebane, R., Kipper, K., Oldekop, M.-L., Evard, H., Herodes, K., et al., 2015. Tutorial review on validation of liquid chromatography–mass spectrometry methods: part I. Anal. Chim. Acta 870, 29–44. Available from: https://doi.org/10.1016/j.aca.2015.02.017.

Kumar, C.G., Mamidyala, S.K., 2011. Extracellular synthesis of silver nanoparticles using culture supernatant of *Pseudomonas aeruginosa*. Colloids Surf. B. Biointerf. 84, 462–466. Available from: https://doi.org/10.1016/J.COLSURFB.2011.01.042.

Linsinger, T., Roebben, G., Gilliland, D., 2012. Requirements on measurements for the implementation of the European Commission definition of the term "nanomaterial.". EUR Report . Available from: https://doi.org/10.2787/63490.

Ma, Y., Prasad, M.N.V., Rajkumar, M., Freitas, H., 2011. Plant growth promoting rhizobacteria and endophytes accelerate phytoremediation of metalliferous soils. Biotechnol. Adv. 29, 248–258. Available from: https://doi.org/10.1016/j.biotechadv.2010.12.001.

Mageswari, A., Subramanian, P., Ravindran, V., Yesodharan, S., Bagavan, A., Rahuman, A. A., et al., 2015. Synthesis and larvicidal activity of low-temperature stable silver nanoparticles from psychrotolerant *Pseudomonas mandelii*. Environ. Sci. Pollut. Res. 22, 5383–5394. Available from: https://doi.org/10.1007/s11356-014-3735-5.

Malarkodi, C., Rajeshkumar, S., Paulkumar, K., Vanaja, M., Jobitha, G.D.G., Annadurai, G., 2013. Bactericidal activity of bio mediated silver nanoparticles synthesized by *Serratia nematodiphila*. Drug Invent. Today 5, 119–125. Available from: https://doi.org/10.1016/J.DIT.2013.05.005.

Malhotra, A., Dolma, K., Kaur, N., Rathore, Y.S., Ashish, Mayilraj, S., et al., 2013. Biosynthesis of gold and silver nanoparticles using a novel marine strain of Stenotrophomonas. Bioresour. Technol. 142, 727–731. Available from: https://doi.org/10.1016/J.BIORTECH.2013.05.109.

Manivasagan, P., Venkatesan, J., Senthilkumar, K., Sivakumar, K., Kim, S.-K., 2013. Biosynthesis, antimicrobial and cytotoxic effect of silver nanoparticles using a novel Nocardiopsis sp. MBRC-1. Biomed. Res. Int. 2013, 287638. Available from: https://doi.org/10.1155/2013/287638.

Mishra, V.K., Kumar, A., 2009. Impact of metal nanoparticles on the plant growth promoting Rhizobacteria. Dig. J. Nanomater. Biostruct. 4, 587–592.

Mma, Y., Mfm, E., 2014. Biofertilizers and their role in management of plant parasitic nematodes. A review. E3 J. Biotechnol. Pharm. Res. 5, 1–6. Available from: https://doi.org/ISSN.2141-7474.

Mocan, T., Matea, C.T., Pop, T., Mosteanu, O., Buzoianu, A.D., Puia, C., et al., 2017. Development of nanoparticle-based optical sensors for pathogenic bacterial detection. J. Nanobiotechnology 15, 1–14. Available from: https://doi.org/10.1186/s12951-017-0260-y.

Mohammed Fayaz, A., Girilal, M., Rahman, M., Venkatesan, R., Kalaichelvan, P.T., 2011. Biosynthesis of silver and gold nanoparticles using thermophilic bacterium *Geobacillus stearothermophilus*. Process Biochem. 46, 1958–1962. Available from: https://doi.org/10.1016/J.PROCBIO.2011.07.003.

Mukhopadhyay, S.S., 2014. Nanotechnology in agriculture: prospects and constraints. Nanotechnol. Sci. Appl. 7, 63–71. Available from: https://doi.org/10.2147/NSA.S39409.

Naik, M.M., Prabhu, M.S., Samant, S.N., Naik, P.M., Shirodkar, S., 2017. Synergistic action of silver nanoparticles synthesized from silver resistant estuarine *Pseudomonas aeruginosa* strain SN5 with antibiotics against antibiotic resistant bacterial human pathogens. Thalass. An Int. J. Mar. Sci. 33, 73–80. Available from: https://doi.org/10.1007/s41208-017-0023-4.

Nair, R., Varghese, S.H., Nair, B.G., Maekawa, T., Yoshida, Y., Kumar, D.S., 2010. Nanoparticulate material delivery to plants. Plant Sci. 179, 154–163. Available from: https://doi.org/10.1016/j.plantsci.2010.04.012.

Najitha Banu, A., Balasubramanian, C., Moorthi, P.V., 2014. Biosynthesis of silver nanoparticles using *Bacillus thuringiensis* against dengue vector, *Aedes aegypti* (Diptera: Culicidae). Parasitol. Res. 113, 311–316. Available from: https://doi.org/10.1007/s00436-013-3656-0.

Nangia, Y., Wangoo, N., Goyal, N., Shekhawat, G., Suri, C.R., 2009. A novel bacterial isolate *Stenotrophomonas maltophilia* as living factory for synthesis of gold nanoparticles. Microb. Cell Fact. 8, 39. Available from: https://doi.org/10.1186/1475-2859-8-39.

Oliveira, J.L.D., Campos, E.V.R., Pereira, A.E.S., Pasquoto, T., Lima, R., Grillo, R., et al., 2018. Zein nanoparticles as eco-friendly carrier systems for botanical repellents aiming sustainable agriculture. J. Agric. Food Chem. 66, 1330–1340. Available from: https://doi.org/10.1021/acs.jafc.7b05552.

Omidi, B., Hashemi, S., Bayat, M., Larijani, K., 2014. Biosynthesis of silver nanoparticles by *Lactobacillus fermentum*. Bull. Environ. Pharmacol. Life Sci. 3, 186–192.

Oves, M., Khan, M.S., Zaidi, A., Ahmed, A.S., Ahmed, F., Ahmad, E., et al., 2013. Antibacterial and cytotoxic efficacy of extracellular silver nanoparticles biofabricated from chromium reducing novel OS4 strain of *Stenotrophomonas maltophilia*. PLoS One 8, e59140. Available from: https://doi.org/10.1371/journal.pone.0059140.

Oza, G., Pandey, S., Shah, R., Sharon, M., 2012. Extracellular fabrication of silver nanoparticles using *Pseudomonas aeruginosa* and its antimicrobial assay. Pelagia Res. Libr. Adv. Appl. Sci. Res. 3, 1776–1783.

Pantidos, N., 2014. Biological synthesis of metallic nanoparticles by bacteria, fungi and plants. J. Nanomed. Nanotechnol. 05. Available from: https://doi.org/10.4172/2157-7439.1000233.

Paul, D., Sinha, S.N., 2014. Extracellular synthesis of silver nanoparticles using *Pseudomonas aeruginosa* KUPSB12 and its antibacterial activity. JJBS Jordan J. Biol. Sci. 7, 245–250.

Paulkumar, K., Rajeshkumar, S., Gnanajobitha, G., Vanaja, M., Malarkodi, C., Annadurai, G., 2013. Biosynthesis of silver chloride nanoparticles using *Bacillus subtilis* MTCC 3053 and assessment of its antifungal activity. ISRN Nanomater. 2013, 1–8. Available from: https://doi.org/10.1155/2013/317963.

Peiris, M.K., Gunasekara, C.P., Jayaweera, P.M., Arachchi, N.D., Fernando, N., 2017. Biosynthesized silver nanoparticles: are they effective antimicrobials? Mem. Inst. Oswaldo Cruz 112, 537–543. Available from: https://doi.org/10.1590/0074-02760170023.

Peiyan, Y., Xin, D., Yan, Y.Y., Qing-Hua, X., 2018. Metal nanoparticles for diagnosis and therapy of bacterial infection. Adv. Healthc. Mater. 7, 1701392. Available from: https://doi.org/10.1002/adhm.201701392.

Policy, Z.D., 2017. Zero draft policy on regulation of nanoproducts in agriculture on regulation of nanoproducts in agriculture. Regulation of Nanoproducts in Agriculture.

Pooja Bansal, J.S.D., Gahlawat, S.K., 2014. Biogenesis of nanoparticles: a review. Afr. J. Biotechnol. 13, 2778–2785. Available from: https://doi.org/10.5897/AJB2013.13458.

Pourali, P., Badiee, S.H., Manafi, S., Noorani, T., Rezaei, A., Yahyaei, B., 2017. Biosynthesis of gold nanoparticles by two bacterial and fungal strains, *Bacillus cereus* and *Fusarium oxysporum*, and assessment and comparison of their nanotoxicity in vitro by direct and indirect assays. Electron. J. Biotechnol. 29, 86–93. Available from: https://doi.org/10.1016/J.EJBT.2017.07.005.

Prasad, R., Bhattacharyya, A., Nguyen, Q.D., 2017. Nanotechnology in sustainable agriculture: recent developments, challenges, and perspectives. Front. Microbiol. 8, 1–13. Available from: https://doi.org/10.3389/fmicb.2017.01014.

Pretty, J., Pervez Bharucha, Z., 2015. Integrated pest management for sustainable intensification of agriculture in Asia and Africa. Insects 6, 152–182. Available from: https://doi.org/10.3390/insects6010152.

Priyadarshini, S., Gopinath, V., Meera Priyadharsshini, N., MubarakAli, D., Velusamy, P., 2013. Synthesis of anisotropic silver nanoparticles using novel strain, *Bacillus flexus* and its biomedical application. Colloids Surf. B. Biointerf. 102, 232–237. Available from: https://doi.org/10.1016/J.COLSURFB.2012.08.018.

Ram, P., Vivek, K., Kumar, S.P., 2014. Nanotechnology in sustainable agriculture: present concerns and future aspects. Afr. J. Biotechnol. 13, 705–713. Available from: https://doi.org/10.5897/AJBX2013.13554.

Reeves, J., 2017. Climate change effects on biological control of invasive plants by insects. CAB Rev. Perspect. Agric. Vet. Sci. Nutr. Nat. Resour. 12. Available from: https://doi.org/10.1079/PAVSNNR201712001.

Roco, M.C., 2003. Nanotechnology: convergence with modern biology and medicine. Curr. Opin. Biotechnol. 14, 337–346. Available from: https://doi.org/10.1016/S0958-1669(03)00068-5.

Rubin, R.L., van Groenigen, K.J., Hungate, B.A., 2017. Plant growth promoting rhizobacteria are more effective under drought: a meta-analysis. Plant Soil 416, 309–323. Available from: https://doi.org/10.1007/s11104-017-3199-8.

Saravanan, M., Vemu, A.K., Barik, S.K., 2011. Rapid biosynthesis of silver nanoparticles from *Bacillus megaterium* (NCIM 2326) and their antibacterial activity on multi drug resistant clinical pathogens. Colloids Surf. B. Biointerf. 88, 325–331. Available from: https://doi.org/10.1016/J.COLSURFB.2011.07.009.

Servin, A.D., White, J.C., 2016. Nanotechnology in agriculture: next steps for understanding engineered nanoparticle exposure and risk. NanoImpact 1, 9–12. Available from: https://doi.org/10.1016/j.impact.2015.12.002.

Siddiqi, K.S., Husen, A., Rao, R.A.K., 2018. A review on biosynthesis of silver nanoparticles and their biocidal properties. J. Nanobiotechnol. 16. Available from: https://doi.org/10.1186/s12951-018-0334-5.

Singh, S., Gupta, G., Khare, E., Behal, K.K., Arora, N.K., 2014. Effect of enrichment material on the shelf life and field efficiency of bioformulation of Rhizobium sp. and P-solubilizing *Pseudomonas fluorescens*. Sci. Res. Rep. 4, 44–50.

Subramanian, K.S., Tarafdar, J.C., 2011. Prospects of nanotechnology in Indian farming. Indian J. Agric. Sci. 81, 887–893.

Sunkar, S., Nachiyar, C.V., 2012. Biogenesis of antibacterial silver nanoparticles using the endophytic bacterium *Bacillus cereus* isolated from *Garcinia xanthochymus*. Asian Pac. J. Trop. Biomed. 2, 953–959. Available from: https://doi.org/10.1016/S2221-1691(13)60006-4.

Thakur, S., Thakur, S.K., Kumar, R., 2018. Bio-nanotechnology and its role in agriculture and food industry. J. Mol. Genet. Med. 12, 1–5. Available from: https://doi.org/10.4172/1747-0862.1000324.

Thompson, M., Ellison, S.L., Wood, R., 2006. The international harmonized protocol for the proficiency testing of analytical chemistry laboratories (IUPAC Technical Report). Pure Appl. Chem 78, 145–196.

Tilman, D., Cassman, K.G., Matson, P.A., Naylor, R., Polasky, S., 2002. Agricultural sustainability and intensive production practices. Nature 418, 671.

Timmusk, S., Seisenbaeva, G., Behers, L., 2018. Titania (TiO2) nanoparticles enhance the performance of growth-promoting rhizobacteria. Sci. Rep. 8, 1–13. Available from: https://doi.org/10.1038/s41598-017-18939-x.

Vacheron, J., Desbrosses, G., Bouffaud, M.-L., Touraine, B., Moënne-Loccoz, Y., Muller, D., et al., 2013. Plant growth-promoting rhizobacteria and root system functioning. Front. Plant Sci. 4.

Vance, M.E., Kuiken, T., Vejerano, E.P., McGinnis, S.P., Hochella, M.F., Rejeski, D., et al., 2015. Nanotechnology in the real world: redeveloping the nanomaterial consumer products inventory. Beilstein J. Nanotechnol. 6, 1769–1780. Available from: https://doi.org/10.3762/bjnano.6.181.

Vejan, P., Abdullah, R., Khadiran, T., Ismail, S., Nasrulhaq Boyce, A., 2016. Role of plant growth promoting rhizobacteria in agricultural sustainability—a review. Molecules 21, 1–17. Available from: https://doi.org/10.3390/molecules21050573.

Wang, P., Lombi, E., Zhao, F.J., Kopittke, P.M., 2016. Nanotechnology: a new opportunity in plant sciences. Trends Plant Sci. 21, 699–712.

Wani, K.A., Kothari, R., 2018. Agricultural Nanotechnology: Applications and Challenges 3, 2146–2148.

Wilson, M.A., Tran, N.H., Milev, A.S., Kannangara, G.S.K., Volk, H., Lu, G.Q.M., 2008. Nanomaterials in soils. Geoderma 146, 291–302. Available from: https://doi.org/10.1016/j.geoderma.2008.06.004.

Yadav, A., Theivasanthi, T., Paul, P.K., Upadhyay, K.C., 2013. Extracellular biosynthesis of silver nanoparticles from plant growth promoting Rhizobacteria Pseudomonas sp.. Int. J. Curr. Microbiol. Appl. Sci. 4, 1057–1068.

Zhang, H., Li, Q., Lu, Y., Sun, D., Lin, X., Deng, X., et al., 2005. Biosorption and bioreduction of diamine silver complex by Corynebacterium. J. Chem. Technol. Biotechnol. 80, 285–290. Available from: https://doi.org/10.1002/jctb.1191.

# The role of fungus in bioactive compound production and nanotechnology

# 9

*Akhileshwar Kumar Srivastava*
The National Institute for Biotechnology, Ben-Gurion University of the Negev, Negev, Israel

## 9.1 Introduction

Plant metabolites have long played a role in human health, as models for synthetic compounds (Masi et al., 2018). These compounds are isolated from living organisms, including polymeric macromolecules such as nucleic acids, proteins, and carbohydrates, as well as low-molecular-weight compounds. These compounds are called secondary metabolites and are defined as organic compounds that do not directly participate in primary metabolic processes like cell growth, cell division, cell respiration, or photosynthesis. In addition, secondary metabolites are mainly obtained from biosynthetic pathways, which are a branch of primary metabolic pathways (Dewick, 2009; Hartmann, 2007). Secondary metabolites have several functions such as (1) alarming signals by volatile pheromones, (2) sex attractants in various insects, (3) defense against predators, and (4) weapons for building ecological niche. However, many of the biological functions are still unknown.

### 9.1.1 *Prevalence of Fungal diversity*

Among living organisms, fungi are one of the richest sources of secondary metabolites and have been used by humans since ancient times (Cimmino et al., 2015; Kornienko et al., 2015). Fungal metabolites have different classes of natural compounds associated with terpenes, phenylpropanoids, polyketides, alkaloids, etc. (Dewick, 2009). These organisms are a valuable source of natural substances that could be used to synthesize of natural biopesticides or antibiotics against human pathogens with alternative modes of action. The structure-activity relationships of some metabolites have been investigated to understand their biological activity, to increase the selectivity and stability of their natural products, as well as to lower their toxicity.

Alexander Fleming discovered the antibiotic penicillin (penicillin F) in 1929, and later, it was developed into a medicine by Florey and Chain in the 1940s

**Role of Plant Growth Promoting Microorganisms in Sustainable Agriculture and Nanotechnology.**
**DOI: https://doi.org/10.1016/B978-0-12-817004-5.00009-9**

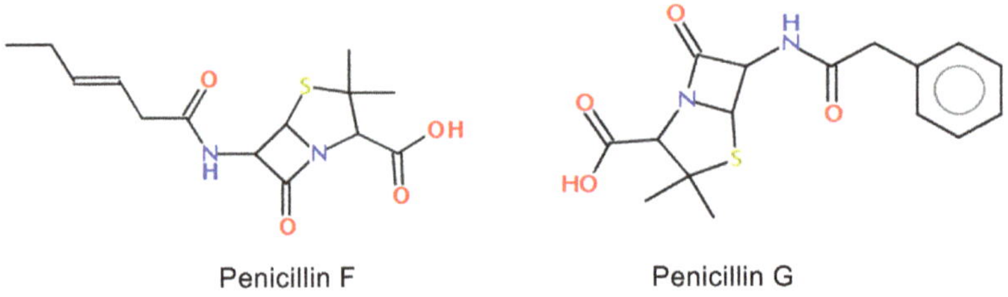

Penicillin F

Penicillin G

**Figure 9.1** Structure of penicillin F and penicillin G.

(penicillin G) (Fig. 9.1). After the discovery of penicillin, extensive research resulted in thousands of new compounds with a broad spectrum of biological activities. More than 1500 fungal secondary metabolites were isolated and characterized between 1993 and 2001, and more than half of these showed antibacterial, antifungal, or antitumor activity (Selim et al., 2012). Fungal secondary metabolites are also disease-causing agents by interacting with plants. These fungi are called phytopathogens and their secondary metabolites are often secreted during infection and are called mycotoxins, phytotoxins, etc. (Meinwald, 2011; Fox & Howlett, 2008).

Diverse groups of fungi have one of the largest natural resources for terrestrial and aquatic systems that play a vital role in ecological balances. It has been estimated that approximately 1.5 million species of fungi are recognized on earth (Hawksworth 2001, 2004). At present, the minimum of 712,285 extant of fungal species are found worldwide according to the ratio of fungal species to plant species in the same region, in which 600,000 fungal species are found in terrestrial plants (Schmit and Mueller, 2007). Unfortunately, only 5%–10% of fungal species have been cultured using current cultivation techniques (Manoharachary et al., 2005). The literature related to fungal collections has exposed the discrepancies in knowledge based on isolation and cultivation methods, fungal physiology, and fungal diversity, and many collected fungal species (more than 20,000 species) still need formal descriptions (Hawksworth and Rossman, 1997). Kirk et al. (2001) reported that about 90% of higher fungi species (including asexual fungi) are recognized worldwide. About 5,000 species of about 1200 genera of higher fungi have been recognized in southern China (Yang and Zang, 2003), and half of these fungal species are of lichenized fungi and macrofungi and the other half are microfungi including aquatic fungi, soil-inhabiting fungi, terrestrial plant-associated fungi, and arthropod-associated fungi (Schmit and Mueller, 2007). Hence, higher fungi are considered as a megadiverse bioresource.

Fungi have a special metabolism and can produce different types of functional secondary metabolites with diverse chemical structures and can stop cell proliferation and differentiation to achieve the self-defense mechanism and thus have potential in drug delivery. The diversification of fungal species also shows their potential in the development of new drugs (Zhong and Xiao, 2009).

Bioactive compounds are considered as a group of compounds having either a harmful or beneficial effect on human health. Antibiotics derived from fungi are one of the important bioactive compounds and many of these compounds are used as dietary supplements. Foods with such bioactive compounds are called "functional foods." Most studies have been done on mushrooms due to their essential amino acids, essential fatty acids, vitamins, and minerals for human health, but they have other advantageous activities for human health including antioxidant, antihypercholesterolemic, and antihyperlipidemic activities; antitumor, antibacterial, and antiviral activities; and antihyperglycemic and antiallergic effects (Díaz-Godínez, 2015).

Nanotechnology has emerged as new area of modern research and is used for the synthesis and application of nanoparticles with size smaller than 100 nm. Nanoparticles are used to create nanostructured materials and devices. Biological synthesis has emerged as an alternative approach to physical and chemical methods of synthesis. This new field of nanotechnology is referred to as "green nanotechnology" or "nanobiotechnology," and it combines biological principles with physical and chemical procedures to develop ecofriendly nanosized particles with specific functions. Higher plants, algae, fungi, yeast, bacteria, and viruses could be sued for the synthesis of nanoparticles. Researchers have focused on fungi as potent bioresources for the synthesis of silver nanoparticles, which possess numerous bioactive properties for applications in biomedicine. This chapter discussed the diversity of fungal metabolites and their role in the development of nanoparticles.

## 9.2 Secondary metabolites of fungal species

The bioactive compounds of fungal species exhibit large structural variations and are created after the active growth phase of the organism and are known as secondary metabolites. Primary metabolites have almost the same functions in all living systems whereas secondary metabolites have specific functions to particular species retrieved from the intermediate pathways of primary metabolism. The secondary metabolites of fungi have different structures associated with terpenoids, alkaloids, quinones, xanthones, peptides, steroids, flavonoids, phenols, and phenolic compounds (Fig. 9.2). The endophytic fungus residing in the plants produces metabolites, which are either same or have higher activity than its corresponding hosts. Many secondary metabolites of fungal origin have been tested for use as pesticides and herbicides in agriculture and in medicine as cholesterol inhibitors, immunosuppressive agents, and anticancer and antitumor agents. Several bioactive compounds from fungi have been explored for use in biomedicine and agrochemistry. The secondary metabolites of fungi are alternative sources of novel bioactive metabolites.

### *9.2.1 Antimicrobial activity*

Antibacterial metabolites have been recognized in fungi. Krohn et al. (1999) reported on some of the antibacterial components in the endophytic fungus

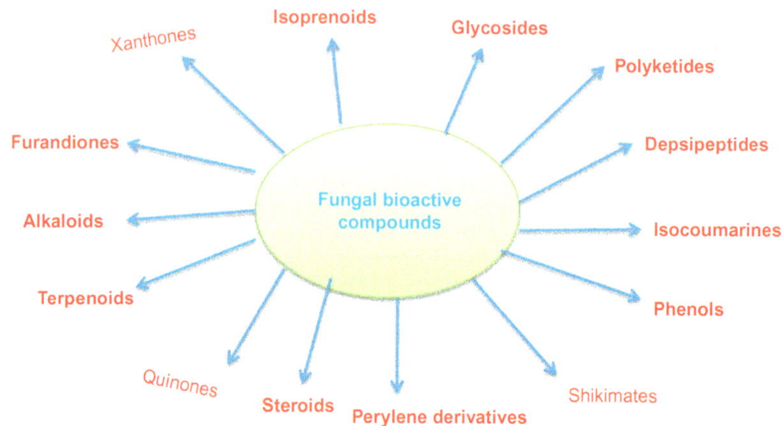


**Figure 9.2** Various groups of fungal metabolites.

*Pleospora herbarum*, namely the ketodivinyllactonic steroid, herbarulide. The isolated ten steroid compounds of *Colletotrichum* sp. showed bactericidal activity (Lu et al., 2000). The tested components of isolated metabolites, 3-oxoergosta-4,6,8 (14),22-tetraene, 3b-hydroxy-ergosta-5-ene, 3b-hydroxy-5a,8a epidioxy-ergosta-6, 3b,5a-dihydroxy-6b-acetoxy-ergosta-7,22-diene, 22-diene 6 isoprenylindole-3-carboxylic acid, and 3b,5a-dihydroxy-6b-phenylacetyloxy-ergosta-7,22-diene showed potential inhibitory effects for pathogens such as *Staphylococcus aureus*, *Bacillus subtilis*, *Pseudomonas* sp., and *Sarcina lutea*. Wagenaar and Clardy (2001) investigated the antibacterial activity of a compound dicerandrols A C isolated from a fungus *Phomopsis longicolla*. Strobel et al. (2001) explained the antibacterial role of five classes of volatile components: esters, alcohols, ketones, lipids, and acids from an important fungus, *Muscodor albus*. The compounds exhibited high efficacy against *E. coli*, *S. aureus*, *Micrococcus luteus*, and *B. subtilis*. The esters of 1-butanol 3-methyl-acetate had a highly inhibitory class of constituents among them. A steroid, ergosterol component of *Penicillium janthinellum* showed antibacterial efficacy for *Leishmania* sp. (Marinho et al., 2005).

The antibacterial compounds 3,4,5-trisubstituted N-methyl-2-pyridone alkaloid and fusapyridons A were isolated from the extract of *Fusarium* sp. (Tsuchinari et al., 2007). A component of YG-45 of *Maackia chinensis* showed inhibitory effect against *S. aureus* and *P. aeruginosa* with the minimum inhibitory concentration (MIC) at 50 and 6.25 μg/mL, respectively. The antibacterial compounds also found in *Penicillium* sp. from the mangrove plant *Cerbera manghas* (Zhuang et al., 2008). Zhuang et al. (2008) also characterized some metabolites 4-(3-hydroxybutan-2-yl)-3,6-dimethylbenzene-1,2-diol, and 3,4,5-trimethyl-1,2-benzenediol and explained about their inhibitory potential on methicillin-resistant *S. aureus* (MRSA) whereas 4-(3-hydroxybutan-2-yl)-3-acetyl-6-methylbenzene-1,2-diol did not show any effect on MRSA. Kjer et al. (2009) studied the antibiotic action of the isolated components of *Alternaria* sp., for inactivation of multidrug-resistant bacteria. Some metabolites such as xanalteric acids I and II had less bactericidal effect against *S. aureus* and other compounds like altenusin showed a wide range of action against several bacteria such as *Enterococcus faecium*, *Enterococcus cloacae*, *Streptococcus pneumonia*, and *Pseudomonas aeruginosa* at MIC ranges 31.5–125 μg/mL.

Fernandes et al. (2009) analyzed the MIC and minimum bactericidal concentration (MBC) for the isolates of *Alternaria alternata*. The ranges of MIC and MBC values of isolates were 50–100 μg/mL for *S. aureus* and 400–800 μg/mL for *E. coli*. Kusari et al. (2009) detected the antibacterial properties for the component deoxypodophyllotoxin of fungus *Aspergillus fumigates* against pathogenic bacteria *S. aureus*, *K. pneumonia*, and *P. aeruginosa*. Lima et al. (2011) reported that the 15 culture filtrates of endophytic fungi associated with *Piper aduncum* L. inhibited about 90% growth of *M. tuberculosis*. Nithya and Muthumary (2010) determined the antibacterial activity of the isolated bioactive components from endophytic fungus *Phomopsis* sp. residing within *Plumeria acutifolia* against the bacterial pathogens *E. coli*, *Pseudomonas* sp., *Klebsilla* sp., *B. subtilis*, *S. aureus*, and *S. typhi* and it had no significant effect on *C. albicans*. Tayung et al. (2011) studied an endophytic fungi *Fusarium solani* found in *Taxus baccata* and reported on its antibacterial compounds 1-tetradecene, 8-pentadecanone, 8-octadecanone, 10-nonadecanone, and octylcyclohexane displaying antibacterial activity against *S. epidermidis*, *S. aureus*, *S. flexneri*, *B. subtilis*, *E. coli*, and *K. pneumonia*. Ho et al. (2012) investigated the endophytes (*Lasmenia* sp., *O. tenuisporum*, *Xylaria cubensis*, and *Cyanodermella* sp.) from the citrus and canthoxylum of *Rutaceae* and cinnamomum of *Lauraceae* against three phytopathogenic bacteria, namely *Erwinia carotovora*, *Xanthomonas campestris*, and *Ralstonia solanacearum*. Among the tested isolates, *Cyanodermella* sp. showed better inhibition for all pathogens. Nath et al. (2012) reported that the isolates of *Xylaria* sp. and *Diaporthe* sp. exhibited antimicrobial properties for clinical pathogens like *Salmonella paratyphi* and *Enteroccocus faecalis*. Senthilmurugan et al. (2013) showed the antibacterial activity of bioactive components of endophytic fungus *Botrytis* sp. related to *Ficus benghalensis* against *E. coli* and *Klebsiella* sp. Pinheiroa et al. (2013) performed on antibacterial activity for two alkaloids, fumigaclavine C and pseurotin A isolated from *Aspergillus* sp. EJC08 associated with medicinal plant *Bauhinia guianensis*. Subban et al. (2013) showed antibacterial activity of the bioactive component 4-(2,4,7-trioxa-bicyclo[4.1.0]-heptan-3-yl) of *Pestalotiopsis mangiferae*, an endophyte of *Mangifera indica* L. The MIC value of this compound against *B. subtilis* and *K. pneumonia* was at 0.039 mg/mL, while for *E. coli* and *M. luteus* it had 1.25 mg/mL following *P. aeruginosa* at 5 mg/mL. At present, about 169 endophytes have been recognized in the medicinal plant, *Aegle marmelos* (Mani et al., 2015). The identified strains were mainly *Curvularia australiensis*, *Alternaria alternate*, *Alternaria citrimacularis*, *Aspergillus niger*, and *Cladosporium cladosporioides*. The compounds of these fungi had potential against the clinical pathogens such as *S. epidermidis*, *S. aureus*, *Shigella* sp., *P. aeruginosa*, *E. faecalis*, *E. coli*, *K. pnuemoniae*, *P. mirabilis*, and *S. typhi*.

## 9.2.2 Antifungal compounds

Fungi also plays an antagonistic role toward pathogenic fungus. Krohn et al. (1999) explored herbarulide, an important compound ketodivinyllactonic steroid from *P. herbarum*. Lu et al. (2000) investigated the antifungal activity of the isolated secondary metabolites from the endophytic fungus *Colletotrichum* sp. of *Artemisia annua*. The metabolites 3b,5a-dihydroxy-6b-phenylacetyloxy-ergosta-7,22-diene, 3-oxo-ergosta-4,6,8(14),22-tetraene, 3b,5a-dihydroxy-6b-acetoxy-ergosta-7,22-diene, and 3b-hydroxy-ergosta-5-ene suppressed the pathogenic action of *A. niger* and *Candida albicans* at MICs ranges 50−100 μg/mL. The other component, 6-isoprenylindole-3-corboxylic acid had the capability to inhibit the growth of the fungal pathogens *Phytophthora capsici*, *Gaeumannomyces graminis var. tritici*, and *Rhizoctonia cerealis*. *Cryptosporiopsis cf. quercina* from the stem of *Tripterygium wilfordii* produced cryptocin, which inhibited the growth of *Pyricularia oryzae* and other plant pathogens (Li et al., 2000).

Strobel et al. (2001) recognized a novel fungi *M. albus* related to *Cinnamomum zeylanicum* and identified five classes of volatile compounds like alcohols, ketones, esters, acids and lipids, which suppressed the activity of pathogenic fungi, namely *Pythium ultimum*, *Rhizoctonia solani*, *Phytophthora cinnamomi*, *Stagnospora nodorum*, *Ustilago hordei*, *Sclerotinia sclerotiorum*, *F. solani*, *A. fumigates*, *Verticillum dahlia*, *Tapesia yallundae*, *Cercospora beticola*, and *C. albicans*. Harper et al. (2003) also showed the antifungal activity of pestacin from fungus *P. microspora* for root-invading fungi *P. ultimum*. Liu et al. (2004) reported 12 bio-active components from fungus *A. fumigates*, which had antifungal action against human pathogens *C. albicans*, *A. niger*, and *T. rubrum*. The compounds, namely asperfumoid, fumitremorgin C, fumigaclavine C, helvolic acid, and physcion, exhibited the inhibitory activity against *C. albicans*.

Silva et al. (2006) investigated the inhibitory effect of five compounds of *Phomopsis cassiae* associated with *Cassia spectabilis* against pathogenic fungi. Among them, the compound 3,12-dihydroxycadalene revealed inhibitory activity for phytopathogens *Cladosporium cladosporioids* and *C. sphaerospermum*. Kjer et al. (2009) reported on a metabolite altenusin from *Alternaria* sp. that exhibited broad spectrum activity against multidrug-resistant fungi *Aspergillus faecalis* and *C. albicans* at MIC 125 and 62.5 μg/mL, respectively. Oliveira et al. (2010) explained dihydroisocoumarin (3R,4R)-3,4-dihydro-4,6-dihydroxy-3-methyl-1-oxo-1H-isochromene-5-carboxylic acid from the endophytic fungus, *Xylaria* sp. residing within *P. aduncum* and showed its antifungal activity against *C. cladosporioides* and *C. sphaerospermum*. Gubiani et al. (2014) identified two other compounds eremophilane-type sesquiterpenes (xylarenones F and G) from the same plant with antiinflammatory properties.

Tayung et al. (2011) reported antifungal compounds from *F. solani* associated with the bark of *T. baccata*, 1-tetradecene, 8-octadecanone, 8-pentadecanone, octylcyclohexane, and 10-nonadecanone, that showed inhibitory action against *C. albicans* and *C. tropicalis*. Sugijanto et al. (2011) explained the fungicidal activity of lecythomycin extracted from fungus *Lecythophora* sp. This compound also played an inhibitory role against fungi like *A. fumigatus* and *Candida kruzei* with MIC of 62.5–125 mg/mL. Nath et al. (2012) reported the compounds from *Xylaria* sp. and *Phomopsis* sp. produced adverse effect for the *C. albicans*. Subbulakshmi et al. (2012) investigated the antifungal activity of endophytic fungi *Alternaria* sp., *Colletotrichum gloeosporioides*, *Pestalotiopsis* sp., *Fusarium* sp., and *Pestalotiopsis* sp. associated with the leaf of *Biota orientalis*, *Pinus excels*, and *Thuja occidentalis*.

Santiago et al. (2012) reported 5-hydroxyramulosin, a polyketide compound from an endophyte *Phoma* sp. associated with the medicinal plant, *Cinnamomum mollissimum*. The isolated components played a higher inhibitory role against *A. niger*. Ho et al. (2012) reported that endophytic fungi *Lasmenia* sp., *O. tenuisporum*, *X. cubensis*, and *Cyanodermella* sp. from the citrus and zanthoxylum of *Rutaceae* and Cinnamomum of *Lauraceae* had different levels of adverse effects against plant pathogens such as *Alternaria solani*, *Botrytis cinerea*, *C. gloeosporioides*, *Colletotrichum higginsianum*, *Cylindrocladiella lageniformis*, *Fusarium oxysporum*, *Monilinia fructicola*, *Penicillium digitatum*, *Pestalotiopsis psidii*, and *Pythium aphanidermatum*. Gherbawy and Gashgari (2013) investigated the antifungal activity of 33 endophytes residing within the leaves of *Calotropis procera* against four phytopathogens: *A. alternata*, *B. cinerea*, *F. oxysporum*, and *P. ultimum*. The isolates of *Chaetomium globosum* and *Myrothecium verrucaria* showed better inhibitory role against the pathogenic fungal strains. Wu et al. (2013) reported the antifungal property of the isolated steroids of *Phomopsis* sp. residing within the plant, *Aconitum carmichaeli*. The components 6-ethoxy-5, 15-dihydroxyergosta-7,22-dien-3-one, 9,14-dihydroxyergosta-4,7,22-triene-3,6-dione, ganodermaside D, and calvasterols had moderate or weak antifungal activities. Subban et al. (2013) reported on the fungicidal activity of the phenolic compound 4-(2,4,7-trioxa-bicyclo[4.1.0]-heptan-3-yl), isolated from the endophytic fungus *P. mangiferae* associated with *M. indica* Linn. against *C. albicans* with a concentration of 0.039 mg/mL.

### 9.2.3 Anticancer compounds

Cancer is one of the most common diseases worldwide. The identification of plant compounds with cytotoxicity has opened paths for the development of anticancer therapy. First, paclitaxel, an anticancerous compound from the endophytic fungus *Taxomyces andreanae* associated with the Yew tree (*Taxus brevifolia*), showed functional diterpenoid occurring in several species of *Taxus* (Strobel et al., 1993; Suffness, 1995). Zhang et al. (2000) reported on the anticancerous component vincristine from the endophytic fungus *Fusarium oxysparum* associated with the phloem (inner bark) of *C. roseus* L. Yang et al. (2004) also reported another anticancerous compound, vincristine, isolated from the endophytic fungi residing in the leaves of *Catharanthus roseus* (L.). The anticancer compound, camptothecin, was explored from the endophytic fungus *Entrophospora infrequens* associated with *Nothapodytes foetida* (Puri et al., 2005). Silva et al. (2006) assessed the antiproliferative properties of five components isolated from *P. cassiae* against HeLa cervical cancer cells; the metabolite 3,12-dihydroxycadalene inhibited the growth of cancerous cell at an IC50 of 20 μM/L whereas 3,12-dihydroxycalamenene and 3,11,12-trihydroxycadalene showed weak inhibition for the same cells. Teles et al. (2006) also isolated secondary metabolites from *Periconia atropurpurea* using ethyl acetate and tested their biological activity against cancerous cells.

Phongpaichit et al. (2007) reported the antiproliferation and cytotoxicity properties of 65 crude extracts of 51 *Garcinia* species (5 species from *Garcinia atroviridis*, 23 from *G. dulcis*, 6 from *G. nigrolineata*, 16 from *G. mangostana*, and 1 from *G. scortechinii*). Approximately 11.1% of the extracts revealed anticancerous properties against the proliferation of NCI-H187 cells and 12.7% against KB cells. About 40% of the extract exhibited cytotoxicity against normal Vero cell lines. Gangadevi and Muthumary (2008) also performed a study on several endophytes of different plants and tested their anticancerous efficacy. Taxol is one of the highly studied fungal metabolites used in the treatment of cancer. The produced taxol amount (163.4 μg/L) from fungus *C. gloeosporioides* (strain JGC-9) exhibited higher cytotoxicity activity against the human cancer cell lines BT 220, Int407, H116, HLK 210, and HL 251. Ge et al. (2009) explained cytotoxic alkaloids from *A. fumigates* found in the stem of *Cynodon dactylon*. The compounds 9-deacetylfumigaclavine C and 9-deacetoxyfumigaclavine C showed cytotoxic activity for the K562 cells (leukemia cancer cell line) with IC50 values of $41.0 \pm 4.6$ and $3.1 \pm 0.9$ μM, respectively. Secalonic acid D, a mycotoxin (ergochrome class) isolated from the mangrove endophyte, had high a cytotoxicity effect against K562 and HL-60 leukemia cells, and hence induced toxicity through apoptosis (Zhang et al., 2009). Nithya and Muthumary (2009) explored anticancerous components using UV, thin-layer chromatography, and Fourier transform infrared spectroscopy and analyzed the taxol from culture filtrate of endophytic fungus *C. gloeosporioides* associated with *P. acutifolia*. The taxol was also produced from the endophytic fungus *Pestalotiopsis* sp. associated with *C. roseus* (Srinivasan and Muthumary, 2009). Zhou et al. (2009) also identified the anticancer compound taxol from *Mucor* sp.

Fernandes et al. (2009) reported on the crude extract of *A. alternate*, an endophyte from *Coffea arabica* L., with antitumor properties that also exhibited cytotoxic activity against HeLa cells with IC50 at 400 μg/mL. Similarly, Kjer et al. (2009) reported the xanalteric acids I and II of endophytic fungus *Alternaria* sp., with a highly growth inhibitory effect against L5178Y cells at concentration 10 μg/mL. An important metabolite (ergoflavin) from endophytic fungi *Claviceps purpurea* (PM0651480) associated with *Mimusops elengi* (*Sapotaceae*) also showed anticancerous properties (Deshmukh et al., 2009). The camptothecin derivatives were isolated from the fungal strains of *F. solani* MTCC 9667 and MTCC 9668 (Shweta et al., 2010). The camptothecin derivatives exhibited as 9-methoxycamptothecin and 10-hydroxycamptothecin. Pandi et al. (2011) also extracted taxol from fungus *Lasiodiplodia theobromae* associated with medicinal plant *Morinda citrifolia* that had cytotoxic effect against MCF-7 cells with an IC50 at 300 μg/mL.

The isolated metabolite *sclerotiorin* from *Cephalotheca faveolata* had the capability to induce apoptosis in cancer cells (Giridharan et al., 2012). Sclerotiorin had apoptotic properties for colon cancer (HCT-116) cells via BAX and inactivated the BCL-2 proteins and further degraded the caspase-3 enzyme promoting apoptosis in cancerous cells. Santiago et al. (2012) extracted 5-hydroxyramulosin (a polyketide compound) from the cultured *Phoma* sp. associated with the plant, *C. mollissimum*. The compound (5-hydroxyramulosin) developed toxicity effect against P388 murine leukemic cells with IC50 value 2.10 μg/mL. Similarly, Lu et al. (2012) reported on the extracts of fungi from *Actinidia macrosperma* with cytotoxic and antitumor activities against brine shrimp and five types of cancer cell lines. The 3-(4,5-dimethylthiazol-2-yl)-2,5-diphenyl tetrazolium bromide (MTT) analysis of fungal metabolites showed approximately 82.4% of fungal isolates with growth inhibitory activity against cancerous cells (50% inhibitory concentration IC50 $<100$ μg/mL). Some of the fungal extract also revealed extensive antitumor activity against all cancer cell lines, suggesting the role of fungi compounds as novel anticancrous agents. Sun and Xia (2002) assessed the toxic effect of seven terpenoid compounds arisugacins B, F, G, I, J, territrem B, and territrem C extracted from fungus *Penicillium* sp. SXH-65 against Hela, HL-60 and K562 cell lines. The metabolites arisugacin B and F had the cytotoxicity at IC50 values (24–60 μM).

## 9.3 Biosynthesis of nanoparticles by fungi

Fungi are unique eukaryotic organisms occurring in several ordinary lodgings and forming decomposer organisms. So far, only about 70,000 species have been explored from the esimated 1.5 million species of fungi on Earth. Recent data on fungus species research has shown that about 5.1 million fungal species have been identified by high throughput sequencing methods (Blackwell, 2011). It is worth noting that the fungi secrete a specific enzyme for hydrolyzing complex molecules into simpler one and digestion of extracellular food, which are absorbed for utilization as energy resources by these organisms (Blackwell, 2011). The implication of fungi in nanobiotechnology is widely accepted as an important way to solve biological challenges. Fungi have gained attention for the development of nanoparticles due to their toleration and metal bioaccumulation capability (Sastry et al., 2003). The handling of fungi are easy at different scale, hence it is used in synthesis of nanoparticles by utilizing a thin solid substrate fermentation technique. Achieving a large amount of enzymes at a commercial level is highly feasible due to extensive secretion of extracellular enzymes by fungi (Castro-Longoria et al., 2012). The economic feasibility and facility of implying biomass for the fungus species is another advantage for the utilization of green approach to synthesize of metallic nanoparticles. In addition, the fungus species are easily cultured and managed in laboratory conditions due to their fast growth rate (Castro-Longoria et al., 2012) and also have high wall-binding and intracellular metal uptake capacity (Volesky and Holan, 1995). Fungi are capable of producing metal nanoparticles/meso and nanostructure through overcoming the enzymatic activity of intracellularly or extracellularly (Table 9.1) and the experimental method for the development of silver

**Table 9.1** Some fungal isolate used for the biosynthesis of metal/metal oxide nanoparticles.

| Fungal species | NPs | Localizations | Size (nm) | Shape | Applications | References |
|---|---|---|---|---|---|---|
| *Candida albicans* | Au | Intracellular | 20–40 | Spherical and nonspherical | Detection of liver cancer | Chuhan et al. (2011) |
| *Helminthosporum solani* | Au | Extracellular | 2–70 | Polydispersed | Anticancer drug | Kumar et al. (2008) |
| *Fusarium oxyporum* | Ag | Extracellular | 20–50 | Spherical | Antibacterial | Durán et al. (2005) |
| *Aspergillus niger* | Ag | Extracellular | 3–30 | Spherical | Antibacterial and antifungal activity | Jaidev and Narasimha (2010) |
| *Aspergillus fumigatus* | Ag | | 15–45 | Mostly spherical | Antiviral against HIV-1 | Alani et al. (2012) |
| *Phoma glomerata* | Ag | | 60–80 | Spherical | Antibiotic | Birla et al. (2009) |

nanoparticles is almost same with slight modification depending on the type of fungi species being implicated (Fig. 9.3) (Ahmad et al., 2003a,b; Durán et al., 2005).

The processing of fungal species in nanotechnology is somewhat new with respect to higher plant metabolites. One area of focus is the biosynthesis of metallic natural products by fungi elucidate the synthesis of silver natural products extracellularly by filamentous fungus *Verticillium* sp. (Mukherjee et al., 2001a,b).

According to the literature on extracellular natural products the biomass of metabolites is exposed to metallic ion solutions (Mukherjee et al., 2011). The first utilized fungus was reported as individual CdS metabolite compounds along with formation of PbS, ZnS, and MoS2 natural products. The sulfate-diminishing enzyme-based procedure for production of natural products was confirmed by the presence of proteins in the aqueous solution. Therefore, fungi can also be transformed into nanoparticles in a size range of 5–50 nm, by changing their morphology (Ahmad et al., 2003a,b).

The spherical silver products (range of 20–50 nm) were developed by utilizing F. oxysporum (Durán et al., 2005) and were compared to the results of studies by Ahmad et al. (2003a,b) and Durán et al. (2005). The variation in morphology and size of silver products depends on temperature not on time (Riddin et al., 2006).

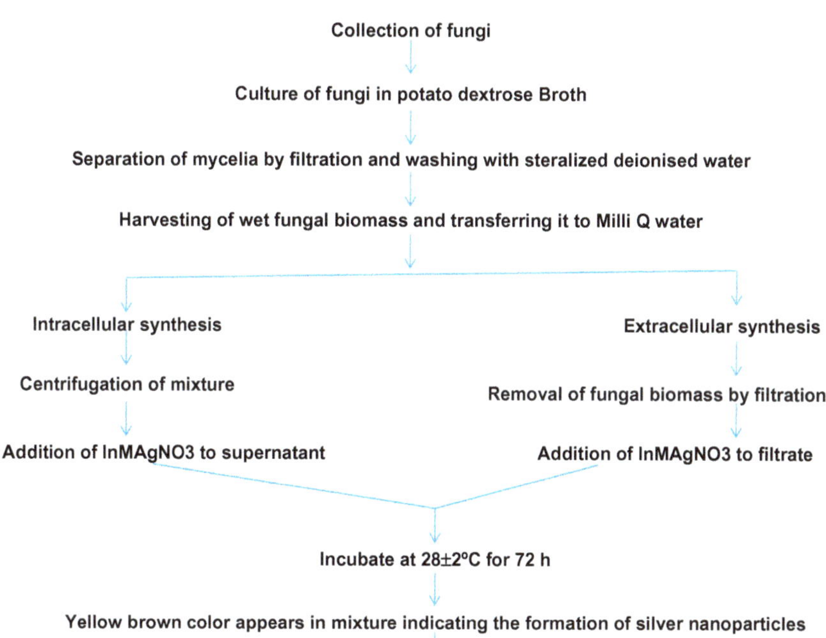


**Figure 9.3** Diagrammatic representation of synthesizing silver nanoparticles using fungi (Sandhu et al., 2017).

Though the most frequently synthesized nanoparticles were quasispherical ones, such morphological variation could be obtained according to the metallic ion solution and incubation condition.

The synthesis of nanoparticles using *F. oxysporum* with particular metals has been performed (Durán et al., 2005). The diminishing metal ions through this fungus were associated with nicotinamide adenine dinucleotide-hydrogen or NAD reduced form (NADH)-based reductases and a shuttle quinone extracellular method (Mukherjee et al., 2001a,b). Moreover, different quantities of NADH made of Au–Ag alloy nanoparticles with different compounds was possible (MubarakAli et al., 2012). *Rhizopus stolonifer* produced α-NADPH-dependent nitrate reductase and phytochelatin that are utilized in formation of silver nanoparticles (range of 10–25 nm) (Binupriya et al., 2010). Govender et al. (2009) have also suggested a mechanism to convert biological $H_2PtCl_6$ and $PtCl_2$ into platinum nanoparticles by hydrogenase enzyme of *F. oxysporum*.

The fungal biomass and/or cell-free extract are used to synthesize nanoparticles with different morphology (Shankar et al., 2004). Even in similar experimental condition, the diverse nanoparticles are produced from different fungal species. For example, particles synthesized from *Verticillium* sp. exhibited cubooctahedral shapes with a range of 100–400 nm magnetite whereas nanoparticles produced by *F. oxysporum* had irregular quasispherical morphology in size range 20–50 nm (Bharde et al., 2006). Such formation of nanoparticles depends on the type and condensation of compounds synthesized by each fungal species, incubation periods, precursor resolutions, and response time. The formation of metallic nanoparticles from the extract of fungus *Rhizopus oryzae*, by regulating its shape, was feasible at room temperature. Nanoparticles were produced by altering the main growth factors such as gold ion concentration, solution pH, and response time (Sun and Xia, 2002; Das et al., 2010).

The pathogenic activity in humans is the main shortcoming to employ such organism for nanoparticles production. The main implication of nanoparticles synthesized by the cultured fungal has been reported in some publications and most of these studies on the assessment of their biological significance has been accomplished. In addition, the studies showed the inhibition of microorganisms like bacteria and fungi after application of silver nanoparticles alone or in combination with antibiotics (Birla et al., 2009). Antimicrobial efficacy of synthesized silver nanoparticles using fungal species was accomplished against bacterial (Jaidev and Narasimha, 2010; Musarrat et al., 2010) and fungal pathogens (Musarrat et al., 2010).

Other metallic nanoparticles synthesized from fungal species as reducing agents have somewhat less impact; however, the conjugate of nano gold-bio formed by *R. oryzae* exhibited high antimicrobial activity for pathogenic bacteria like *P. aeruginosa*, *E. coli*, *B. subtilis*, *S. aureus*, *Salmonella* sp., and the yeasts *S. cerevisiae* and *C. albicans* (Das et al., 2010). In recent years, it has been shown that antimicrobial activity of fungus synthesized of TiO2 nanoparticles could be a new potent antibacterial remedy (Rajakumar et al., 2012).

Hydrated metal ions like $AuCl_4$ or Ag + could synthesis metal nanoparticles through fungi like *Verticillium*, *F. oxysporum* extracellularly (Ahmad et al., 2003a,b) or intracellularly (Mukherjee et al., 2001a,b) metal nanoparticles. Using fungus like *Colletotrichum* sp. (Shankar et al., 2003), Au nanoparticles with morphological variations were produced like rods, flat sheets, and triangles. The formation of nanoparticles depends on characteristics of nanomaterial's differing in shapes. Alteration in the molar proportion of metal ions in synthetic solutions, alloy nanoparticles with several components could be achieved by nanotechnology and also be implicated for the synthesis of additional alloy or composite systems like Au-CdS, Ag-CdS, and CdS-PbS (Senapati et al., 2005).

In addition, fungi can also be used to synthesize semiconductors instead of metallic nanoparticles. For example, when *F. oxysporum* is imparted to the hydrated CdSO4 solution that patters to sulfate-decreasing enzyme for synthesizing of extracellular CdS nanoparticles (Ahmad et al., 2002). Using the microbial methods, metal ions and its toxic ions can be scuttled from ores to purify water and nanoparticle synthesis. The cells of microorganisms provide the size and space of constrained synthesis. Moreover, if not all the minerals are synthesized inside the cells of microorganisms and the cells are influenced by the growth of such minerals inside the cells, the monodispersity of the minerals may be reduced. Hence, it is required more studies to produce higher-monodispersity nanomaterials and different variety of compounds through microorganism synthesis.

## 9.4 Future prospectives

Nanotechnology for the production of nanoparticles using fungal cells is a recent phenomenon. Fungus like *Colletotrichum* sp., *A. clavatus*, and *Pestalotia* sp. have been used for development of nanoparticles against pathogenic microbes. The advantages of nanotechnology are highly reproductive, comfortable processing methods, financial viability, sustainability against elevated concentrations of metal, easy handling at large-scale production of nanoparticles, and good dispersion of nanoparticles in higher amounts. Therefore, fungi are preferred for production of large amounts of nanoparticles over other alternative methods. While a number of papers have been published on the biological synthesis of nanoparticles, the potential of fungi is still vastly unexplored.

Silver nanoparticles are widely implicated for synthesis of nanomaterials, and they have shown promise in biomedicine for wound healing, antimicrobial and antiinflammatory action, etc. The application of silver nanoparticles in drug delivery and diagnostic systems has resulted in a new branch of nanotechnology using biological agents called nanobiotechnology, which combines biological principles with physical and chemical procedures to develop nanosized particles for specific functions. The biosynthesis of nanoparticles is also an attractive to the harmful chemical and physical methods of nanoparticle synthesis.

As discussed in this chapter, the bioactive compounds of fungi exhibit antimicrobial, anticancerous, antiviral, and anitifungal properties that could be used in nanomedicine for the treatment of disease. Further research is needed to explore fungi as a prospective bioresource for the synthesis of nanoparticles.

## References


Ahmad, A., Mukherjee, P., Mandal, D., Senapati, S., Khan, M.I., Kumar, R., et al., 2002. Enzyme mediated extracellular synthesis of CdS nanoparticles by the fungus *Fusarium oxysporum*. J. Am. Chem. Soc. 124, 12108–12109.

Ahmad, A., Mukherjee, P., Senapati, S., Mandal, D., Khan, M.I., Kumar, R., et al., 2003a. Extracellular biosynthesis of silver nanoparticles using the fungus *Fusarium oxysporum*. Colloids Surf. B. Biointerf. 28, 313–318.

Ahmad, A., Senapati, S., Khan, M.I., Kumar, R., Ramani, R., Srinivas, V., et al., 2003b. Intracellular synthesis of gold nanoparticles by a novel alkalotolerant actinomycete, *Rhodococcus* species. Nanotechnology 14, 824–828.

Alani, F., Moo-Young, M., Anderson, W., 2012. Biosynthesis of silver nanoparticles by a new strain of Streptomyces sp. compared with *Aspergillus fumigatus*. World J. Microbiol. Biotechnol. 28, 1081–1086.

Bharde, A., Rautaray, D., Bansal, V., Ahmad, A., Sarkar, I., Yusuf, S.M., et al., 2006. Extracellular biosynthesis of magnetite using fungi. Small 2, 135–141.

Binupriya, A.R., Sathishkumar, M., Yun, S.I., 2010. Biocrystallization of silver and gold ions by inactive cell filtrate of *Rhizopus stolonifer*. Colloids Surf. B. Biointerf. 79, 531–534.

Birla, S.S., Tiwari, V.V., Gade, A.K., Ingle, A.P., Yadav, A.P., Rai, M.K., 2009. Fabrication of silver nanoparticles by *Phoma glomerata* and its combined effect against *Escherichia coli, Pseudomonas aeruginosa* and *Staphylococcus aureus*. Lett. Appl. Microbiol. 48, 173–179.

Blackwell, M., 2011. The fungi: 1, 2, 3 ... 5.1 million species? Am. J. Bot. 98, 426–438.

Castro-Longoria, E., Moreno-Velásquez, S.D., Vilchis-Nestor, A.R., Arenas-Berumen, E., Avalos-Borja, M., 2012. Production of platinum nanoparticles and nanoaggregates using *Neurospora crassa*. J. Microbiol. Biotechnol. 22, 1000–1004.

Chuhan, A., Zubair, S., Tufail, S., Sherwani, A., Sajid, M., Raman, S.C., et al., 2011. Fungus-mediated biological synthesis of gold nanoparticles: potential in detection of liver cancer. Int. J. Nanomed. 6, 2305–2319.

Cimmino, A., Masi, M., Evidente, M., Superchi, S., Evidente, A., 2015. Fungal phytotoxins with potential herbicidal activity: chemical and biological characterization. Nat. Prod. Rep. 32, 1629–1653.

Das, S.K., Das, A.R., Guha, A.K., 2010. Microbial synthesis of multishaped gold nanostructures. Small 6, 1012–1021.

Dewick, P.M., 2009. Medicinal Natural Products: A Biosynthetic Approach, 3rd. ed. John Wiley & Sons, Ltd, Chichester, UK9780470741689.

Díaz-Godínez, G., 2015. Fungal bioactive compounds: An overview. Wiley Online Library. Available from: http://doi.ogr/10.1002/9781118733103.

Durán, N., Marcato, P.D., Alves, O.L., de Souza, G.I.H., Esposito, E., 2005. Mechanistic aspects of biosynthesis of silver nanoparticles by several *Fusarium oxysporum* strains. J. Nanobiotechnol. 3. Available from: https://doi.org/10.1186/1477-3155-3-8.

Fernandes, M.R.V., Costa, S.T.A., Pfenning, L.H., Costa-Neto, C.M., Heinrich, T.A., Alencar, S.M., et al., 2009. Biological activities of the fermentation extract of the endophytic fungus *Alternaria alternate* isolated from *Coffea arabica* L. Braz. J. Pharm. Sci. 45, 678–685.

Fox, E.M., Howlett, B.J., 2008. Secondary metabolism: regulation and role in fungal biology. Curr. Opin. Microbiol. 11, 481-48.

Gangadevi, V., Muthumary, J., 2008. Isolation of *Colletotrichum gloeosporioides*, a novel endophytic taxol-producing fungus from the leaves of a medicinal plant, *Justicia gendarussa*. Mycol. Balc. 5, 1–4.

Ge, H.M., Yu, Z.G., Zhang, J., Wu, J.H., Tan, R.X., 2009. Bioactive alkaloids from endophytic *Aspergillus fumigatus*. J. Nat. Prod. 72, 753–755.

Gherbawy, Y.A., Gashgari, R.M., 2013. Molecular characterization of fungal endophytes from *Calotropis procera* plants in Taif region (Saudi Arabia) and their antifungal activities. Plant Biosyst. 148, 1085–1092.

Govender, Y., Riddin, T., Gericke, M., Whiteley, C.G., 2009. Bioreduction of platinum salts into nanoparticles: a mechanistic perspective. Biotechnol. Lett. 31, 95–100.

Gubiani, J.R., Zeraik, M.L., Oliveira, C.M., Ximenes, V.F., Nogueira, C.R., Fonseca, L.M., et al., 2014. Biologically active eremophilane-type sesquiterpenes from *Camarops* sp., an endophytic fungus isolated from *Alibertia macrophylla*. J. Nat. Prod. 77, 668–672.

Harper, J.K., Ford, E.J., Strobel, G.A., Arif, A., Grant, D.M., Porco, J., et al., 2003. Pestacin: a 1,3,-dihydro isobenofuran from *Pestalotipsis microspora* possessing antioxidant and antimycotic activities. Tetrahedron 59, 2471–2476.

Hartmann, T., 2007. From waste products to ecochemicals: fifty years research of plant secondary metabolism. Phytochemistry 68, 2831–2846.

Hawksworth, D.L., 2001. The magnitude of fungal diversity: the 1.5 million species estimate revisited. Mycol. Res. 105, 1422–1432.

Hawksworth, D.L., 2004. Fungal diversity and its implication for genetic resource collections. Stud. Mycol. 50, 9–18.

Hawksworth, D.L., Rossman, A.Y., 1997. Where are all the undescribed fungi? Phytopathology 87, 888–891.

Ho, M.Y., Wen-Chuan, C., Hung-Chang, H., Wen-Hsia, C., Wen-Hsin, C., 2012. Identification of endophytic fungi of medicinal herbs of *Lauraceae* and *Rutaceae* with antimicrobial property. Taiwania 57, 229–241.

Jaidev, L.R., Narasimha, G., 2010. Fungal mediated biosynthesis of silver nanoparticles, characterization and antimicrobial activity. Colloids Surf. B. Biointerf. 81, 430–433.

Kirk, P.M., Cannon, P.F., David, J.C., Stalpers, J.A., 2001. Ainsworth and Bisby's Dictionary of the Fungi, 9th edn CABI, Wallingford, UK, pp. 1–655.

Kjer, J., Wray, V., Edrada-Ebel, R., Ebel, R., Pretsch, A., Lin, W., et al., 2009. Xanalteric acids I and II and related phenolic compounds from an endophytic *Alternaria* sp. isolated from the mangrove plant *Sonneratia alba*. J. Nat. Prod. 72, 2053–2057.

Kornienko, A., Evidente, A., Vurro, M., Mathieu, V., Cimmino, A., Evidente, M., et al., 2015. Toward a cancer drug of fungal origin. Med. Res. Rev. 35, 937–967.

Krohn, K., Biele, C., Aust, H.J., Draeger, S., Schulz, B., 1999. Herbarulide, a ketodivinyllactone steroid with an unprecedented Homo-6-oxaergostane skeleton from the endophytic fungus *Pleospora herbarum*. J. Nat. Prod. 62, 629–630.

Kumar, S.A., Peter, Y.A., Nadeau, J.L., 2008. Facile biosynthesis, separation and conjugation of gold nanoparticles to doxorubicin. Nanotechnology 19. Available from: https://doi.org/10.1088/0957-4484/19/49/495101.

Kusari, S., Lamshoft, M., Spiteller, M., 2009. *Aspergillus fumigatus* Fresenius, an endophytic fungus from *Juniperus communis* L. Horstmann as a novel source of the anticancer prodrug deoxypodophyllotoxin. J. Appl. Microbiol. 107, 1019–1030.

Li, J.Y., Strobel, G., Harper, J., Lobkovsky, E., Clardy, J., 2000. Cryptocin, a potent tetramic acid antimycotic from the endophytic fungus *Cryptosporiopsis cf. quercina*. Org. Lett. 2, 767–770.

Lima, A.M.D., Julia, I.S., Joao, V.B.S., Ana, C.A.C., Clarice, M.C., Francisco, C.M.C., et al., 2011. Effects of culture filtrates of endophytic fungi obtained from *Piper aduncum* L. on the growth of *Mycobacterium tuberculosis*. Electron. J. Biotechnol. 14, 1–6.

Liu, J.Y., Song, Y.C., Zhanga, Z., Wang, L., Guo, Z.J., Zou, W.X., et al., 2004. *Aspergillus fumigates* CY018, an endophytic fungus in *Cynodon dactylon* as a versatile producer of new and bioactive metabolites. J. Biotechnol. 114, 279–287.

Lu, H., Wen, X.Z., Jun, C.M., Jun, H., Ren, X.T., 2000. New bioactive metabolites produced by *Colletotrichum* sp., an endophytic fungus in *Artemisia annua*. Plant Sci. 151, 67–73.

Mani, V.M., Soundari, A.P.G., Karthiyaini, D., Preeth, K., 2015. Bioprospecting endophytic fungi and their metabolites from medicinal tree *Aegle marmelos* in Western Ghats, India. Mycobiology 43, 303–310.

Manoharachary, C., Sridhar, K., Singh, R., Adholeya, A., Suryanarayanan, T.S., Rawat, S., et al., 2005. Fungal biodiversity: distribution, conservation and prospecting of fungi from India. Curr. Sci. 89, 58–71.

Marinho, A.M.R., Rodrigues-Filho, E., Antonio, G.F., Lourivaldo, S.S., 2005. C25 steroid epimers produced by *Penicillium janthinellum*, a fungus isolated from fruits *Melia azedarach*. J. Braz. Chem. Soc. 6, 1342–1346.

Masi, M., Nocera, P., Reveglia, P., Cimmino, A., Evidente, A., 2018. Fungal metabolite antagonists of plant pests and human pathogens: structure-activity relationship studies. Molecules 23, 834.

Meinwald, J., 2011. Natural products as molecular messengers. J. Nat. Prod. 74, 305–309.

MubarakAli, D., Gopinath, V., Rameshbabu, N., Thajuddin, N., 2012. Synthesis and characterization of CdS nanoparticles using C-phycoerythrin from the marine cyanobacteria. Mater. Lett. 74, 8–11.

Mukherjee, P., Ahmad, A., Mandal, D., Senapati, S., Sainkar, S.R., Khan, M.I., et al., 2001a. Bioreduction of $AuCl_4$-ions by the fungus, *Verticillium* sp. and surface trapping of the gold nanoparticles formed. Angew. Chem. Int. Ed. 40, 3585–3588.

Mukherjee, P., Ahmad, A., Mandal, D., Senapati, S., Sainkar, S.R., Khan, M.I., et al., 2001b. Fungus-mediated synthesis of silver nanoparticles and their immobilization in the mycelial matrix: a novel biological approach to nanoparticle synthesis. Nano Lett. 1, 515–519.

Musarrat, J., Dwivedi, S., Singh, B.R., Al-Khedhairy, A.A., Azam, A., Naqvi, A., 2010. Production of antimicrobial silver nanoparticles in water extracts of the fungus *Amylomyces rouxii* strain KSU-09. Bioresour. Technol. 101, 8772–8776.

Nath, A., Raghunatha, P., Joshi, S.R., 2012. Diversity and biological activities of endophytic fungi of *Emblica officinalis*, an ethnomedicinal plant of India. Mycobiology 40, 8–13.

Nithya, K., Muthumary, J., 2009. Growth studies of *Colletotrichum gloeosporioides* (Penz.) Sacc. – a taxol producing endophytic fungus from *Plumeria acutifolia*. Indian J. Sci. Technol 2, 14–19.

Nithya, K., Muthumary, J., 2010. Secondary metabolite from *Phomopsis* sp. isolated from *Plumeria acutifolia* Poiret. Recent Res. Sci. Technol. 2, 99–103.

Oliveira, C.M., Luis, O.R., Geraldo, H.S., Ludwig, H.P., Maria, C.M.Y., Roberto, G.S.B., et al., 2010. Dihydroisocoumarins produced by *Xylaria* sp. and *Penicillium* sp., endophytic fungi associated with *Alibertia macrophylla* and *Piper aduncum*. Phytochem. Lett. 179, 1–4.

Phongpaichit, S., Nikom, J., Rungjindamai, N., Sakayaroj, J., Hutadilok-Towatana, N., Rukachaisirikul, V., et al., 2007. Biological activities of extracts from endophytic fungi isolated from *Garcinia* plants. FEMS Immunol. Med. Microbiol. 51, 517–525.

Pinheiroa, E.A.A., Josiwander, M.C., Diellem, C.P.S., de Andre, O.F., Patri, S.B.M., Giselle, M.S.P.G., et al., 2013. Antibacterial activity of alkaloids produced by endophytic fungus *Aspergillus* sp. EJC08 isolated from medical plant *Bauhinia guianensis*. Nat. Prod. Res. 27, 1633–1638.

Puri, S.C., Verma, V., Amna, T., Qazi, G.N., Spiteller, M., 2005. An endophytic fungus from *Nothapodytes foetida* that produces camptothecin. J. Nat. Prod. 68, 1717–1719.

Rajakumar, G., Rahuman, A., Roopan, S.M., Khanna, V.G., Elango, G., Kamaraj, C., et al., 2012. Fungus-mediated biosynthesis and characterization of TiO2 nanoparticles and their activity against pathogenic bacteria. Spectrochim. Acta A. Mol. Biomol. Spectrosc. 91, 23–29.

Riddin, T.L., Gericke, M., Whiteley, C.G., 2006. Analysis of the inter- and extracellular formation of platinum nanoparticles by *Fusarium oxysporum* f. sp. lycopersici using response surface methodology. Nanotechnology 17, 3482–3489.

Sandhu, S.S., Shukla, H., Shukla, S., 2017. Biosynthesis of silver nanoparticles by endophytic fungi: its mechanism, characterization techniques and antimicrobial potential. Afr. J. Biotechnol. 16, 683–698.

Santiago, C., Chris, F., Murray, H.G.M., Juriyati, J., Jacinta, S., 2012. Cytotoxic and antifungal activities of 5-hydroxyramulosin, a compound produced by an endophytic fungus isolated from *Cinnamomum mollissimum*. Evid-Based Compl Alt. Med. 1–6. Available from: https://doi.org/10.1155/2012/689310.

Sastry, M., Ahmad, A., Islam Khan, M., Kumar, R., 2003. Biosynthesis of metal nanoparticles using fungi and actinomycete. Curr. Sci. 85, 162–170.

Schmit, J.P., Mueller, G.M., 2007. An estimate of the lower limit of global fungal diversity. Biodivers. Conserv. 16, 99–111.

Selim, K.A., El-Beih, A.A., Abdel-Rahman, T.M., El-Diwany, A.I., 2012. Biology of endophytic fungi. Curr. Res. Environ. Appl. Mycol. 2, 31–82.

Senapati, S., Ahmad, A., Khan, M.I., Sastry, M., Kumar, R., 2005. Extracellular biosynthesis of bimetallic Au–Ag alloy nanoparticles. Small 1, 517–520.

Senthilmurugan, V., Sekar, G.R., Kuru, S., Balamurugan, S., 2013. Phytochemical screening, enzyme and antibacterial activity analysis of endophytic fungi *Botrytis* sp. isolated from *Ficus benghalensis* (L.). Int. J. Pharm. Biosci. 2, 264–273.

Shankar, S.S., Ahmad, A., Pasricha, R., Sastry, M., 2003. Bioreduction of chloroaurate ions by geranium leaves and its endophytic fungus yields gold nanoparticles of different shapes. J. Mater. Chem. 13, 1822–1826.

Shankar, S.S., Ahmad, A., Pasricha, R., Khan, M.I., Kumar, R., Sastry, M., 2004. Immobilization of biogenic gold nanoparticles in thermally evaporated fatty acid and amine thin films. J. Colloid. Interf. Sci. 274, 69–75.

Silva, G.H., Teles, H.L., Zanardi, L.M., Young, M.C.M., Eberlin, M.N., Hadad, R., et al., 2006. Cadinane sesquiterpenoids of *Phomopsis cassia*, an endophytic fungus associated with *Cassia spectabilis* (*Leguminoseae*). Phytochemistry 67, 1964–1969.

Srinivasan, K., Muthumary, J., 2009. Taxol production from *Pestalotiopsis* sp an endophytic fungus isolated from *Catharanthus roseus*. J. Ecobiotechnol. 1, 028–031.

Strobel, G., Stierle, A., Stierle, D., Hess, W.M., 1993. *Taxomyces andreanae* a proposed new taxon for a bulbilliferous hyphomycete associated with Pacific yew. Mycotaxon 47, 71–78.

Strobel, G.A., Emilie, D., Joe, S., Chris, M., 2001. Volatile antimicrobials from *Muscodor albus*, a novel endophytic fungus. Microbiology 147, 2943–2950.

Subban, K., Ramesh, S., Muthumary, J., 2013. A novel antibacterial and antifungal phenolic compound from the endophytic fungus *Pestalotiopsis mangiferae*. Nat. Prod. Res. 27, 1445–1449.

Suffness, M., 1995. Taxol Science and Applications. CRC Press, Boca Raton.

Sugijanto, N.E., Diesel, A., Rateb, M., Pretsch, A., Gogalic, S., Zaini, N.C., et al., 2011. Lecythomycin, a new macrolactone glycoside from the endophytic fungus *Lecythophora* sp. Nat. Prod. Commun. 6, 677–678.

Sun, Y., Xia, Y., 2002. Shape-controlled synthesis of gold and silver nanoparticles. Science 298, 2176–2179.

Tayung, K., Barik, B.P., Jha, D.K., Deka, D.C., 2011. Identification and characterization of antimicrobial metabolite from an endophytic fungus, *Fusarium solani* isolated from bark of Himalayan yew. Mycosphere 2, 203–213.

Teles, H.L., Sordi, R., Silva, G.H., Castro-Gamboa, I., Bolzani, V.S., Pfenning, L.H., et al., 2006. Aromatic compounds produced by *Periconia atropurpurea*, an endophytic fungus associated with *Xylopia aromatica*. Phytochemistry 67, 2686–2690.

Tsuchinari, M., Keiko, S., Fuminori, H., Tetsuya, M., Takuya, K., Yoshihito, S., 2007. Fusapyridons A and B, novel pyridone alkaloids from an endophytic fungus, *Fusarium* sp. YG-45. Z. Naturforsch. 62, 1203–1207.

Volesky, B., Holan, Z.R., 1995. Biosorption of heavy metals. Biotechnol. Prog. 11, 235–250.

Wagenaar, M.M., Clardy, J., 2001. Dicerandrols, new antibiotic and cytotoxic dimers produced by the fungus *Phomopsis longicolla* isolated from an endangered mint. J. Nat. Prod. 64, 1006–1009.

Wu, S.H., Rong, H., Cui-Ping, M., You-Wei, C., 2013. Two new steroids from an endophytic fungus *Phomopsis* sp. Chem. Biodivers. 10, 1276–1283.

Yang, Z.L., Zang, M., 2003. Tropical affinities of higher fungi in southern China. Acta Bot. Yunnanica 25, 129–144.

Yang, X., Zhang, L., Guo, B., Guo, S., 2004. Preliminary study of a vincristine-producing endophytic fungus isolated from leaves of *Catharanthus roseus*. Chin. Tradit. Herb. Drug 35, 79–81.

Zhang, W., Draeger, S., Schulz, B., Krohn, K., 2009. Ring B aromatic steroids from an endophytic fungus, *Colletotrichum* sp. Nat. Prod. Commun. 4, 1449–1454.

Zhang, L., Guo, B., Li, H., Zeng, S., Shao, H., Gu, S., et al., 2000. Preliminary study on the isolation of endophytic fungus of *Catharanthus roseus* and its fermentation to produce products of therapeutic value. Chin. Tradit. Herb. Drug 31, 805–807.

Zhong, J.J., Xiao, J.H., 2009. Secondary metabolites from higher fungi: discovery, bioactivity, and bioproduction. Adv. Biochem. Eng. Biotechnol. 113, 79–150.

Zhuang, H., Wen-Li, M., Hai-Bin, C., Yan-Bo, Z., Hai-Peng, L., Kui, H., 2008. Antibacterial constituents from the endophytic fungus *Penicillium* sp. of mangrove plant *Cerbera manghas*. Chem. J. Chin. Univ. Chin. 29, 749–752.

## Further reading

Bérdy, J., 2005. Bioactive microbial metabolites: a personal view. J. Antibiotics 58, 1–26.

Nithya, K., Muthumary, J., 2011. Bioactive metabolite produced by *Phomopsis* sp., an endophytic fungus in *Allamanda cathartica* Linn. Recent Res. Sci. Technol. 3, 44–48.

# Role of actinomycetes in bioactive and nanoparticle synthesis

# 10

*Akanksha Gupta*[1], *Divya Singh*[2], *Sandeep Kumar Singh*[1], *Vipin Kumar Singh*[1], *Anand Vikram Singh*[1] *and Ajay Kumar*[1]
[1]Center of Advanced Study in Botany Institute of Science, Banaras Hindu University, Varanasi, India, [2]Department of Chemical Engineering & Technology, Indian Institute of Technology, Banaras Hindu University, Varanasi, India

## 10.1 Introduction

The continuous and excessive use of chemical fertilizers and pesticides in traditional agriculture causes severe concern for the health and productivity of plant, soil, human beings, and the surrounding environments from last few decades. In this regard, beneficial microbes appear as better alternative of chemical pesticides, due to ecofriendly nature, less expense, and suitable for maintaining nutrient availability of both soils and plants. (Jeffries et al., 2003). Soil microbiota is composed of many beneficial microbes such as bacteria, fungi, actinomycetes, etc. Actinomycetes are a group of aerobic, Gram-positive bacteria that form thread-like structures in soil (Ames et al., 1984; Nonomura, 1989; Halder et al., 1991; Elliott and Lynch, 1995).

The word "actinomycetes" is derived from the Greek words "aktis" (meaning ray) and "mykes" (meaning fungus) (Das et al., 2008). Although actinomycetes have characteristics of both bacteria and fungi but considered as true bacteria with apparent resemblance to fungi (Bhatti et al., 2017). Actinomycetes are omnipresent (Srinivasan et al., 1991) but are the dominant microbiome of dry alkaline soil. Actinomycetes are an exceptionally versatile group of organisms that play an important role in the soil nutrient cycle as well as produce a wide range of biological active substances like antimicrobial products, amino acids, enzymes, and vitamins (Veiga et al., 1983; Boer et al., 2005). They even produce anticancerous metabolites that are active against cholesterol or have immunosuppressive properties. They produce substances responsible for the characteristic odor of wet soil (Rowbotham and Cross, 1977). Actinomycetes are also used as biocontrol agents to many plant pathogens (Jeffrey et al., 2007; Sprusansky et al., 2005). Oskay et al. (2004) reported that actinomycetes inhibit the plant pathogen *Agrobacterium tumefaciens* and *Erwinia amylovora*, which cause crown gall and fire blight of apple, respectively. An overview of actinomyctes is given in Fig. 10.1.

Actinomycetes can be autotrophous or heterotrophous based on their nature. They use various substances as their source of carbon and nitrogen, and can live in

**Role of Plant Growth Promoting Microorganisms in Sustainable Agriculture and Nanotechnology.**
**DOI: https://doi.org/10.1016/B978-0-12-817004-5.00010-5**

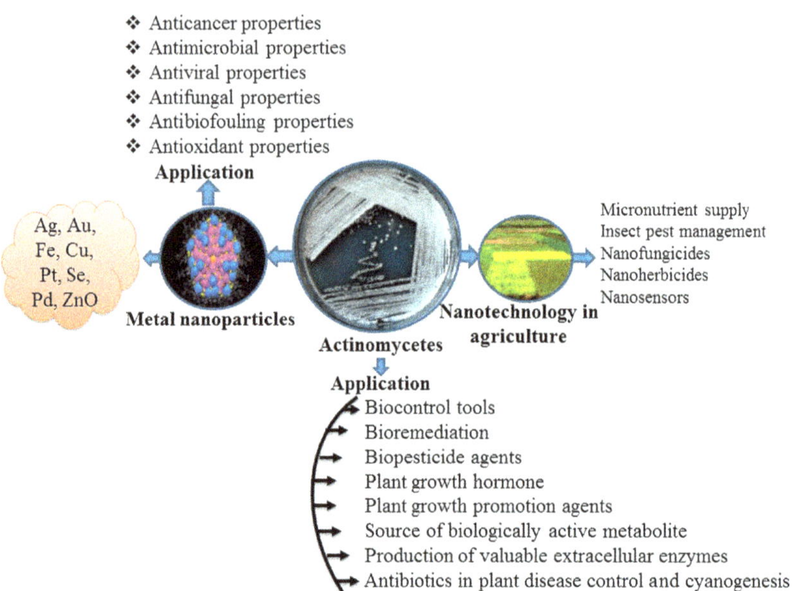


**Figure 10.1** An overview of actinomycte functioning.

a wide range of temperatures (from 7°C to 70°C) and places (George et al., 2012). Actinomycetes have been found in the most uninhabitable places on earth like the Mariana Trench (Pathom-Aree et al., 2006; Takami et al., 1997), soil of desert (Diraviyam et al., 2011), Antarctica (Moncheva et al., 2002; Raja et al., 2010) and even from the sediments of the ocean (Colquhoun et al., 1998; Walker and Colwell, 1975). Bacterial population in the soil is far abundant among all the microbial population whereas actinomycetes stand out to be the second most abundant organism with the population of $10^4-10^8$ $g^{-1}$ of soil. They are found to be less sensitive to low pH and can be found in every layer of the soil, but their population declines with depth (Takahashi and Omura, 2003).

Cross (1981) isolated different kinds of actinomycetes from different kinds of freshwater sources, such as various species of *Micromonospora*, *Streptomyces*, *Actinoplanes*, and *Thermoactinomycetes*, which produce endospores and *Rhodococcus*. According to Goodfellow and Haynes (1984), most of the species of actinomycetes present in freshwater sources are the result of surface water runoff to the freshwater bodies. Makkar and Cross (1982) reported that *Actinoplanes* can tolerate water stress in the form of mature sporangia and spread their spores when water is available as these spores are motile in nature and need water for movement. *Actinoplanes* are isolated from allochthonous leaves that fall and end up on the shore of the lake. According to Rowbotham and Cross (1977), many species of *Micromonospora* are endemic to water bodies and can only be isolated from the deposits of lakes; their spores serve as the dormant phase and propagate through the currents in the water sources. Human activities have led to the presence of a coprophyle *Rhodococcus coprophilus* in water bodies as a result of washoff of the animal dung of herbivores.

Continuous surface runoff of water-containing spores of *Streptomyces* to water bodies made them the most dominant species in these sources (Jiang and Xu, 1996; Terkina et al., 2002).

Besides being abundant in freshwater sources, actinomycetes are also found in the marine habitat due to dumping of various substances in the sea. According to Grein and Meyers (1958), species found from the marine environment are the ecoclines of the species present on the land. Okami and Okazaki (1974) suggested that currents of the river or rain may carry spores of actinomycetes to the marine habitat. Weyland (1969) isolated species of actinomycetes from sediments present in the deep ocean. Species isolated from the sea are found to be more metabolically active than species found on land and have developed tolerance against the saline environment of the sea (Jensen et al., 1991).

The ubiquitous nature of actinomycetes is well portrayed by various report that showed, actinomycetes had been frequently isolated from extreme environments, for example, Nocardiopsis and Streptomycetes isolated from soil of higher pH up to the range of 10–12 (Jiang et al., 1993). Groth et al. (1997) isolated a new species called *Bogoriella caseilytica* from the soil of Soda Lake with a pH of about 10. A halophyle called *Saccharomonospora halophile* was isolated from marsh soil by Al-Zarban et al. (2002); this strain showed an optimum growth at 10% concentration of NaCl. Mevs et al. (2000) isolated *Modestobactermultiseptatus* from the soils of the Transantarctic Mountain. Similarly, Suzuki et al. (1994) isolated a strain of actinomycetes called *Cryobacterium psychrophilum* from the soil of Antarctica; later on this strain was found as an obligate psychrophyle, which can't grow at temperature higher than 18°C. Some researchers have also isolated strains of *Micromonospora* and *Streptomyces* from peat soil and forest soil, which has very low pH. Takahashi et al. (1996) isolated strains (e.g., *Nocardia, Microbispora, Saccharothrix, Amycolaptosis, Microtetraspora*, and *Actinomadura*) from the soil of Mojave Desert, California, which is thermotolerant in nature. Chaphalkar and Deys (1998) reported strains of *Streptomyces* species from water and soil samples of a crater created by a meteor.

Due to the presence of true aerial hyphae, actinomycetes resemble fungi but ended up being considered as bacteria as they possess characteristics that put them in the bacterial group such as the composition of their cell wall (Davenport et al., 2000; Manulis et al., 1994). Morphologically the cell wall composition is similar to gram-positive bacteria (De Schrijver and De Mot, 1999; Das et al., 2008). But before being separated as a bacterial group they were called ray fungi due to the appearance of a radiating sun in the tissue.

Actinomycetes are known to present in soil with big populations (e.g., many millions in each gram of soil) (Lechevalier, 1981), and around 20 genera of actinomycetes can be isolated only from soil itself (Williams and Wellington, 1982a). According to Goodfellow and Williams (1983), if the soil is calcareous and more humic or dryer in nature, the population count of actinomycetes can reach up to $10^6$ $g^{-1}$ dry weight soil and become the dominant microbiota of that soil. But when it comes to soil with low pH and anaerobic or water-logged conditions, the actinomycete population count declines to $10^2$–$10^3$ $g^{-1}$ dry weight soil (Williams and Wellington, 1982; El-Tarabily and Sivasithamparam, 2006).

## 10.2 Role of actinomycetes in soil and plant health

Soil is normally counted as a biotic factor in the environment because microbes living in the soil made it dynamic system where large number of biological process occurring at the same time. Plants take nutrients from the soil and derive energy by which other organisms live. Microbes not only make soil viable but also provide minerals and other nutrients to plants by degrading the dead parts and debris of the plants. Microbes not only clean up the mess but also release beneficial products to the soil, which help plants grow better. Bacteria, fungi, actinomycetes, and algae make the main population of the soil microbiota, in which *Streptomyces* and *Micromonospora* species dominate the actinomycetes group. These filamentous fungi-like group species produce spores (one to two spores) on the tip of the branches of their filaments, and they are aerobic and use proteins, cellulose, lignin, and hemicellulose-like complex compounds as their source of carbon and nitrogen. Some of the species of actinomycetes show the ability to reduce nitrates, and they are slow in their activity so the organic matter added to soil is first degraded by bacteria and fungi later on actinomycetes act on it. Bacteria and fungi degrade simpler compounds of organic matter. The complex organic compounds degraded by the actinomycete, releases dark pigment humus to the soil. Humus is known as the most resistant matter in soil and takes more time to be degraded; it works as a substrate for actinomycetes (Bhatti et al., 2017).

Actinomycetes not only produce humus but also many metabolites known for their antimicrobial activity (Lazzarini et al., 2000; Basilio et al., 2003; Terkina et al., 2006) such as gentamicin, streptomycin, erythromycin, and rifamycin. *Streptomyces* and *Micromonospora* are the most studied genera of this group (Franco-Correa et al., 2010). *Streptomyces* alone is known to produce two-thirds of the antibiotics produced by the actinomycetes group; chloramphenicol, tetracyclines, aminoglycoside, β-lactams, anthracyclines, and macrolides are some of them (Bull et al., 1992; Basilio et al., 2003; Terkina et al., 2006). About 75% of the total antibiotics obtained from microbes are only produced by members of the group actinomycetes (Newman et al., 2003; Jiménez-Esquilín and Roane, 2005).

## 10.3 Actinomycetes as plant growth-promoting agents

Actinomycetes are a bacteria group and around 1000 species of 100 genera are known to exists in different habitats including agricultural soil. They have many processes that are beneficial to soil health including decomposition of cellulose matter, ammonium fixation, and synthesis as well as degradation of humus (Bhatti et al., 2017). They also offer many positive effects that help plants to grow well; for example, like act against many pathological agents and act as biocontrol agents (Doumbou et al., 2001), produce around 61% of biologically active molecules (Moncheva et al., 2002; Lam, 2006), and act as a good tool for genetic improvement of species (Bhattacharyya and Sen, 2004), as well as for bioremediation of lignin, cellulose (Crawford, 1978), petroleum, and heavy metal contaminant (Baniasadi et al., 2009). Tan et al. (2009) reported that actinomycetes can be used as producers of probiotics to complete the nutritional requirements of plants. Bressan (2003) documented species of the actinomycetes as biological pesticides. Many researchers have documented species of actinomycetes that have plant growth-promoting potential, and some of them have been successfully isolated from various soils, rhizospheres of crop plants, and endophytic species as well. However, studies are needed in this field before declaring any species of actinomycetes or their products as plant growth-promoting agents, as some aspects like interaction of species and their products to the host plant and rhizosphere need to be explored (Singh et al., 2018). The species of actinomycetes produce many metabolites; some of them work as plant growth promoting properties that helps in sustainable agriculture.

## 10.4 Actinomycetes in the synthesis of bioactive compounds

Microbial secondary metabolites have received significant attention in recent years. Thousands of microbial secondary metabolites have been isolated, identified, and characterized; most of them have been used in pharmacological, veterinary, and agriculture sectors. Actinomycetes are the largest producers of distinct and novel bioactive compounds accounting for 70% of all microbial secondary metabolites (Janardhan et al., 2014; Subramani and Aalbersberg, 2012). It has been estimated that each strain of actinomycetes can produce 10–20 secondary metabolites (Sosio et al., 2000). A list of bioactive compounds synthesized by actinomycetes is given in Table 10.1. *Amycolatopsis*, *Actinoplanes*, *Streptomyces*, *Saccharopolyspora*, and *Micromonospora* are the main group of actinomycetes that produce commercially important metabolites with antimicrobial, antitumor, insecticidal, immunosuppressive, and enzyme inhibition activities (Manivasagan et al., 2013a,b).

Bacterial growth is inhibited by antibacterial substances. Microbes resistant to antibiotics are one of the main reasons for death in infectious diseases. The frequency of drug resistance in microbes is increasing continuously at alarming rates. Decreased efficiency and antibiotic resistance in microbes has created the need to develop alternatives (Manivasagan et al., 2013a,b).

Actinomycetes have been comprehensively studied for antimicrobial activities. Abyssomicin C is functionally a polycyclic polyketide antibiotic produced by the *Verrucosispora* strain effective against Gram-positive bacteria. It inhibits synthesis of para-aminobenzoic acid and thus effectively prevents biosynthesis of folic acid at the early stages as compared to sulfa drugs (Bister et al., 2004). Lomofungin is a natural phenazine antibiotic isolated from *Streptomyces lomondensis* (Gomes et al., 2018). It has a wide range of antifungal and antibacterial activities. Lomofungin is effective against pathogenic fungi and Gram-positive and Gram-negative bacteria (Zhang et al., 2015). A novel antibiotic compound bonactin obtained from *Streptomyces* sp. BD21-2 possesses antimicrobial (for Gram-positive and Gram-negative bacteria) and

**Table 10.1** Role of actinomycetes in bioactive compound production.

| Bioactive compounds | Source | Isolated from | Role of bioactive compounds | Reference |
|---|---|---|---|---|
| Glaciapyrroles A, B and C | *Streptomyces* sp. NPS008187 | Alaskan marine sediment | Antibacterial | Macherla et al. (2005) |
| Resistoflavine | *Streptomyces chibaensis* AUBN | Marine sediment samples of Bay of Bengal, India | Cytotoxic and antibacterial activity | Gorajana et al. (2007) |
| Daryamides | *Streptomyces* sp. CNQ-085 | Marine sediment of San Diego, California | Cytotoxic, antifungal | Asolkar et al. (2006) |
| Lomofungin | *Streptomyces lomondensis* | Rhizosphere soil in Shanghai, China | Antifungal, antibacterial | Gomes et al. (2018) |
| Marinomycins A-D | *Marinispora* sp. | Sediment of La Jolla, California | Antifungal; anticancer | Kwon et al. (2006) |
| Bonactin | *Streptomyces* sp. BD21-2 | Kailua Beach, Oahu, Hawaii | Antimicrobial, antifungal | Schumacher et al. (2003) |
| Streptokordin | *Streptomyces* sp. KORDI-3238 | Deep-sea sediment at Ayu Trough (Pacific Ocean) | Anticancer | Jeong et al. (2006) |
| Proximicins | *Verrucosispora* sp. | Raune Fjord (Norway) | Cytostatic | Schneider et al. (2008) |
| Cyanosporasides | *Salinispora pacifica* | Deep-sea Sediments from Palau | Cytotoxic | Oh et al. (2006) |
| Cyclomarin A | *Streptomyces* sp. | Marine bacterium collected in the vicinity of San Diego | Antiinflammatory | Renner et al. (1999) |
| Pyrizinostatin | *Streptomyces* sp. SA-2289 | Marine sediment | Pyroglutamyl peptidase Inhibition | Aoyagi et al. (1992) |
| Lodopyridone | *Saccharomonospora* sp. | Mouth of the La Jolla Submarine Canyon | Anticancer | Maloney et al. (2009) |
| Dehydroxynocardamine; Desmethylenylnocardamine | *Streptomyces* sp. | Jaeju Island, Korea | Enzyme sortase B Inhibitor | Lee et al. (2005) |
| Arenicolides | *Salinispora arenicola* | Sediment sample, Guam | Antitumor | Williams et al. (2007) |
| Bisanthraquinone | *Streptomyces* sp. | La Paragua, Puerto Rico | Antibacterial | Socha et al. (2006) |
| Komodoquinone A | *Streptomyces* sp. KS3 | Marine sediment, Komodo Island, Indonesia | Neuritogenic | Itoh et al. (2003) |

| | | | | |
|---|---|---|---|---|
| Ammosamides | *Streptomyces* sp. CNR-698 | Bottom sediments, Bahamas Islands | Cytotoxic | Hughes et al. (2009) |
| Dermacozines A-G | *Dermacoccus* abyssi sp. MT 1.1 and MT 1.2 | Mariana Trench sediment | Cytotoxic, radial scavenging activity | Abdel-Mageed et al. (2010) |
| Pyridinium | *Amycolatopsis alba* | Marine sediments of coastal Visakhapatnm, India | Antimicrobial | Dasari et al. (2012) |
| Helquinoline | *Janibacter limosus* | Marine sediment | Antibacterial | Asolkar et al. (2004) |
| Mechercharmycins A | *Thermoactinomyces* sp. YM3-251 | Sea mud collected at Mecherchar in the Republic of Palau (North Pacific Ocean) | Antitumor | Kanoh et al. (2005) |
| Streptochlorin | *Streptomyces* sp. | Shallow water sediment from Ayajin Bay | Antiproliferative | Shin et al. (2007) |
| Actinofuranones | *Streptomyces* sp. CNQ766 | Marine sediment from island of Guam | Cytotoxic activity | Cho et al. (2006) |

antifungal activities (Schumacher et al., 2003). Glaciapyrroles A, B, and C are novel pyrrolosesquiterpes produced by *Streptomyces* sp. (NPSOO8187). These metabolites possess antibacterial properties (Macherla et al., 2005). *Streptomyces* are the major producer of therapeutic important metabolites such as tetracyclins, amphotericin, tacrolimus and adriamycin having antibacterial, antifungal, immunosuppressant and anticancerous activity (Hopwood, 2007). Bisanthraquinone is produced by *Streptomyces* sp. possessing bactericidal activity (Socha et al., 2006). A secondary metabolite called pyridinium, isolated from *Amycolatopsis alba*, possesses cytotoxicity against brain (U87MG), breast (MCF-7), and cervix cell lines (HeLa), and also has antibacterial activity specifically for both Gram-positive and Gram-negative bacteria (Dasari et al., 2012). Moreover, novel antibiotic helquinolines have been isolated from *Janibacter limosus*. It shows moderate activity against *Bacillus subtilis*, *Staphylococcus aureus*, and *Streptomyces virdochromogenes* Tu57 (Asolkar et al., 2004).

Numerous antibiotics have been successfully applied in the treatment of tumors and microbial diseases in recent decades. Therefore, tumor and bacterial cells have developed multidrug resistance or antimicrobial resistance creating challengs in medicine. The solution to this problem primarily relied on the development and discovery of new bio-active compounds from different sources such as microbes or plants. Various bioactive microbial secondary metabolites have been studied and most of them are used by the pharmaceutical industry as strong antibiotics. However, from last few decades, exhaustion of new bioactive secondary metabolites from actinomycetes has opened the door for various researchers (Manivasagan et al., 2014). A new antitumor–antibiotic metabolite called marinomycins is produced by *Marinispora* sp. Marinomycins exhibit antimicrobial activity against many pathogens resistant to antibiotics and also show remarkable and discriminating cytotoxicity toward melanoma cell lines (Kwon et al., 2006). Mechercharmycins are produced by *Thermoactinomyces* sp. Mechercharmycin A is a strong antitumor compound compared to mechercharmycin B (Kanoh et al., 2005). Novel antibiotic daryamides are antitumor compounds produced by *Streptomyces* strain CNQ-085. It has been estimated that daryamide A is more cytotoxic to human colon carcinoma cell line HCT-116 ($IC_{50}$ 3.15 μg/mL) than daryamides B and C (Asolkar et al., 2006). However, cyanosporasides are novel 3-keto-pyranohexose sugar-containing cyano- and chlorinated cyclopenta[a]indene ring isolated from *Salinispora pacifica* (Oh et al., 2006). They exhibit antitumor activity (Olano et al., 2009). Arenicolides, 26-membered polyunsaturated macrolactones obtained from *Salinispora arenicola*, show moderate cytotoxicity for human colon adenocarcinoma cell line HCT-116 ($IC_{50}$ 3.6 μg/mL) (Williams et al., 2007).

Cancer is one of the most serious health problems in the world. Radiotherapy, surgery, chemotherapy, and immunotherapy are therapeutic some of the available cancer treatments. Each method is effective for specific conditions, but a combination of these techniques is more efficient in tumor treatment. Most of the secondary metabolites produced by marine actinomycetes possess antitumor property (Manivasagan et al., 2014). Streptokordin, a methylpyridine derivative, has been isolated from *Streptomyces* sp. KORDI-3238. Although it possesses cytotoxic activity against numerous human cancer cell lines, it exhibits no significant microbial growth inhibition (Jeong et al., 2006). In addition, lodopyridone obtained from

*Saccharomonospora* sp. is effective against human colon adenocarcinoma cell line HCT-116 ($IC_{50}$ 3.6 μM) (Maloney et al., 2009). Moreover streptochlorin, an indole alkaloid, isolated from *Streptomyces* sp., inhibits angiogenesis, which is a crucial step in the progress of tumor and metastasis (Choi et al., 2007; Shin et al., 2007).

Many metabolites produced by actinomycetes inhibit cell growth and development accounting for cytostatic activity. Proximicins are obtained from *Verrucosispora maris* AB-18-032 and *Verrucosispora* strain MG-37. Proximicins A, B, and C are weak antibacterial agents with strong cytostatic effects on human tumors. These proximicins significantly inhibit human gastric adenocarcinoma AGS cells by arresting the G0/G1 phase of the cell cycle as well as increasing p53 and p21 levels (Fiedler et al., 2008; Schneider et al., 2008). However, a unique anthracycline komodoquinone A was obtained from *Streptomyces* sp. KS3. It can induce cell differentiation in neuroblastoma (Itoh et al., 2003).

Some bioactive secondary metabolites are toxic to cells. Resistoflavine, a cytotoxic compound, is produced by *Streptomyces chibaensis* $AUBN_1/7$. It possesses toxicity against hepatic carcinoma HePG2 cell lines and human erythroleukemia cell line K-562. It also shows weak antibacterial activity (Kock et al., 2005; Gorajana et al., 2007). Actinofuranones are polyketide compounds isolated from streptomyce strain CNQ766. Although it exhibits weak cytotoxicity toward macrophages and mouse splenocyte T-cells ($IC_{50}$ 20 μg/mL), it had no effect on human colon carcinoma HCT-116 cells (Cho et al., 2006).

Antioxidants refer to the substances that can impede or inhibit oxidation. Various organic substances are classified as antioxidants including carotenoids, vitamins A, C, E, etc. (Manivasagan et al., 2014). Dermacozines A-G are novel phenazine type metabolites produced by *Dermacoccus abyssi* sp. They possess antiprotozoal, antitumor, and free radical scavenging activities. Dermacozine C exhibited the highest antioxidant activity while dermacozines F and G show moderate toxicity toward leukemia cell line K562 (Abdel-Mageed et al., 2010). Lipocarbazoles, a novel compound isolated from *Tsukamurella pseudospumae* strain Acta 1857, exhibit strong antioxidant activity.

Some secondary metabolites secreted by actinomycetes can inhibit enzyme activities. One such compound is pyrizinostatin synthesized by *Streptomyces* sp. SA-2289 (Aoyagi et al., 1992). It can inhibit pyroglutamyl peptidase enzyme activity. Dehydroxynocardamine and desmethylenylnocardamine are cyclic peptides, produced by *Streptomyces* sp. They show weak activity against enzyme sortase B (Lee et al., 2005). However, cyclomarin A isolated from *Streptomyces* sp. reduces inflammation in both in vitro and in vivo conditions (Renner et al., 1999).

## 10.5 Role of actinomycetes in nanotechnology

Generally, the term nanotechnology refers to fabrication and utilization of materials within nanometer range. Due to their small size, the nature and properties of materials within nanorange are quite different from the parent substance. The rapidly rising field of nanotechnology has revolutionized the different disciplines of science because of its influence on energy production, electronics, drug synthesis, and gene transfer technology (Rao and Cheetham, 2001). Improvement in existing methodologies together with the development of novel techniques is thus essential to produce nanomaterials with important applications in human health, food, cosmetics, agriculture, etc. Presently, different physical, chemical, and biological methods are being used to design and synthesize nanomaterials (Iravani et al., 2014). Most physicochemical methods are expensive and hazardous. In contrast, the application of biological processes for generation of nanomaterials (bionanotechnology) is cheap, nontoxic, and environmentally friendly. Therefore, the application of different biological agents including algae, fungi, bacteria, actinomycetes, and plants for fabrication of different nanomaterials is gaining momentum among researchers worldwide.

Actinomycetes were previously considered as ray fungi but with later advances in microbiological techniques and instrumentation they were placed in different groups. Actinomycetes, the gram-positive filamentous bacteria showing several attributes similar to fungi and bacteria, are one of the potential microorganisms with potential applications in agricultural, medical, and environmental sectors. Apart from their common occurrences in a given environmental conditions, they have also been reported from extreme environments of temperature, pressure, heavy metals, organic environment, salinity, and pH (Sastry et al., 2003). The existence of diverse actinomycetes in harsh environments relies on unique biochemical and physiological mechanisms supporting the vital biological processes. The specific life-sustaining mechanisms involve in maintaining the structure and function of enzymes, cell membrane and intracellular milieu.

Metal nanoparticles are gaining increasing importance in different sectors. There have been many reports on extracellular and intracellular production of nanoparticles of different metals including copper, iron, zinc, gold, silver, palladium, platinum, selenium, titanium, and cadmium based on biological entities. The search for a new biological route for nanoparticle synthesis is challenging.

Ahmad et al. (2003b) first demonstrated gold nanoparticle biosynthesis by extremophilic actinomycete *Thermomonospora* sp. The nanoparticles generated had an average size of 8 nm with good dispersion. The absolute extracellular reduction of $AuCl_4$ solution leading to formation of nanoparticles was achieved in 120 hours.

Recently, Nabila and Kannabiran (2018) demonstrated the synthesis of CuO nanoparticles for the first time by employing actinomycetes and tested their efficacy against seven bacterial pathogens affecting humans and fishes. The mean size of the nanoparticles as revealed by TEM analysis was 61.7 nm. The biologically fabricated CuO nanoparticles exhibited better activity against the bacterial pathogens as compared to cell-free extract. Analytical investigation of nanoparticles through dynamic light-scattering technique resulted in zeta potential equivalent to 31.1 mV. The crystalline nature of the nanoparticles was confirmed by XRD. Among the tested bacteria, *Bacillus cereus* was the most susceptible pathogen.

The application of actinomycete (*Streptomyces griseoruber*)-mediated extracellular formation of selenium nanoparticle was presented by Ranjitha and Ravishankar

(2018). The biomolecules present in the culture media filtrate produced by actinomycete were responsible for reduction of selenium to generate nanoparticles. The selenium nanoparticles had absorption maxima at 575 nm wavelength with size ranging from 100–250 nm as revealed through UV–Vis spectrophotometry and TEM investigation. Interestingly, the biologically fabricated selenium nanoparticles had cytotoxic nature against the colon cancer line HT-29. The decreased cellular viability of cell lines with increase in concentration of nanoparticle was documented indicating their exploitation as chemotherapeutic agents.

The biogenic synthesis of nanoparticle of different metals such as gold, silver, selenium, iron, copper, zinc, and manganese as well as their application as antimicrobial agent against pathogenic bacteria, fungi, and parasites is rapidly increasing throughout the world. Saad et al. (2018) demonstrated the synthesis of copper nanoparticle using endophytic actinomycete *Streptomyces capillispiralis* Ca-1 recovered from *Convolvulus arvensis*. Modifications in color of culture media in presence of copper indicated the formation of nanoparticles, which was confirmed by UV–Vis spectroscopy. The well-dispersed size of synthesized nanoparticles as elucidated through TEM ranged from 3.6–59 nm indicating stability. The investigation of copper nanoparticles by XRD depicted the face-centered cubic form. The presence of different functional groups associated with actinomycete-induced nanoparticle biosynthesis as shown by FTIR analysis was considered responsible for copper nanoparticle formation. The biologically fabricated copper nanoparticles showed possible applications in control of human pathogens and management of plant diseases.

Al-Dhabi et al. (2018) demonstrated the antibacterial property of silver nanoparticles synthesized by extract of *Streptomyces* sp. Al-Dhabi-87 culture. The size of spherical nanoparticles ranged from 10 to 17 nm. The minimum inhibitory concentration of nanoparticles against the pathogenic bacteria *Pseudomonas aeruginosa*, *Klebsiella pneumoniae*, and *Escherichia coli* was determined to be 39, 78, and 152 μg/mL, respectively. Moreover, nanoparticles with good inhibitory action against drug-resistant bacteria including *Acinetobacter baumannii*, and *Staphylococcus aureus* have paved the way for promising applications in medical sciences.

Recently, Vijayabharathi et al. (2018) described the extracellular synthesis of silver nanoparticles using rhizospheric actinomycete *Streptomyces griseoplanus* SAI-25 isolated from rice. The externally produced silver nanoparticles exhibited absorption maxima between 413 and 417 nm. Analytical investigation of cell-free extract through FTIR spectroscopy revealed the presence of biologically important groups such as amine, alcohol, and phenol catalyzing the reduction and thus formation of silver nanoparticles. The actinomycete-induced nanoparticle formation displayed antifungal activity against the plant pathogen *Macrophomina phaseolina* suggesting the application of silver nanoparticles as biopesticides to control the loss of crop productivity.

Application of actinomycetes for extracellular as well as intracellular generation of metal-based nanoparticle is ecofriendly, green, simple, nontoxic, and cost-effective method. Nanoparticles may be good candidates for the treatment of number of life-threatening diseases due to altered morphology, high activity, larger surface area, novel composition, and useful properties (Manimaran and Kannabiran 2017).

**Table 10.2** Synthesis of nanoparticles by actinomycetes and their applications.

| Actinomycetes | Types of nanoparticle | Application | References |
|---|---|---|---|
| *Streptomyces* sp. VITPK1 | Silver | Excellent microbial resource for the extracellular synthesis of AgNPs | Sanjenbam et al. (2014) |
| *Streptomyces parvuus* SSNP11 | Silver | Antimicrobial property against gram-negative as well as gram-positive bacterial strains | Prakasham et al. (2014) |
| *Streptomyces* sp. BDUKAS10 | Silver | Antimicrobial activity was evaluated between cell-free supernatant and microbe | Sivalingam et al. (2012) |
| *Streptomycetes viridogens* HM10 | Gold | Antibacterial activity against *S. aureus* and *E. coli* in well-diffusion method | Balagurunathan et al. (2011) |
| *Nocardiopsis* sp. MBRC-1 | Silver | Strong antimicrobial activity against bacteria and fungi | Manivasagan et al. (2013a) |
| *Streptomyces* sp. | Zinc | Antibacterial activity | Rajamanickam et al. (2012) |
| *Rhodococcus* sp. | Gold | Metal ions were not toxic to the cells and the cells continued to multiply after biosynthesis of the gold nanoparticles | Ahmad et al. (2003a) |
| *Streptomyces* sp. | Silver | Antimicrobial activity against disease-causing microorganism | Saminathan (2015) |
| *Streptomyces* sp JAR1 | Silver | Antibacterial activity of antibiotics increased in presence of the biologically synthesized silver nanoparticles against the pathogens | Chauhan et al. (2013) |
| *Streptomyces* sp. VITBT7 | Silver | Antimicrobial activity found against fungal and bacterial pathogens | Subashini and Kannabiran (2013) |
| Actinomycetes isolate VITBN4 | Copper | Antibacterial activity exhibited by copper oxide nanoparticles against both human and fish bacterial pathogens | Nabila and Kannabiran (2018) |
| *Streptomyces* sp. ES2-5 | Selenium | Applied on contaminated site because have ability of spore reproduction, Se(IV) reduction, and adaptation in soil | Tan et al. (2016) |
| *Streptomyces griseobrunneus* FSHH12 | Selenium | Photocatalytic degradation | Ameri et al. (2016) |

Actinomycetes possess the fascinating potential to generate the nanoparticles of various metals exhibiting antibacterial (Adiguzel et al., 2018), antifungal (Wypij et al., 2018), antioxidant, anticancer, antibiofouling, and antiparasitic activity (Manivasagan et al., 2016). Some actinomycetes and synthesis of different metal nanoparticles and their applications are listed in Table 10.2

## 10.6 Future perspectives

In the last few decades, there has been growing interest in actinomycetes-based metal nanoparticle biosynthesis but extensive study is still needed for identification of potential actinomycetes. Since the biological agent-driven induction of nanoparticle formation is very slow often taking many hours in contrast to very fast and quick process of physicochemical methods, the research should be carried out in order to enhance the rate of synthesis so as to overcome the limitations associated. For nanoparticles, the shape, surface area, and size are important factors determining their activity. Control of different biological processes to control the size and shape of nanoparticles has still not been investigated in much detail. Studies have shown that biologically (actinomycete-mediated) produced nanoparticles are amenable to changes in their structural and functional properties due to lack of stability, necessitating further study in this direction. More emphasis should be paid on extracellular mechanisms governing nanoparticle biosynthesis because intracellularly produced nanoparticles require further processing, which is expensive and time consuming. Factors affecting the synthesis of nanoranged materials are also important; thorough optimization would be helpful in fabricating nanoparticles with desired properties. With the elimination of these limitations, actinomycete-mediated nanoparticle biosynthesis may play a significant role in the medical, food, agricultural, environmental, and energy sectors.

## References

Abdel-Mageed, W.M., Milne, B.F., Wagner, M., Schumacher, M., Sandor, P., Pathom-aree, W., et al., 2010. Dermacozines, a new phenazine family from deep-sea dermacocci isolated from a Mariana Trench sediment. Org. Biomol. Chem. 8 (10), 2352–2362.

Adiguzel, A.O., Adiguzel, S.K., Mazmanci, B., Tunçer, M., Mazmanci, M.A., 2018. Silver nanoparticle biosynthesis from newly isolated *Streptomyces* genus from soil. Mater. Res. Express. 5 (4), 045402.

Ahmad, A., Senapati, S., Khan, M.I., Kumar, R., Sastry, M., 2003a. Extracellular biosynthesis of monodisperse gold nanoparticles by a novel extremophilic actinomycete, *Thermomonospora* sp.. Langmuir 19 (8), 3550–3553.

Ahmad, A., Senapati, S., Khan, M.I., Kumar, R., Ramani, R., Srinivas, V., et al., 2003b. Intracellular synthesis of gold nanoparticles by a novel alkalotolerant actinomycete, *Rhodococcus* species. Nanotechnology 14 (7), 824.

Al-Dhabi, N.A., Ghilan, A.K.M., Arasu, M.V., 2018. Characterization of silver nanomaterials derived from marine *Streptomyces* sp. Al-Dhabi-87 and its in vitro application against multi-drug resistant and extended-spectrum beta-lactamase clinical pathogens. Nanomaterials 8 (5), 279.

Al-Zarban, S.S., Abbas, I., Al-Musallam, A.A., Steiner, U., Stackebrandt, E., Kroppenstedt, R.M., 2002. *Nocardiopsis halotolerans* sp. nov., isolated from salt marsh soil in Kuwait. Int. J. Syst. Evol. Microbiol. 52 (2), 525–529.

Ameri, A., Shakibaie, M., Ameri, A., Faramarzi, M.A., Amir-Heidari, B., Forootanfar, H., 2016. Photocatalytic decolorization of bromothymol blue using biogenic selenium nano-particles synthesized by terrestrial actinomycete *Streptomyces griseobrunneus* strain FSHH12. Desalin. Water Treat. 57 (45), 21552–21563.

Ames, R.N., Reid, C.P.P., Ingham, E.R., 1984. Rhizosphere bacterial population responses to root colonization by a vesicular-arbuscular mycorrh1zal fungus. New Phytol. 96 (4), 555–563.

Aoyagi, T., Hatsu, M., Imada, C., Naganawa, H., Okami, Y., Takeuchi, T., 1992. Pyrizinostatin: a new inhibitor of pyroglutamyl peptidase. J. Antibiot. (Tokyo) 45 (11), 1795–1796.

Asolkar, R.N., Schroeder, D., Heckmann, R., Lang, S., Wagner-Doebler, I., Laatsch, H., 2004. Helquinoline, a new tetrahydroquinoline antibiotic from *Janibacter limosus* Hel1. J. Antibiot. (Tokyo) 57 (1), 17–23.

Asolkar, R.N., Jensen, P.R., Kauffman, C.A., Fenical, W., 2006. Daryamides A – C, weakly cytotoxic polyketides from a marine-derived actinomycete of the genus *Streptomyces* strain CNQ-085. J. Nat. Prod. 69 (12), 1756–1759.

Balagurunathan, R., Radhakrishnan, M., Rajendran, R.B., Velmurugan, D., 2011. Biosynthesis of gold nanoparticles by actinomycete *Streptomyces viridogens* strain HM10. Indian J. Biochem. Biophys. 48 (5), 331–335.

Baniasadi, F., Shahidi, G.H., Nik, A.K., 2009. In vitro petroleum decomposition by actino-mycetes isolated from petroleum contaminated soils. Am. Eurasian J. Agric. Environ. Sci. 6 (3), 268–270.

Basilio, A., Gonzalez, I., Vicente, M.F., Gorrochategui, J., Cabello, A., Gonzalez, A., et al., 2003. Patterns of antimicrobial activities from soil actinomycetes isolated under different conditions of pH and salinity. J. Appl. Microbiol. 95 (4), 814–823.

Bhattacharyya, B.K., Sen, S.K., 2004. Gene manipulation in *Streptomyces*. Indian J. Biotechnol., 3, 22–28.

Bhatti, A.A., Haq, S., Bhat, R.A., 2017. Actinomycetes benefaction role in soil and plant health. Microb. Pathog. 111, 458–467.

Bister, B., Bischoff, D., Ströbele, M., Riedlinger, J., Reicke, A., Wolter, F., et al., 2004. Abyssomicin C—a polycyclic antibiotic from a marine *Verrucosispora* strain as an inhibitor of the p-aminobenzoic acid/tetrahydrofolate biosynthesis pathway. Angew. Chem. Int. Ed. 43 (19), 2574–2576.

Boer, W.D., Folman, L.B., Summerbell, R.C., Boddy, L., 2005. Living in a fungal world: impact of fungi on soil bacterial niche development. FEMS Microbiol. Rev. 29 (4), 795–811.

Bressan, W., 2003. Biological control of maize seed pathogenic fungi by use of actinomy-cetes. Biocontrol. 48 (2), 233–240.

Bull, A.T., Goodfellow, M., Slater, J.H., 1992. Biodiversity as a source of innovation in bio-technology. Annu. Rev. Microbiol. 46 (1), 219–246.

Chaphalkar, S.R., Dey, S., 1998. Thermostable alkaline metalloprotease from newly isolated alkalophilic *Streptomyces diastaticus* strain SS1. Indian J. Biochem. Biophys. 35 (1), 34–40.

Chauhan, R., Kumar, A., Abraham, J., 2013. *Streptomyces* A biological approach to the synthesis of silver nanoparticles with sp JAR1 and its antimicrobial activity. Sci. Pharm. 81 (2), 607.

Cho, J.Y., Kwon, H.C., Williams, P.G., Kauffman, C.A., Jensen, P.R., Fenical, W., 2006. Actinofuranones A and B, polyketides from a marine-derived bacterium related to the genus *Streptomyces* (Actinomycetales). J. Nat. Prod. 69 (3), 425–428.

Choi, I.K., Shin, H.J., Lee, H.S., Kwon, H.J., 2007. Streptochlorin, a marine natural product, inhibits NF-kappaB activation and suppresses angiogenesis in vitro. J. Microbiol. Biotechnol. 17 (8), 1338–1343.

Colquhoun, J.A., Mexson, J., Goodfellow, M., Ward, A.C., Horikoshi, K., Bull, A.T., 1998. Novel rhodococci and other mycolate actinomycetes from the deep sea. Antonie van Leeuwenhoek 74 (1–3), 27–40.

Crawford, D.L., 1978. Lignocellulose decomposition by selected *Streptomyces* strains. J. Appl. Environ. Microbiol. 35 (6), 1041–1045.

Cross, T., 1981. Aquatic actinomycetes: a critical survey of the occurrence, growth and role of actinomycetes in aquatic habitats. J. Appl. Bacteriol. 50 (3), 397–423.

Das, S., Lyla, P.S., Khan, S.A., 2008. Distribution and generic composition of culturable marine actinomycetes from the sediments of Indian continental slope of Bay of Bengal. Chin. J. Oceanol. Limnol. 26 (2), 166–177.

Dasari, V.R.R.K., Muthyala, M.K.K., Nikku, M.Y., Donthireddy, S.R.R., 2012. Novel Pyridinium compound from marine actinomycete, Amycolatopsis alba var. nov. DVR D4 showing antimicrobial and cytotoxic activities in vitro. Microbiological research 167 (6), 346–351.

Davenport, R.J., Curtis, T.P., Goodfellow, M., Stainsby, F.M., Bingley, M., 2000. Quantitative use of fluorescent in situ hybridization to examine relationships between mycolic acid-containing actinomycetes and foaming in activated sludge plants. Appl. Environ. Microbiol. 66 (3), 1158–1166.

De Schrijver, A., De Mot, R., 1999. A subfamily of MalT-related ATP-dependent regulators in the LuxR family. Microbiology 145 (6), 1287–1288.

Diraviyam, T., Radhakrishnan, M., Balagurunathan, R., 2011. Antioxidant activity of melanin pigment from *Streptomyces* species D5 isolated from Desert soil, Rajasthan, India. Drug Invent. 3 (3), 12–13.

Doumbou, C.L., Hamby Salove, M.K., Crawford, D.L., Beaulieu, C., 2001. Actinomycetes, promising tools to control plant diseases and to promote plant growth. Phytoprotection 82 (3), 85–102.

Elliott, L.F., Lynch, J.M., 1995. The international workshop on establishment of microbial inocula in soils: cooperative research project on biological resource management of the Organization for Economic Cooperation and Development (OECD). Am. J. Altern. Agric. 10 (2), 50–73.

El-Tarabily, K.A., Sivasithamparam, K., 2006. Non-streptomyceteactinomycetes as biocontrol agents of soil-borne fungal plant pathogens and as plant growth promoters. Soil Biol. Biochem. 38 (7), 1505–1520.

Fiedler, H.P., Bruntner, C., Riedlinger, J., Bull, A.T., Knutsen, G., Goodfellow, M., et al., 2008. Proximicin A, B and C, novel aminofuran antibiotic and anticancer compounds isolated from marine strains of the actinomycete *Verrucosispora*. J Antibiot. 61 (3), 158.

Franco-Correa, M., Quintana, A., Duque, C., Suarez, C., Rodríguez, M.X., Barea, J.M., 2010. Evaluation of actinomycete strains for key traits related with plant growth promotion and mycorrhiza helping activities. Appl. Soil Ecol. 45 (3), 209–217.

George, M., Anjumol, A., George, G., Hatha, A.M., 2012. Distribution and bioactive potential of soil actinomycetes from different ecological habitats. Afr. J. Microbiol. Res. 6 (10), 2265–2271.

Gomes, E.D.B., Dias, L.R.L., Rita de Cassia, M., 2018. Actinomycetes bioactive compounds: biological control of fungi and phytopathogenic insect. Afr. J. Biotechnol. 17 (17), 552–559.

Goodfellow, M., Williams, S.T., 1983. Ecology of actinomycetes. Annu. Rev. Microbiol. 37 (1), 189–216.

Goodfellow, M., Haynes, J.A., 1984. Actinomycetes in marine sediments. Biological, Biochemical and Biomedical Aspects of Actinomycetes. Academic Press, New York, pp. 453–472.

Gorajana, A., Venkatesan, M.V., Vinjamuri, S., Kurada, B.V., Peela, S., Jangam, P., et al., 2007. Resistoflavine cytotoxic compound from a marine actinomycete, *Streptomyces chibaensis* AUBN(1)/7. Microbiol. Res. 162 (4), 322–327.

Grein, A., Meyers, S.P., 1958. Growth characteristics and antibiotic production of actinomycetes isolated from littoral sediments and materials suspended in sea water. J. Bacteriol. 76 (5), 457.

Groth, I., Schumann, P., Rainey, F.A., Martin, K., Schuetze, B., Augsten, K., 1997. *Bogoriella caseilytica* gen. nov., sp. nov., a new alkaliphilic actinomycete from a soda lake in Africa. Int. J. Syst. Evol. Microbiol. 47 (3), 788–794.

Halder, A.K., Mishra, A.K., Chakrabarty, P.K., 1991. Solubilization of inorganic phosphates by *Bradyrhizobium*. Indian J. Exp. Biol. 29 (1), 28–31.

Hopwood, D.A., 2007. Therapeutic treasures from the deep. Nat. Chem. Biol. 3 (8), 457.

Hughes, C.C., MacMillan, J.B., Gaudêncio, S.P., Jensen, P.R., Fenical, W., 2009. The ammosamides: structures of cell cycle modulators from a marine-derived *Streptomyces* species. Angew. Chem. Int. Ed. 48, 725–727.

Iravani, S., Korbekandi, H., Mirmohammadi, S.V., Zolfaghari, B., 2014. Synthesis of silver nanoparticles: chemical, physical and biological methods. Res. Pharm. Sci. 9 (6), 385.

Itoh, T., Kinoshita, M., Aoki, S., Kobayashi, M., 2003. Komodoquinone A, a novel neuritogenic anthracycline, from marine *Streptomyces* sp. KS3. J. Nat. Prod. 66 (10), 1373–1377.

Janardhan, A., Kumar, A.P., Viswanath, B., Saigopal, D.V.R., Narasimha, G., 2014. Production of bioactive compounds by actinomycetes and their antioxidant properties. Biotechnol. Res. Int. .

Jeffrey, L.S.H., Sahilah, A.M., Son, R., Tosiah, S., 2007. Isolation and screening of actinomycetes from Malaysian soil for their enzymatic and antimicrobial activities. J. Trop. Agric. Fd. Sci. 35 (1), 159.

Jeffries, P., Gianinazzi, S., Perotto, S., Turnau, K., Barea, J.M., 2003. The contribution of arbuscular mycorrhizal fungi in sustainable maintenance of plant health and soil fertility. Biol. Fertil. Soils 37 (1), 1–16.

Jensen, P.R., Dwight, R.Y.A.N., Fenical, W., 1991. Distribution of actinomycetes in near-shore tropical marine sediments. Appl. Environ. Microbiol. 57 (4), 1102–1108.

Jeong, S.-Y., Shin, H.J., Kim, T.S., Lee, H.-S., Park, S.-K., Kim, H.M., 2006. Streptokordin, a new cytotoxic compound of the methylpyridine class from a marine-derived *Streptomyces* sp. KORDI-3238. J Antibiot. 59, 234–240.

Jiang, C.L., Xu, L.H., 1996. Diversity of aquatic actinomycetes in lakes of the middle plateau, yunnan, china. Appl. Environ. Microbiol. 62 (1), 249–253.

Jiang, W., Lechner, J., Carbon, J., 1993. Isolation and characterization of a gene (CBF2) specifying a protein component of the budding yeast kinetochore. J. Cell Biol. 121 (3), 513–519.

Jiménez-Esquilín, A.E., Roane, T.M., 2005. Antifungal activities of actinomycete strains associated with high-altitude sagebrush rhizosphere. J. Ind. Microbiol. Biotechnol. 32 (8), 378–381.

Kanoh, K., Matsuo, Y., Adachi, K., Imagawa, H., Nishizawa, M., Shizuri, Y., 2005. Mechercharmycins A and B, cytotoxic substances from marine-derived *Thermoactinomyces* sp. YM3- 251. J. Antibiot. (Tokyo) 58 (4), 289–292.

Kock, I., Maskey, R.P., Biabani, M.F., Helmke, E., Laatsch, H., 2005. 1-Hydroxy-1-norresistomycin and resistoflavin methyl ether: new antibiotics from marine-derived *Streptomycetes*. J. Antibiot. 58 (8), 530.

Kwon, H.C., Kauffman, C.A., Jensen, P.R., Fenical, W., 2006. Marinomycins AD. antitumor-antibiotics of a new structure class from a marine actinomycete of the recently discovered genus "*Marinispora*". J. Am. Chem. Soc. 128 (5), 1622–1632.

Lam, K.S., 2006. Discovery of novel metabolites from marine actinomycetes. Curr. Opin. Microbiol. 9 (3), 245–251.

Lazzarini, A., Cavaletti, L., Toppo, G., Marinelli, F., 2000. Rare genera of actinomycetes as potential producers of new antibiotics. Antonie van Leeuwenhoek 78 (3–4), 399–405.

Lechevalier, M.P., 1981. Ecological associations involving actinomycetes. Zentralbl. Bakteriol. Hyg. Abt I Suppl. 11, 159–166.

Lee, H.S., Shin, H.J., Jang, K.H., Kim, T.S., Oh, K.B., Shin, J., 2005. Cyclic peptides of the Nocardamine class from a marine derived bacterium of the genus *Streptomyces*. J. Nat. Prod. 68 (4), 623–625.

Macherla, V.R., Liu, J., Bellows, C., Teisan, S., Nicholson, B., Lam, K.S., et al., 2005. Glaciapyrroles A, B, and C, pyrrolosesquiterpenes from a *Streptomyces* sp. isolated from an Alaskan marine sediment. J. Nat. Prod. 68 (5), 780–783.

Makkar, N.S., Cross, T., 1982. Actinoplanetes in soil and on plant litter from freshwater habitats. J. Appl. Bacteriol. 52 (2), 209–218.

Maloney, K.N., MacMillan, J.B., Kauffman, C.A., Jensen, P.R., DiPasquale, A.G., Rheingold, A.L., et al., 2009. Lodopyridone, a structurally unprecedented alkaloid from a marine actinomycete. Org. Lett. 11 (23), 5422–5424.

Manimaran, M., Kannabiran, K., 2017. Actinomycetes-mediated biogenic synthesis of metal and metal oxide nanoparticles: progress and challenges. Lett. Appl. Microbiol 64 (6), 401–408.

Manivasagan, P., Venkatesan, J., Senthilkumar, K., Sivakumar, K., Kim, S.K., 2013a. Biosynthesis, antimicrobial and cytotoxic effect of silver nanoparticles using a novel *Nocardiopsis* sp. MBRC-1. BioMed Res. Int. . Available from: https://doi.org/10.1155/2013/287638.

Manivasagan, P., Venkatesan, J., Sivakumar, K., Kim, S.K., 2013b. RETRACTED: Marine actinobacterial metabolites: current status and future perspectives. Microbiol. Res. 168, 311–332.

Manivasagan, P., Venkatesan, J., Sivakumar, K., Kim, S.K., 2014. Pharmaceutically active secondary metabolites of marine actinobacteria. Microbiol. Res. 169 (4), 262–278.

Manivasagan, P., Venkatesan, J., Sivakumar, K., Kim, S.K., 2016. Actinobacteria mediated synthesis of nanoparticles and their biological properties: a review. Crit. Rev. Microbiol. 42 (2), 209–221.

Manulis, S., Shafrir, H., Epstein, E., Lichter, A., Barash, I., 1994. Biosynthesis of indole-3-acetic acid via the indole-3-acetamide pathway in *Streptomyces* spp. Microbiology 140 (5), 1045–1050.

Mevs, U., Stackebrandt, E., Schumann, P., Gallikowski, C.A., Hirsch, P., 2000. *Modestobacter multiseptatus* gen. nov., sp. nov., a budding actinomycete from soils of the Asgard Range (Transantarctic Mountains). Int. J. Syst. Evol. Microbiol. 50 (1), 337–346.

Moncheva, P., Tishkov, S., Dimitrova, N., Chipeva, V., Antonova-Nikolova, S., Bogatzevska, N., 2002. Characteristics of soil actinomycetes from Antarctica, 3, 3–14.
Nabila, M.I., Kannabiran, K., 2018. Biosynthesis, characterization and antibacterial activity of copper oxide nanoparticles (CuO NPs) from actinomycetes. Biocatal. Agric. Biotechnol. 15, 56–62.
Newman, D.J., Cragg, G.M., Snader, K.M., 2003. Natural products as sources of new drugs over the period 1981 – 2002. J. Nat. Prod. 66 (7), 1022–1037.
Nonomura, H., 1989. Genus *Streptosporangium*, Bergey's Manual of Systematic Bacteriology., vol. 4. Williams & Wilkins, Baltimore, pp. 2545–2551.
Oh, D.C., Williams, P.G., Kauffman, C.A., Jensen, P.R., Fenical, W., 2006. Cyanosporasides A and B, chloro-and cyano-cyclopenta [a] indene glycosides from the marine actinomycete "*Salinispora pacifica*". Org. Lett. 8 (6), 1021–1024.
Okami, Y., Okazaki, T., 1974. Studies on marine microorganisms. III. J. Antibiot. 27 (4), 240–247.
Olano, C., Méndez, C., Salas, J.A., 2009. Antitumor compounds from marine actinomycetes. Mar. Drugs 7 (2), 210–248.
Oskay, A.M., Üsame, T., Cem, A., 2004. Antibacterial activity of some actinomycetes isolated from farming soils of Turkey. Afr. J. Biotechnol. 3 (9), 441–446.
Pathom-Aree, W., Stach, J.E., Ward, A.C., Horikoshi, K., Bull, A.T., Goodfellow, M., 2006. Diversity of actinomycetes isolated from challenger deep sediment (10,898 m) from the Mariana Trench. Extremophiles 10 (3), 181–189.
Prakasham, R.S., Kumar, B.S., Kumar, Y.S., Kumar, K.P., 2014. Production and characterization of protein encapsulated silver nanoparticles by marine isolate *Streptomyces parvulus* SSNP11. Indian J. Microbiol. 54 (3), 329–336.
Raja, A., Prabakaran, P., Gajalakshmi, P., Rahman, A.H., 2010. A population study of psychrophilic actinomycetes isolated from Rothang Hill-Manali soil sample. J. Pure Appl. Microbiol. 4 (2), 847–851.
Rajamanickam, U., Mylsamy, P., Viswanathan, S., Muthusamy, P., 2012. Biosynthesis of zinc nanoparticles using actinomycetes for antibacterial food packaging. In: International Conference on Nutrition and Food Sciences, Singapore.
Ranjitha, V.R., Ravishankar, V.R., 2018. Extracellular synthesis of selenium nanoparticles from an actinomycetes *Streptomyces griseoruber* and evaluation of its cytotoxicity on HT-29 cell line. Pharm. Nanotechnol. 6 (1), 61–68.
Rao, C.N.R., Cheetham, A.K., 2001. Science and technology of nanomaterials: current status and future prospects. J. Mater. Chem. 11 (12), 2887–2894.
Renner, M.K., Shen, Y.-C., Cheng, X.-C., Jensen, P.R., Frankmoelle, W., Kauffman, C.A., et al., 1999. Cyclomarins AC, new anti-inflammatory cyclic peptides produced by a marine bacterium (*Streptomyces* sp.). J. Am. Chem. Soc. 121, 11273–11276.
Rowbotham, T.J., Cross, T., 1977. *Rhodococcus coprophilus* sp. nov.: an aerobic nocardioform actinomycete belonging to the 'rhodochrous' complex. Microbiology 100 (1), 123–138.
Saad, E.L., Salem, S.S., Fouda, A., Awad, M.A., El-Gamal, M.S., Abdo, A.M., 2018. New approach for antimicrobial activity and bio-control of various pathogens by biosynthesized copper nanoparticles using endophytic actinomycetes. J. Radiat. Res. Appl. Sci.
Saminathan, K., 2015. Biosynthesis of silver nanoparticles using soil actinomycetes *Streptomyces* sp. Int. J. Curr. Microbiol. Appl. Sci. 4 (3), 1073–1083.
Sanjenbam, P., Gopal, J.V., Kannabiran, K., 2014. Anticandidal activity of silver nanoparticles synthesized using *Streptomyces* sp. VITPK1. J. Mycol. Med. 24 (3), 211–219.

Sastry, M., Ahmad, A., Khan, M.I., Kumar, R., 2003. Biosynthesis of metal nanoparticles using fungi and actinomycete. Curr. Sci. 85 (2), 162–170.

Schneider, K., Keller, S., Wolter, F.E., Röglin, L., Beil, W., Seitz, O., et al., 2008. Proximicins A, B, and C—antitumor furan analogues of netropsin from the marine actinomycete *Verrucosispora* induce upregulation of p53 and the cyclin kinase inhibitor p21. Angew. Chem. Int. Ed. 47, 3258–3261.

Schumacher, R.W., Talmage, S.C., Miller, S.A., Sarris, K.E., Davidson, B.S., Goldberg, A., 2003. Isolation and structure determination of an antimicrobial ester from a marine sediment-derived bacterium. J. Nat. Prod. 66 (9), 1291–1293.

Shin, H.J., Jeong, H.S., Lee, H.S., Park, S.K., Kim, H.M., Kwon, H.J., 2007. Isolation and structure determination of streptochlorin, an antiproliferative agent from a marine-derived *Streptomyces* sp. 04DH110. J. Microbiol. Biotechnol. 17 (8), 1403–1406.

Singh, D.P., Patil, H.J., Prabha, R., Yandigeri, M.S., Prasad, S.R., 2018. Actinomycetes as potential plant growth-promoting microbial communities. In Crop Improvement Through Microbial Biotechnology. 27–38.

Sivalingam, P., Antony, J.J., Siva, D., Achiraman, S., Anbarasu, K., 2012. Mangrove *Streptomyces* sp. BDUKAS10 as nanofactory for fabrication of bactericidal silver nanoparticles. Colloids Surf. B: Biointerfaces 98, 12–17.

Socha, A.M., LaPlante, K.L., Rowley, D.C., 2006. New bisanthraquinone antibiotics and semisynthetic derivatives with potent activity against clinical *Staphylococcus aureus* and *Enterococcus faecium* isolates. Bioorg. Med. Chem. 14, 8446–8454.

Sosio, M., Bossi, E., Bianchi, A., Donadio, S., 2000. Multiple peptide synthetase gene clusters in actinomycetes. Mol. Gen. Genet. 264 (3), 213–221.

Sprusansky, O., Stirrett, K., Skinner, D., Denoya, C., Westpheling, J., 2005. The bkdR gene of *Streptomyces* coelicolor is required for morphogenesis and antibiotic production and encodes a transcriptional regulator of a branched-chain amino acid dehydrogenase complex. J Bacteriol. 187 (2), 664–671.

Srinivasan, M.C., Laxman, R.S., Deshpande, M.V., 1991. Physiology and nutritional aspects of actinomycetes: an overview. World J. Microbiol. Biotechnol. 7 (2), 171–184.

Subashini, J., Kannabiran, K., 2013. Antimicrobial activity of *Streptomyces* sp. VITBT7 and its synthesized silver nanoparticles against medically important fungal and bacterial pathogens. Der Pharm. Lett. 5, 192–200.

Subramani, R., Aalbersberg, W., 2012. Marine actinomycetes: an ongoing source of novel bioactive metabolites. Microbiol. Res. 167 (10), 571–580.

Suzuki, K., Nagai, K., Shimizu, Y., Suzuki, Y., 1994. Search for actinomycetes in screening for new bioactive compounds. Actinomycetologica 8 (2), 122–127.

Takahashi, Y., Omura, S., 2003. Isolation of new actinomycete strains for the screening of new bioactive compounds. J Gen Appl Microbiol. 49 (3), 141–154.

Takahashi, Y., Itami, T., Maeda, M., Suzuki, N., Kasornchandra, J., Supamattaya, K., et al., 1996. Polymerase chain reaction (PCR) amplification of bacilliform virus (RV-PJ) DNA in *Penaeus japonicus* Bate and systemic ectodermal and mesodermal baculovirus (SEMBV) DNA in Penaeus monodon Fabricius. J. Fish Dis. 19 (5), 399–403.

Takami, H., Inoue, A., Fuji, F., Horikoshi, K., 1997. Microbial flora in the deepest sea mud of the Mariana Trench. FEMS Microbiol. Lett. 152 (2), 279–285.

Tan, H., Deng, Z., Cao, L., 2009. Isolation and characterization of actinomycetes from healthy goat faeces. Lett. Appl. Microbiol. 49 (2), 248–253.

Tan, Y., Yao, R., Wang, R., Wang, D., Wang, G., Zheng, S., 2016. Reduction of selenite to Se (0) nanoparticles by filamentous bacterium *Streptomyces* sp. ES2-5 isolated from a selenium mining soil. Microb. Cell Fact. 15 (1), 157.

Terkina, I.A., Drukker, V.V., Parfenova, V.V., Kostornova, T.Y., 2002. The biodiversity of actinomycetes in Lake Baikal. Microbiology 71 (3), 346–349.

Terkina, I.A., Parfenova, V.V., Ahn, T.S., 2006. Antagonistic activity of actinomycetes of Lake Baikal. Appl. Biochem. Microbiol. 42 (2), 173–176.

Veiga, M., Esparis, A., Fabregas, J., 1983. Isolation of cellulolytic actinomycetes from marine sediments. Appl. Environ. Microbiol. 46 (1), 286.

Vijayabharathi, R., Sathya, A., Gopalakrishnan, S., 2018. Extracellular biosynthesis of silver nanoparticles using *Streptomyces griseoplanus* SAI-25 and its antifungal activity against *Macrophomina phaseolina*, the charcoal rot pathogen of sorghum. Biocatal. Agric. Biotechnol. 14, 166–171.

Walker, J.D., Colwell, R.R., 1975. Factors affecting enumeration and isolation of actinomycetes from Chesapeake Bay and Southeastern Atlantic Ocean sediments. Mar. Biol. 30 (3), 193–201.

Weyland, H., 1969. Actinomycetes in North Sea and Atlantic ocean sediments. Nature 223 (5208), 858.

Williams, S.T., Wellington, E.M.H., 1982. Actinomycetes. Methods of soil analysis. Part 2. Chemical and microbiological properties, (methodsofsoilan 2), pp. 969–987.

Williams, P.G., Miller, E.D., Asolkar, R.N., Jensen, P.R., Fenical, W., 2007. Arenicolides A – C, 26-membered ring macrolides from the marine actinomycete *Salinispora arenicola*. J. Org. Chem. 72 (14), 5025–5034.

Wypij, M., Świecimska, M., Czarnecka, J., Dahm, H., Rai, M., Golinska, P., 2018. Antimicrobial and cytotoxic activity of silver nanoparticles synthesized from two haloalkaliphilic actinobacterial strains alone and in combination with antibiotics. J. Appl. Microbiol. 124 (6), 1411–1424.

Zhang, C., Sheng, C., Wang, W., Hu, H., Peng, H., Zhang, X., 2015. Identification of the lomofungin biosynthesis gene cluster and associated flavin-dependent monooxygenase gene in Streptomyces lomondensis S015. PLoS One 10 (8), e0136228.

## Further reading

Kumar, D.V.R.R., Donthireddy, S.R.R., 2012. *Amycolatopsis alba* var. nov DVR D4, a bioactive actinomycete isolated from Indian marine environment. J. Biochem. Technol. 3 (2), 251–256.

# Cyanobacteria as a source of nanoparticles and their applications

*Snigdha Rai[1], Wang Wenjing[2,3], Alok Kumar Shrivastava[4] and Prashant Kumar Singh[5,6]*

[1]Molecular Biology Section, Centre for Advanced Study in Botany, Department of Botany, Banaras Hindu University, Varanasi, India, [2]State Key Laboratory of Cotton Biology, Henan Key Laboratory of Plant Stress Biology, School of Life Science, Henan University, Kaifeng, China, [3]Department of Life Science, School of Biology and Food Science, Shangqiu Normal University, Shangqiu, P. R. China, [4]Department of Botany, Mahatma Gandhi Central University, Motihari, India, [5]Department of Vegetable and Fruit Science, Institute of Plant Science, Agriculture Research Organization (ARO), The Volcani Center, Rishon LeZion, Israel, [6]Department of Vegetables and Field Crops, Institute of Plant Sciences, Agricultural Research Organization, The Volcani Center, Rishon LeZion, Israel

## 11.1 Introduction

Nanoparticles (NPs) are ultrafine particles in the range of 1–100 nm size. Nobel Laureate Richard Feynman in his visionary lecture "There is Plenty of Room at the Bottom" (Feynman, 1960) inspired revolutionary developments in manipulating matter at the atomic scale. Reduction in size results in a larger surface-to-volume ratio reflecting the unique physical and chemical properties of NPs and allowing them a wide range of applications. The past few decades have witnessed an exponential growth of activities in the field of nanometer-scale science and technology. In fact, more than 1000 products containing NPs were commercially available by 2010 (Ostiguy et al., 2010; Sebastian et al., 2014). Some nanomaterials occur naturally, and some nanomaterials are engineered, both of which are used in commercial products and processes from cosmetics, soap, and food packaging to diagnostics and drug delivery. NPs are now widely considered to have potential in areas as diverse as drug development, water decontamination, information and communication technologies, and in the production of stronger and lighter materials (Salata, 2004). A higher surface-to-volume ratio increases the possibility of the substrate to bind with NPs in chemical reactions (Pareek et al., 2017). NPs in biology or medicine are used as fluorescent biological labels, drug and gene delivery, biodetection of pathogens, detection of proteins, probing of DNA structure, tissue engineering, tumor destruction via heating (hyperthermia), separation and purification of biological molecules and cells, MRI contrast enhancement, and in phagokinetic studies (Salata, 2004).

**Role of Plant Growth Promoting Microorganisms in Sustainable Agriculture and Nanotechnology.**
**DOI: https://doi.org/10.1016/B978-0-12-817004-5.00011-7**

### 11.1.1 Nanoparticles currently in use

NPs are divided categorized by their morphology, size, and chemical properties. Based on physical and chemical characteristics, some of the well-known classes of NPs are as follows.

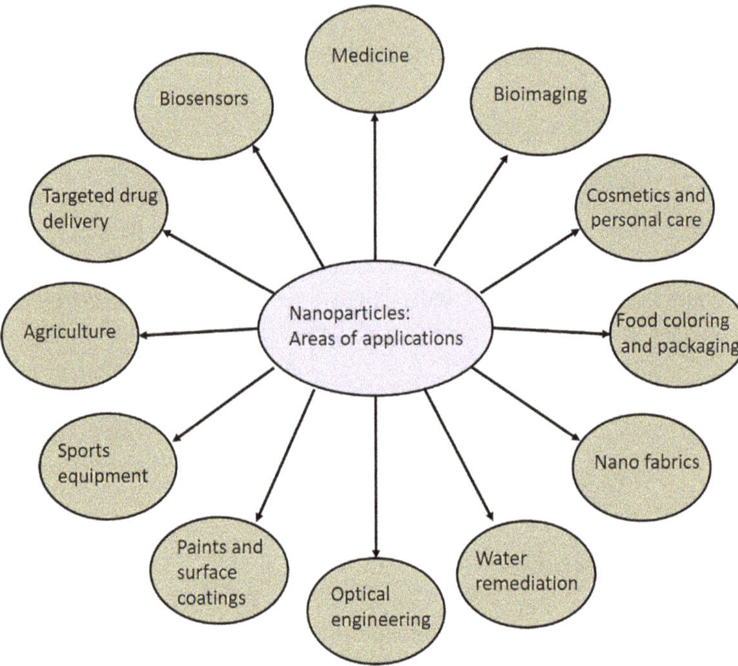


**Figure 11.1** Diagrammatic presentation of the major areas of nanoparticle applications.

**Table 11.1** Potential applications of nanoparticles and expected impact of nanotechnology.

| Product areas with end-products containing nanoparticles | Considerable impact of nanoparticles in various sectors |
|---|---|
| Cosmetics and personal care products | Medical and pharmaceutical sector |
| Paints and coatings | Bionanotechnology and biosensors |
| Consumer electronics | Composite materials |
| Household products | Environment sector including water |
| Sports products and textiles | Automotive sectors |
| Medical and healthcare products | Aeronautics |
| Food and nutritional ingredients | The energy sector, including fuel cells, batteries, and photovoltaics |
| Food packaging and agrochemicals | Remediation |
| Veterinary medicines | Electronics and optoelectronics, photonics |
| Construction materials | Construction sector including reinforcement of |
| Catalysts and lubricants | materials |

*Source*: Adapted and modified from Lovestam G., Rauscher, H, Gert, R., Sokull K.B., Niel G., Jean-Philippe P., and Hermann S., (2010). Considerations on a Definition of Nanomaterial for Regulatory Purposes. European Commission, Joint Research Centre, Luxembourg. ISBN NO. 978-92-79-16014-1.40 p.

1. *Carbon-based NPs*

   Fullerenes and carbon nanotubes are two major classes of NPs used for commercial applications such as fillers (Saeed and Khan, 2016), efficient gas adsorbents for environmental remediation (Ngoy et al., 2014), and as a support medium for different inorganic and organic catalysts (Mabena et al., 2011). Carbon NPs have electrical conductivity, heat conductivity, and mechanical properties. Carbon NPs are composed of carbon and thus have high stability, good conductivity, low toxicity, and environmental friendliness. Applications of carbon-based NPs are in the early stages, but are expected to be used in drug and gene delivery, bioimaging (Jing et al., 2016), and energy storage (Qi-Long et al., 2016).
2. *Metallic NPs*

   Metal NPs are submicron-scale entities made of pure metals (e.g., gold, platinum, silver, titanium, zinc, cerium, iron, thallium) or their compounds (e.g., oxides, hydroxides, sulfides, phosphates, fluorides, chlorides). Size and aspect ratio of metallic NPs (such as gold NPs) determine the unique properties, particularly optical properties like dispersion color. A large diversity of nanoscale oxides also known as metal oxide NPs is because of their metallic nature. An important effect of reduction in size is the electronic properties of these metal and metal oxide NPs. The strong influence of the conductivity and chemical reactivity never affect the bandgap because of their small size (Hoffmann, 1988; Albright et al., 1985). Metallic NPs are used in molecular diagnostics, electronics, catalysis, drug delivery, and sensing (Castro et al., 2014). Metallic NPs can be synthesized and modified with various chemical functional groups, which allows them to be conjugated with antibodies, ligands, and drugs of interest and thus opens a wide range of potential applications in biotechnology, magnetic separation, preconcentration of target analytes, targeted drug delivery, vehicles for gene and drug delivery, and more importantly diagnostic imaging (Mody et al., 2010).
3. *Ceramics NPs*

   These NPs (also known as nanoceramics or nanopowders) are classified as inorganic, heat-resistant, nonmetallic solids made of oxides, carbides, phosphates, and carbonates of both metallic and nonmetallic compounds. The material offers unique properties such as such as high heat resistance and chemical inertness (Thomas et al., 2015). The biomedical field is the most explored field of ceramic NPs. Nanoceramics have great potential as drug carriers to deliver and target the active pharmaceutical ingredient to the desired site in a controlled manner, resulting in achievement of therapeutic concentration of the drug at the target site (Nissan, 2004). Additionally, nanoceramics are suggested for areas such in energy supply and storage and communication and transportation systems.
4. *Semiconductor NPs*

   A wide bandgap is a characteristic of semiconductor NPs, and therefore these NPs showed a significant alteration in their properties with bandgap tuning. They are essential materials in photocatalysis, photooptics, and electronic devices (Sun et al., 2000). Semiconductor nanocrystals have broad applications in solar energy conversion, optoelectronic devices, molecular and cellular imaging, and in ultrasensitive detection (Smith and Nie, 2010).
5. *Polymeric NPs*

   Polymeric NPs (PNPs) are the particles that are prepared from polymers. A polymer is a class of natural or synthetic substances composed of macromolecules that are multiples of monomers. The most traditional field of application is waterborne paints, adhesives, and coatings. They are mostly nanosphere or nanocapsule shaped (Mansha et al., 2017). PNPs are readily functionalized and thus find applications ranging from photonics, electronics, sensors, medicine, pollution control, and environmental technology. More recently, they have found applications in biomedical fields such as bioimaging, drug delivery, and diagnostics (Mallakpour and Behranvand, 2016).

6. *Lipid-based NPs*

These NPs contain lipid moieties and are effectively used in many biomedical applications. Like PNPs, lipid NPs are extensively used in the pharmaceutical industry and used to improve the oral bioavailability of the poorly water-soluble drugs. Lipid NPs possess a solid core made of lipid and a matrix contains soluble lipophilic molecules. Surfactants or emulsifiers stabilize the outer core of these NPs (Rawat et al., 2011). Lipid-based NPs are now extensively used in biomedical fields such as bioimaging, drug delivery, and diagnostics (Mallakpour and Behranvand, 2016; Puri et al., 2009). Lipid NPs enhance the absorption of drugs in the gastrointestinal tract because of improved mucosal adhesion and enhanced residence time. Lipid NPs may also protect loaded drugs from chemical and enzymatic degradation and gradually release drug molecules from the lipid matrix into blood, resulting in improved therapeutic profiles.

## 11.1.2 Nanoparticle synthesis

### 11.1.2.1 Industrial synthesis

There are two primary methods of industrial synthesis of NPs: (1) top down and (2) bottom up.

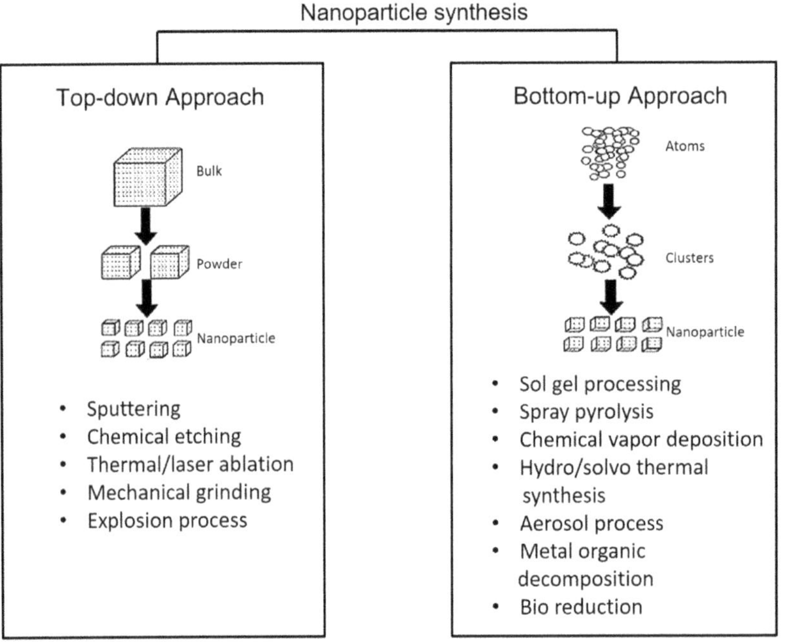


**Figure 11 2** Top-down and bottom-up approaches for nanoparticle synthesis.

1. *Top-down approach*

   This approach includes preparation of NPs by cutting more significant pieces of material until only a NP remains. This method is commonly achieved by using lithographic or etching techniques; however, grinding in a ball mill can also be used in some cases. However, it poses some difficulties in generating uniform NPs and is more likely to introduce internal stress, structural defects, and contamination. For example, nanowires made by lithography are not smooth and may contain a lot of impurities and structural defects on the surface. These imperfections affect the physical properties and surface chemistry of nanomaterials, because of a large surface-to-volume ratio. Due to inelastic surface scattering of the imperfect surfaces, reduced conductivity and excessive heat generation take place, which impose extra challenges on the device design and fabrication processes (Guozhong, 2004).
2. *Bottom-up approach*

The more convenient and faster method for producing NPs on a commercial scale is the "bottom-up" approach. This method refers to the buildup of material from the bottom: atom-by-atom, molecule-by-molecule, or cluster-by-cluster, where a NP is "grown" from simple molecules. By limiting the concentration, functionalizing the surface of the particle, or using a micelle to template the growth, the size of the NP can be controlled. The bottom-up approach relies on the principle of supersaturation to control particle size (Christian et al., 2008). Bottom-up methods to produce NPs from atoms are chemical processes based on transformations in solution such as sol-gel processing, hydro or solvothermal syntheses, metal-organic decomposition, or in vapor phase chemical vapor deposition. To control the formation and growth of the NPs by the bottom-up method, two approaches are used:

**(a)** Arrested precipitation requires either the introduction of a chemical or exhaustion of one of the reactants to block the reaction.
**(b)** Reduction in growth volume of NPs.

The bottom-up approach also promises a better chance to obtain nanostructures with fewer defects, more homogeneous chemical composition, and better short- and long-range order. These can be achieved because the bottom-up approach is driven mainly by the reduction of Gibbs-free energy so that nanostructures and nanomaterials produced are in a state closer to a thermodynamic equilibrium state. In contrast, the top-down approach often introduces internal stress, in addition to surface defects and contamination.

### 11.1.3 Why biosynthesize nanoparticles? Disadvantages of conventional synthesis methods

Physical and chemical techniques have been extensively used in recent years to produce NPs. These methods can provide large quantities of NPs with defined size and shape in a relatively short time, but

1. The processes are complicated, outdated, and costly.
2. These methods create a considerable amount of hazardous toxic waste (Li et al., 2011).

The NPs synthesis methods currently in use massively rely on the use of organic solvents with high toxicities (Duan et al., 2014). Physical approaches to NP production require higher utilization of energy, are more expensive, offer less precision in the size of the NPs generated, and sometimes impurity problems are present. While chemical synthesis is faster it also has disadvantages. Chemical synthesis of NPs requires a capping agent, a reducing agent, and a reaction solvent. Capping agents are widely employed in the colloidal synthesis and stabilization of NPs to impose constraints on the size evolution, to control the particle morphology, and to protect or passivate the surface from aggregation (Duan et al., 2014). Many surfactants such as heteroatom-functionalized long-chain hydrocarbons, polymers, dendrimers, etc., are used as capping agents in NP synthesis, but these are difficult to remove and do not readily degrade and require complex removal processing before the particles can be used as industrial products, which increases energy costs (Duan et al., 2014). Some highly reactive reducing agents such as sodium borohydride ($NaBH_4$), hydrazine ($N_2H_4$), and formaldehyde are used in large volumes in the synthesis process, requiring safe handling and discarding a significant amount, as well as these chemicals, have to be removed entirely from the synthesized NPs for the use in biological applications. Similarly, reaction solvents used in the process, which provide a medium for the dissolution of precursors, transferring heat and reactants and dispersing resulting NPs, pose a threat to the environment.

With the evolution of techniques in physical and chemical NP synthesis, concerns regarding the environment, health, and safety have increased. Thus, the concept of "green nanotechnology" started gaining attention. The sustainable approach of green nanotechnology has a single goal—a better environment—but there are at least two paths to that goal:

1. Environmentally friendly method of NP production
2. Production of nanomaterials without harming the environment or human health by the process or by its outcome.

The biological approach to NP synthesis uses microorganisms, algae, fungi, and plants to deliver high-yield, low-cost, and environmentally friendly NPs (Narayanan and Sakthivel, 2010). NPs produced by the abiogenic enzymatic process are safe for pharmaceutical uses and are environmentally friendly. Furthermore, these NPs have a longer shelf-life and stability as natural capping takes place. It is cost effective and a single-step method of NPs production with secure downstream processing as well as purification. Another advantage over chemical synthesis is that the biological process can be easily scaled up; produced. Among the advantages mentioned above for biosynthesis of NPs, natural capping of the NPs has been of great importance. With an enzymatic process, the use of expensive chemicals is eliminated, less energy is utilized, and the more acceptable "green" route is adopted. For example, extracellular synthesis of silver NPs using microbes is primarily found to be nitrate reductase-mediated synthesis. The enzyme nicotinamide adenine dinucleotide (NADH)-dependent reductase is associated with a reduction of $Ag^+$ to Ag in the case of bacteria, cyanobacteria, and fungi (Hulkoti and Taranath, 2014).

In NP biosynthesis, organisms take up the metal ions supplied and convert them into the elemental metal of desired shape and size through enzymes generated by the cell activities. Many microorganisms produce inorganic materials either intracellularly or extracellularly. The intracellular method includes the transportation of ions into the microbial cell to form NPs in the presence of enzymes. The extracellular synthesis of NPs involves trapping the metal ions on the surface of the cells and reducing ions in the presence of enzymes (Zhang et al., 2011). Synthesis of NPs may be triggered by several compounds such as carbonyl groups, terpenoids, phenolics, flavonones, amines, amides, proteins, pigments, alkaloids, and other reducing agents present in the plant extracts and microbial cells (Asmathunisha and Kathiresan, 2013). However, the exact mechanism of NP synthesis by natural extracts is not yet fully understood.

## 11.2 Role of microbes (cyanobacteria) in nanoparticle synthesis and contribution

There is a need for ecofriendly, safe, reliable, and clean methods for the preparation of NPs. Various biological routes such as the use of plant extracts, bacteria, fungi, and algae are considered safe and nontoxic and provide more environmentally sound synthesis of NPs. These emerging and expanding bottom-up methodologies are based on the exploitation of biomass in the form of aqueous extracts or whole cells. Among the biological systems used for NP synthesis, particular attention has been paid to microalgae as they are involved in toxic metal bioremediation. Microalgae have been shown to produce silver, gold, cadmium, and platinum NPs (Brayner et al., 2007; Parial et al., 2012). Among the microalgae, cyanobacteria have unique features such as high biomass yield, oxygenic photosynthesis, habitat diversity (contaminated and polluted environments) and growth on nonarable lands, useful byproducts and biofuel generation, enhance soil fertility, and reduced greenhouse gas emissions. A characteristic feature of converting $CO_2$ to other forms of carbon catalyzed by sunlight has implications on reduced cost for growth medium. Hence cheaper production costs, and potentially reduced carbon footprint of the process makes cyanobacteria a sustainable source of NPs biosynthesis (Pasula and Lim, 2017).

The capacity of cyanobacteria to adsorb metals is a function of their cell surface, which consists of complex structures in distinct layers each with unique molecular functional groups and metal-binding properties (Yee et al., 2004). The external cell wall in cyanobacteria is covered by S-layers consisting of glycoprotein that comprises the entire cell surface, beneath which lies the outer membrane, the peptidoglycan layer, and the cytoplasmic membrane (Hoiczyk and Hansel, 2000). The rapid binding of metals on to the cyanobacterial cell surface occurs as a result of negatively charged groups on membrane and in extracellular polymeric substances (EPS) while the intracellular levels are maintained via (1) metal chelation by phytochelatin and metallothionein and (2) sequestration in polyphosphate bodies

(Hudek et al., 2012). The delicate balance between metal requirements and toxic overexposure is tightly regulated through the uptake and trafficking via exopolysaccharides (EPS) layers and cell membrane (Hudek et al., 2012; Mehta and Gaur, 2005). For metals to be internalized (imported), cyanobacteria have evolved unique metal transport systems (Ma et al., 2009). Porins of outer membrane facilitate the nonselective passive diffusion of metal ions across the outer membrane, while active transport systems operate both in the outer and inner membranes (Ma et al., 2009). Yee et al. (2004) showed that *Calothrix*, a filamentous cyanobacterium, has heterogeneously distributed metal-binding sites on its surface. Proton-active surface carboxyl, phosphoryl, hydroxyl, and amine functional groups located on the cell wall and exopolymer sheath interact with metal ions (Yee et al., 2004).

Cyanobacteria is considered one of the best biological systems for NP synthesis intracellularly and extracellularly. However, there are only a few reports on the biological synthesis of noble metal NPs by using cyanobacteria (El-Naggar et al., 2017). *Spirulina* has been utilized for the extracellular synthesis of gold, silver, and Au/Ag bimetallic NPs (Chakraborty et al., 2009). Cyanobacteria such as *Anabaena*, *Calothrix*, and *Leptolyngbya* have been shown to form Au, Ag, Pd, and Pt NPs of well-controlled size (Brayner et al., 2007). These NPs are synthesized intracellularly and naturally released in the culture medium, where they are stabilized by algal polysaccharides, allowing their easy recovery. The size of the recovered particles as well as the reaction yield has been shown to depend on the cyanobacterial genus. Investigations of NP formation indicate that the intracellular enzyme "nitrogenase" is responsible for the metal reduction but that the cellular environment is involved in the colloid growth process (Brayner et al., 2007). The cyanobacterium *Nostoc linckia* was used to synthesize selenium NPs from selenite; both extracellular and intracellular formation of amorphous NPs, in size from 10 to 80 nm were observed (Cepoi et al., 2015). Selenite can be reduced to elemental selenium by reaction with reactive thiol groups of proteins/peptides in the so-called "paintertype" reaction (Zinicovscaia et al., 2016). Silver and gold are the most commonly reported NPs synthesized using cyanobacteria as the host system as discussed in the following.

## 11.2.1 Synthesis of gold nanoparticles from cyanobacteria

The formation of gold NPs within the cell wall of cyanobacteria may be due to the polyphosphates, polysaccharides, and carboxyl groups present on the cell membrane, which catalyze the reduction of gold ions (Pasula and Lim, 2017). Synthesis of nanogold has been reported using *Plectonema boryanum* (Lengket et al., 2006a,b). A controlled sized Nanogold NPs has been reported to produce by using *Plectonema boryanum* (Lengket et al., 2006a,b). The cyanobacteria in aqueous gold (III)−chloride solution caused the precipitation of amorphous gold (I)−sulfide NPs at the cell walls. The metallic gold was finally deposited in the form of octahedral platelets ($\sim$10 nm to 6 μm) near cell surfaces and solutions. The X-ray absorption spectroscopy analysis showed that the reduction of gold (III)−chloride to metallic gold by cyanobacteria results in the formation of an intermediate Au (I) species and gold

(I)−sulfide (Lengket et al., 2006a). Intracellular gold NPs have been made using *Synechocystis* sp. PCC 6803 (Focsan et al., 2011). The NPs of average size $13 \pm 2$ nm were found to localize at the cell wall, plasma membrane, and inside the cytoplasm. The study compared the gold NP synthesis to the metabolic activity of cyanobacteria namely, photosynthesis and respiration.

### *11.2.2 Synthesis of silver nanoparticles from cyanobacteria*

Cyanobacterial extracts like phycocyanin and polysaccharides can extracellularly reduce silver ions. According to Patel et al. (2015), the active factor involved in NP formation appears to be the extracellular polysaccharide, pigment phycocyanin, and some other biomolecules that act as natural surfactants (capping agents) on specific facets of the forming crystal. The exopolysaccharide layer plays a vital role in metal ion accumulation and reduction (Zinicovscaia et al., 2016). Zhang et al. (2011) showed that in the molecular structure of polysaccharides, there are reactive amino, hydroxyl, or carboxyl groups that have a significant effect on the formation, stabilization, and growth of selenium NPs. Sixteen different strains of cyanobacteria and microalgae were tested for their ability to produce silver NPs of which 14 were successful (Patel et al., 2015). Both cell extracts and extracellular medium were capable of producing NPs indicating that the extracellular medium contains excreted compounds responsible for the synthesis of NPs of sizes 13−31 nm. Experiments showed that extracellular polysaccharides released by cyanobacteria and algae act as reducing agents in the NP synthesis. Interestingly, the extracellular medium failed to produce silver NPs in the dark suggesting that light plays a role in the process (Birla et al., 2013). The study also demonstrated that C-phycocyanin, which is the blue-colored accessory pigment produced by cyanobacteria, can reduce silver to form silver NPs.

## 11.3 Commercial applications of cyanobacterial nanoparticles

Synthesizing NPs via biological entities offers a clean, nontoxic. and environmentally friendly method. NPs with a wide range of sizes, shapes, compositions, and physicochemical properties can be synthesized (Shah et al., 2015). The biological production of highly stable and well-characterized NPs can be obtained by optimizing vital aspects such as the type of organism, cell growth rate, and enzyme activity. The enzymes and proteins in the biological system provide extractability for the synthesized NPs. Thus chemical stabilizers are avoided, another step toward "green synthesis" (Das et al., 2017). Several strains of microalgae carry out the process of noble metal NP biosynthesis intracellularly. Also, in some cases, the produced NPs are released into culture media to form stable colloids, thus easing their recovery. One significant advantage of biosynthesized NPs is that they are nontoxic and can be used for clinical and biomedical uses including drug carriers for targeted delivery, cancer treatment, gene therapy and DNA analysis, as antibacterial agents, biosensors, to enhance reaction rates, and in MRI. These bioproducts are also better suited for cosmetics usage since they are less allergic. NPs through biomethods have led to conservation of natural and nonrenewable resources along with a reduction in environmental pollution (Shankar et al., 2016); these methods are equal and sometimes better than other commercial approaches for everyday items such as stain-resistant clothing and electronics. Different metal NPs such as Ag, Au, Pt, Pd, and Cu have been successfully biosynthesized using green reducing agents.

For example, biosynthesized NPs such as AuNP can passively accumulate in tumors where its unique optical and chemical properties can be used in tumors thermal treatments (Hirsch et al., 2003; Zheng and Sache, 2009). Studies have shown that biocompatible AuNPs can be successfully used as carrier platforms for the targeted delivery of anticancer drugs, thus improving delivery and minimizing treatment durations and side effects (Paciotti et al., 2006; Cheng et al., 2010). Some AuNP-based diagnostic kits for fast diagnoses of cancer, HIV and other diseases are under clinical trials (Kumar et al., 2011). Green synthesized AuNPs have also been used in the development of biosensors, quantification of blood glucose, disease markers, toxic metals, and insecticides (Dykman and Khlebtsov, 2011; Liu and Lu, 2003). AuNPs have antimicrobial activity and primarily react with sulfur or phosphorus-holding bases in proteins (Nadeem et al., 2017). When NPs attach to thiol functional groups of enzymes, such as NADH dehydrogenases, they interrupt the respiratory chains by the generation of a high number of free radicles, leading to metabolic stress. Also, AuNPs may inhibit the binding of tRNA to ribosomal subunit (Cui et al., 2012).

According to the report by El-Naggar et al. (2017), biosynthesized silver NPs (AgNPs) demonstrate the antibacterial effect on both Gram classes of bacteria. Silver ions released from silver NPs when coming in contact with bacterial cells may deactivate the production of some enzymes and cellular proteins necessary for adenosine triphosphate synthesis or influence the bacterial DNA replication functions. Silver ions may also disrupt the working of membrane-bound enzymes of the respiratory chain (Agnihotri et al., 2014). Like Au and Ag, ZnONPs display good antibacterial activity and have been used in food packaging and wastewater treatments (Shah et al., 2015); biosynthesized ZnONPs are also more suitable and cause less allergic symptoms when used in cosmetic products like sunscreens. According to Parchi et al. (2013), biosynthesized NPs are more suitable for orthopedic implants due to their better compatibility with bone-forming cells (osteoblasts), can colonize as well as synthesize new bone matrix and thus increasing the average lifetime of the implant. Biosynthesized bone fillers might have less of a risk of rejection and toxicity. Likewise, iron oxide NPs (IONPs) have several biological uses. Among the various IONPs, $Fe_3O_4$ NPs are the most typical and widely used. One promising application is in cancer diagnostics and therapy, using magnetic resonance imaging, magnetic hyperthermia, magnetic targeting, and cell tracking (Yu et al., 2014). Recently, nanoscale copper oxide has gained much attention due to its remarkable antibacterial activity (Chatterjee et al., 2014), and products with copper-containing surfaces have been used for sterilization processes in hospitals (Mikolay et al., 2010).

## 11.4 Concluding remarks and prospects

The important challenges frequently encountered in the biosynthesis of NPs are controlling the shape and size of particles and achieving the monodispersity in solution phase. Cyanobacteria, due to (1) lower energy inputs required in culturing and (2) shorter generation time, are a favorite organism host for NP synthesis. Both intracellular and extracellular synthesis have been reported in these organisms. However, the challenges that remain to realize the full potential of cyanobacteria are scaling up for production-level processing, minimizing the time required while choosing a suitable strain, and selection of a suitable biocatalyst state. Arguably, the majority of cyanobacteria-mediated NP synthesis is laboratory scale and has not been proven beyond the bench. It is, therefore, important to adapt the small-scale protocols to procedures that are amenable to standardized and robust scale-up to fulfill the promise of biogenic NPs as a viable industrial activity. There is a need for comprehensive understanding of the complex underlying mechanism in cyanobacteria for NP synthesis. Future studies must concentrate on the enzymatic mechanisms and proteins responsible for NP synthesis. Due to the vast cyanobacterial diversity, there is a need to tap the potential of several untouched local cyanobacterial isolates for NPs biosynthesis. Most of the genetic manipulations are restricted to a few model systems. There is also the need to focus on improving the available genetic tools for systemic manipulation of the genetic circuits in these microorganisms. Cyanobacterial biosynthesis of NPs is a comparatively new idea and in the developmental stage. Though there are several parameters that need to be optimized for the biosynthesis of NPs, the need for a green and futuristic approach is the main driving force for exploring potential organisms such as cyanobacteria. The large-scale synthesis of NPs using cyanobacteria is interesting because it can be cost effective (lower input requirements for batch culture since light can be used for energy synthesis) and does not require any hazardous, toxic, and expensive

**Table 11.2** Nanoparticle produced by cyanobacterial biosynthesis.

| Name of the species | Type of nanoparticle | Size and shape | Author and year |
|---|---|---|---|
| *Spirulina platensis* | Ag, Au | 7–16 nm<br>6–10 nm | Govindaraju et al. (2008) |
| *Oscillatoria willei* | Ag | 100–200 nm | Ali et al. (2011) |
| *Phormidium tenue* | Cd | 5 nm | Ali et al. (2012) |
| *Plectonema boryanum* | Au | 10–6 μm<br>Octahedral | Lengket et al. (2006a,b) |
| *Limnothrix* sp. | Ag | 31.86 nm<br>elongated | Patel et al. (2015) |
| *Synechocystis* sp. *PCC 6803* | Au | 13 ± 2 nm | Focsan et al. (2011) |
| *Nostoc linckia* | Se | 10–80 nm<br>Amorphous shape | Zinicovscaia et al. (2016) |

chemical materials for synthesis and stabilization processes of NPs biosynthesis (Iravani, 2014). Procedures to develop specific size and shape of the particles by the use of a specific strain of cyanobacteria are still needed (Figs. 11.1 and 11.2) (Tables 11.1 and 11.2).

## Acknowledgments

Prashant Kumar Singh is thankful to Agriculture Research Organisation (ARO), Israel for a postdoctoral fellowship. Alok Kumar Shrivastava is grateful for the Department of Science and Technology (DST)-Science and Engineering Research Board (SERB) for the Young Scientist Award.

## References

Agnihotri, S., Mukherji, S., Mukherji, S., 2014. Size-controlled silver nanoparticles synthesized over the range 5-100 nm using the same protocol and their antibacterial efficacy. RSC Adv. 4, 3974–3983.

Albright, T.A., Burdett, J.K., Whangbo, M.H., 1985. Orbital Interactions in Chemistry. Wiley-Interscience, New York.

Ali, D.M., Sasikala, M., Gunasekaran, M., Thajuddin, N., 2011. Biosynthesis and characterization of silver nanoparticles using marine cyanobacterium, *Oscillatoria willei* NTDM01. Dig. J. Nanomater. Biostruct. 6 (2), 385–390.

Ali, D.M., Gopinath, V., Gopinath, N., Rameshbabu, N., Thajuddin, N., 2012. Synthesis and characterization of CdS nanoparticles using C-phycoerythrin from the marine cyanobacteria. Mater. Let. 74. Available from: https://doi.org/10.1016/j.matlet.2012.01.026.

Asmathunisha, N., Kathiresan, K., 2013. A review on biosynthesis of nanoparticles by marine organisms. Coll. Surf. B: Biointerf. 103, 283–287.

Birla, S.S., Gaikwad, S.C., Gade, A.K., Rai, M.K., 2013. Rapid synthesis of silver nanoparticles from *Fusarium oxysporum* by optimizing physiocultural conditions. Sci. World J. 2013, 796018. Available from: https://doi.org/10.1155/2013/796018.

Brayner, R., Barberousse, H., Hernadi, M., Djedjat, C., Yepremian, C., Coradin, T., 2007. Cyanobacteria as bioreactors for the synthesis of Au, Ag, Pd, and Pt nanoparticles via an enzyme-mediated route. J. Nanosci. Nanotechnol. 7 (8), 2696–2708. Available from: https://doi.org/10.1166/jnn.2007.600.

Castro, L., Blázquez, M.L., Muñoz, J.A., González, F.G., Ballester, A., 2014. Mechanism and applications of metal nanoparticles prepared by biomediated process. Rev. Adv. Sci. Eng. 3 (3), 192–216.

Cepoi, L., Rudi, L., Chiriac, T., Valuta, A., Zinicovscaia, I., Duca, G., 2015. Biochemical changes in cyanobacteria during the synthesis of silver nanoparticles. Can. J. Microbiol. 61, 13–21. Available from: https://doi.org/10.1139/cjm-2014-0450.

Chakraborty, N., Banerjee, A., Lahiri, S., Panda, A., Ghosh, A.N., Pal, R., 2009. Biorecovery of gold using cyanobacteria and an eukaryotic alga with special reference to nanogold formation-a novel phenomenon. J. Appl. Phycol. 21, 145–152.

Chatterjee, A.K., Chakraborty, R., Basu, T., 2014. Mechanism of antibacterial activity of copper nanoparticles. Nanotechnology 25, 135101. Available from: https://doi.org/10.1088/0957-4484/25/13/135101.

Cheng, Y., Samia, A.C., Li, J., Kenney, M.E., Resnick, A., Burda, C., 2010. Delivery and efficacy of a cancer drug as a function of the bond to the gold nanoparticle surface. Langmuir 26, 2248–2255.

Christian, P., Von der Kammer, F., Baalousha, M., Hofmann, T., 2008. Nanoparticles: structure, properties, preparation and behaviour in environmental media. Ecotoxicology 17 (5), 326–343. Available from: https://doi.org/10.1007/s10646-008-0213-1.

Cui, Y., Zhao, Y., Tian, Y., Zhang, W., Lu, X., Jiang, X., 2012. The molecular mechanism of action of bactericidal gold nanoparticles on *Escherichia coli*. Biomaterials 33, 2327–2333. Available from: https://doi.org/10.1016/j.biomaterials.2011.11.057.

Das, R.K., Pachapur, V.L., Lonappan, L., Naghdi, M., Pulicharla, R., Maiti, S., et al., 2017. Biological synthesis of metallic nanoparticles: plants, animals and microbial aspects. Nanotechnol. Environ. Eng. 2 (18). Available from: https://doi.org/10.1007/s41204-017-0029-4.

Duan, H., Wang, D., Li, Y., 2014. Green chemistry for nanoparticle synthesis. Chem. Soc. Rev. 2015 (44), 5778–5792. Available from: https://doi.org/10.1039/c4cs00363b.

Dykman, L., Khlebtsov, N., 2011. Gold nanoparticles in biology and medicine: recent advances and prospects. Acta Nat. 3 (2), 34–55.

El-Naggar, N.E., Hussein, M.H., El-Sawah, A.A., 2017. Biofabrication of silver nanoparticles by phycocyanin, characterization, in vitro anticancer activity against breast cancer cell line and in vivo cytotxicity. Sci Rep. 7 (1), 10844. Available from: https://doi.org/10.1038/s41598-017-11121-3.

Feynman, R.P., 1960. There's plenty of room at the bottom. Eng. Sci 23, 22–36.

Focsan, M., Ardelean, I.I., Craciun, C., Astileanet, S., 2011. Interplay between gold nanoparticle biosynthesis and metabolic activity of cyanobacterium *Synechocystis* Sp. PCC.6803. Nanotechnology 22 (48), 485101. Available from: https://doi.org/10.1088/0957-4484/22/48/485101.

Govindaraju, K., Khaleel, S., Vijayakumar, B., Kumar, G., Singaravelu, G., 2008. Silver, gold and bimetallic nanoparticles production using single-cell protein (*Spirulina platensis*) Geitler. J. Mater. Sci. 43 (15), 5115–5122.

Guozhong, C., 2004. Nanostructures & Nanomaterials Synthesis, Properties & Applications, 1st Edition Imperial College Press1-86094-4159.

Hirsch, L.R., Stafford, R.J., Bankson, J.A., Sershen, S.R., Rivera, B., Price, R.E., et al., 2003. Nanoshell-mediated near-infrared thermal therapy of tumours under magnetic resonance guidance. Proc. Natl. Acad. Sci. 100, 13549–13554.

Hoffmann, R., 1988. Solids and Surfaces: A Chemist's View of Bonding in Extended Structures. VCH, New York.

Hoiczyk, E., Hansel, A., 2000. Cyanobacterial cell walls: news from an unusual prokaryotic envelope. J. Bacteriol. 182 (5), 1191–1199.

Hudek, L., Rai, S., Michalczyk, A., Rai, L.C., Neilan, B.A., Ackland, M.L., 2012. Physiological metal uptake by *Nostoc punctiforme*. Biometals. 25 (5), 893–903. Available from: https://doi.org/10.1007/s10534-012-9556-4.

Hulkoti, N.I., Taranath, T.C., 2014. Biosynthesis of nanoparticles using microbes—a review. Coll. Surf. B: Biointerf. 121, 474–483.

Iravani, S., 2014. Bacteria in nanoparticle synthesis: current status and future prospects. Int. Sch. Res. Notic. Article ID 359316, 18 pages.

Jing, L., Chen, X., Huang, P., 2016. Graphene-based nanomaterials for bioimaging. Adv. Drug Deliv. Rev. 105 (Pt B), 242–254. Available from: https://doi.org/10.1016/j.addr.2016.05.013.

Kumar, A., Boruah, B.M., Liang, X.J., 2011. Gold nanoparticles: promising nanomaterials for the diagnosis of cancer and HIV/AIDS. J. Nanomater. 2011, Available from: https://doi.org/10.1155/2011/202187. Article ID 202187.

Lengket, M., Fleet, M.E., Southam, G., 2006a. Morphology of gold nanoparticles synthesized by filamentous cyanobacteria from gold(1)-thiosulfate and gold (111)-chloride complexes. Langmuir 22, 2780–2787.

Lengket, M.F., Ravel, B., Fleet, M.E., Wanger, G., Gordon, R.A., Southam, G., 2006b. Mechanisms of gold bioaccumulation by filamentous cyanobacteria from gold(111)-chloride complex. Envi. Sci. Technol. 40, 6304–6309.

Li, X., Xu, H., Chen, Z., Chen, G., 2011. Biosynthesis of nanoparticles by microorganisms and their applications. J. Nanomater 2011, 16. Available from: https://doi.org/10.1155/2011/270974. Article ID: 270974.

Liu, J., Lu, Y., 2003. A colorimetric lead biosensor using DNAzyme-directed assembly of gold nanoparticles. J. Am. Chem. Soc. 125 (22), 6642–6643. Available from: https://doi.org/10.1021/ja034775u.

Lovestam, G., Rauscher, H., Gert, R., Sokull, K.B., Niel, G., Jean-Philippe, P., 2010. Considerations on a Definition of Nanomaterial for Regulatory Purposes. European Commission. Joint Research Centre, Luxembourg, ISBN NO. 978-92-79-16014-1.40 p.

Ma, Z., Jacobsen, F.E., Giedroc, D.P., 2009. Coordination chemistry of bacterial metal transport and sensing. Chem. Rev. 109, 4644–4681.

Mabena, L.F., Ray, S.S., Mhlanga, S.D., Coville, N.J., 2011. Nitrogen-doped carbon nanotubes as a metal catalyst support. Appl. Nanosci. 1, 67–77. Available from: https://doi.org/10.1007/s13204-011-0013-4.

Mallakpour, S., Behranvand, V., 2016. Polymeric nanoparticles: recent development in synthesis and application. eXPRESS Polym. Lett. 10 (11), 895–913.

Mansha, M., Khan, I., Ullah, N., Qurashi, A., 2017. Synthesis, characterization and visible-light-driven photoelectrochemical hydrogen evolution reaction of carbazole-containing conjugated polymers. Int. J. Hydrogen Energy . Available from: https://doi.org/10.1016/j.ijhydene.2017.02.053.

Mehta, S.K., Gaur, J.P., 2005. Use of algae for removing heavy metal ions from wastewater: progress and prospects. Crit. Rev. Biotechnol. 25, 113–152.

Mikolay, S., Huggett, L., Tikana, G., Braun, G.J., Nies, D.H., 2010. Survival of bacteria on metallic copper surfaces in a hospital trial. Appl. Microbiol. Biotechnol. 87 (5), 1875–1879.

Mody, V.V., Siwale, R., Singh, A., Mody, H.R., 2010. Introduction to metallic nanoparticles. J. Pharm. Bioall. Sci. 2 (4), 282–289. Available from: https://doi.org/10.4103/0975-7406.72127.

Nadeem, M., Abbasi, B.H., Younas, M., Ahmad, W., Khan, T., 2017. A review of the green syntheses and anti-microbial applications of gold nanoparticles. Green Chem. Lett. Rev. 10 (4), 216–227. Available from: https://doi.org/10.1080/17518253.2017.1349192.

Narayanan, K.B., Sakthivel, N., 2010. Biological synthesis of metal nanoparticles by microbes. Adv. Coll. Interf. Sci. 156 (1), 1–13.

Ngoy, J.M., Wagner, N., Riboldi, L., Bolland, O., 2014. A $CO_2$ capture technology using multi-walled carbon nanotubes with polyaspartamide surfactant. Energy Proc. 63, 2230–2248. Available from: https://doi.org/10.1016/j.egypro.2014.11.242.

Nissan, B.B., 2004. Nanoceramics in biomedical applications. Opportunities and uncertainties MRS Bulletin 29 (1), 28–32. ISBN 0 854036040.

Ostiguy, C., Roberge, B., Woods, C., Soucy, B., 2010. Engineered nanoparticles: current knowledge about OHS risks and prevention measures. IRSST Studies and Research Projects Report 656, second ed.

Paciotti, G.F., Mayer, L., Weinreich, D., Goia, D., Pavel, N., McLaughlin, R.E., et al., 2006. Colloidal gold: a novel nanoparticle vector for tumour directed drug delivery. Drug Deliv. 11, 169–183.

Parchi, P.D., Vittorio, O., Andreani, L., Piolanti, N., Cirillo, G., 2013. How nanotechnology can really improve the future of orthopedic implants and scaffolds for bone and cartilage defects. J Nanomed. Biotherap. Discov. 3, 114. Available from: https://doi.org/10.4172/2155-983X.1000114.

Pareek, V., Bhargava, A., Gupta, R., Jain, N., Panwar, J., 2017. Synthesis and applications of noble metal nanoparticles: a review. Adv. Sci. Eng. Med. 9, 527–544. Available from: https://doi.org/10.1166/asem.2017.2027.

Parial, D., Patra, H.K., Dasgupta, A.K.R., Pal, R., 2012. Screening of different algae for green synthesis of gold nanoparticles. Eur. J. Phycol. 47, 22–29.

Pasula, R.R., Lim, S., 2017. Engineering nanoparticle synthesis using microbial factories. Eng. Biol. 1 (1), 12–17.

Patel, V., Berthold, D., Puranik, P., Gantarb, M., 2015. Screening of cyanobacteria and microalgae for their ability to synthesize silver nanoparticles with antibacterial activity. Biotechnol. Rep. 5, 112–119. Available from: https://doi.org/10.1016/j.btre.2014.12.001.

Puri, A., Loomis, K., Smith, B., Lee, J.H., Yavlovich, A., Heldman, E., et al., 2009. Lipid-based nanoparticles as pharmaceutical drug carriers: from concepts to clinic. Crit. Rev. Ther. Drug Carrier Syst 26 (6), 523–580. 2009.

Qi-Long, Y., Gozin, M., Zhao, F.Q., Cohena, A., Panget, S.P., 2016. Highly energetic compositions based on functionalized carbon nanomaterials. Nanoscale 8 (9), 4799–4851. Available from: https://doi.org/10.1039/C5NR0755E.

Rawat, M.K., Jain, A., Singh, S., Mehnert, W., Thunemann, A.F., Souto, E.B., et al., 2011. Studies on binary lipid matrix based solid lipid nanoparticles of repaglinide: in vitro and in vivo evaluation. J. Pharm. Sci. 100, 2366–2378. Available from: https://doi.org/10.1002/jps.22435.

Saeed, K., Khan, I., 2016. Preparation and characterization of single-walled carbon nanotube/nylon 6,6 nanocomposites. Instrum. Sci. Technol. 44, 435–444. Available from: https://doi.org/10.1080/10739149.2015.1127256.

Salata, O.V., 2004. Applications of nanoparticles in biology and medicine. J. Nanobiotechnol. 2, 3. Available from: https://doi.org/10.1186/1477-3155-2-3.

Sebastian, V., Arruebo, M., Santamaria, J., 2014. Reaction engineering strategies for the production of inorganic nanomaterials. Small 10, 835–853.

Shah, M., Fawcett, D., Sharma, S., Tripathi, S.K., Poinern, J.G.E., 2015. Green synthesis of metallic nanoparticles via biological entities. Materials 8 (11), 7278–7308.

Shankar, P.D., Shobana, S., Karuppusamy, I., Pugazhendhi, A., Ramkumar, V.S., Arvindnarayan, S., et al., 2016. A review on the biosynthesis of metallic nanoparticles (gold and silver) using biocomponents of microalgae: formation mechanism and applications. Enzyme Microb. Technol. 95, 28–44. Available from: https://doi.org/10.1016/j.enzmictec.2016.10.015.

Sharma, D., Kanchi, S., Bisetty, K., 2015. Biogenic synthesis of nanoparticles: a review. Arabian J. Chem. Available from: https://doi.org/10.1016/j.arabjc.2015.11.002 (online).

Smith, A.M., Nie, S., 2010. Semiconductor nanocrystals: structure, properties, and band gap engineering. Acc. Chem. Res. 43 (2), 190–200. Available from: https://doi.org/10.1021/ar9001069.

Sun, S., Murray, C.B., Weller, D., Folks, L., Moser, A., 2000. Monodisperse FePt nanoparticles and ferromagnetic FePt nanocrystal superlattices. Science 80 (287), 1989–1992. Available from: https://doi.org/10.1126/science.287.5460.1989.

Thomas, S.C., Harshita, Mishra, P.K., Talegaonkar, S., 2015. Ceramic nanoparticles: fabrication methods and applications in drug delivery. Curr. Pharm. Design. 21, 6165–6188.

Yee, N., Benning, L.G., Phoenix, V.R., Ferris, F.G., 2004. Characterization of metal-cyanobacteria sorption reactions: a combined macroscopic and infrared spectroscopic investigation. Environ. Sci. Technol 38 (3), 775–782.

Yu, L., Liu, J., Wu, K., Klein, T., Jiang, Y., Wang, J.P., 2014. Evaluation of hyperthermia of magnetic nanoparticles by dehydrating DNA. Sci. Rep. 4, 7216.

Zhang, X., Yan, S., Tyagi, R.D., Surampalli, R.Y., 2011. Synthesis of nanoparticles by microorganisms and their application in enhancing microbiological reaction rates. Chemosphere 82 (4), 489–494. 2011.

Zheng, Y., Sache, L., 2009. Gold nanoparticles enhance DNA damage induced by anticancer drugs and radiation. Radiat. Res. 172, 114–119.

Zinicovscaia, I., Rudi, L., Valuta, A., Cepoi, L., Vergel, K., Frontasyev, M.V., et al., 2016. Biochemical changes in *Nostoc linckia* associated with selenium nanoparticles biosynthesis. Ecol. Chem. Eng. 23 (4), 559–569.

# Biosynthesis of nanoparticles and applications in agriculture

# 12

*Monika Singh, Meenakshi Srivastava, Ajay Kumar and K.D Pandey*
Centre of Advanced Study in Botany, Banaras Hindu University, Varanasi, India

## 12.1 Introduction

Nanotechnology is a promising field of interdisciplinary science research and is opening up opportunities in the fields of medicine, electronics, physics, chemistry, pharmaceutical science, material science, etc. Nanotechnology involves nanoparticles with size in the order of 100 nm or less than 100 nm (Auffan et al., 2009). According to reports by the United Nations, the world's population is projected to reach 8.5 billion by 2030 and we will need to produce at least 50% more food. Thus, increasing agricultural productivity and improving postharvest processing are vital to feed the world's growing population. Farmers have been attempting to increase agricultural yields in conventional ways for many decades. In this context, nanotechnology emerges as a promising tool in the various sectors of agri and food technology such as sustainable agriculture, food processing, water industry, forestry, environmental problems, and sustainable utilization of food resources. Many successful attempts have been made for the synthesis of metal nanoparticles using microorganisms, including bacteria, fungi, algae, yeast, cyanobacteria, and actinomycetes (Golinska et al., 2014).

The biosynthesis of metallic nanoparticles by microbes is a green and eco-friendly technology. This chapter focuses on the use of different microbes for the synthesis of metallic nanoparticles such as silver, gold, platinum, palladium, cadmium, selenium, titanium, and tellurium. During the last few years, an array of experiments has been conducted to measure the potential impact of nanotechnology on sustainable agriculture and crop improvement. The biomedical applications of metallic nanoparticles are promising with their tremendous effects in the fields of medicine, drug delivery, and agriculture (Sahoo et al., 2009; Gurunathan et al., 2009; Hulkoti and Taranath, 2014).

Selenium (Se) is an essential element for the food crops in which elemental selenium is one of the dominant species in the soil, but the mechanism of its uptake by the plants is still unknown. Selenium nanoparticles (SeNPs) have been rapidly oxidized to Se (IV) and converted to organic forms selenocystine, se-methyl-selenocysteine, and selenomethionine, readily absorbed by *Triticum aestivum* roots (Hu et al., 2018). SeNPs have many biotechnological applications in different fields. *Azoarcus* sp. is a facultative anaerobe that combines the ability to degrade under aerobic or anaerobic conditions and also the ability to produce SeNPs of

**Role of Plant Growth Promoting Microorganisms in Sustainable Agriculture and Nanotechnology.**
**DOI: https://doi.org/10.1016/B978-0-12-817004-5.00012-9**

agricultural importance (Fernandez-Llamosas et al., 2016). In a similar manner, *Pseudomonas putida* KT2440 also has the ability to synthesize nanoparticles of elemental selenium (SeNPs) from selenite (Avendano et al., 2016).

Titanium is also an important element for growth of plants. Ti applied via roots or leaves at low concentrations can improve crop performance through stimulating the activity of certain enzymes, enhancing chlorophyll content and photosynthesis, promoting nutrient uptake, strengthening stress tolerance, and improving crop yield and quality (Lyu et al., 2017). A titanium metal oxide is one of the most important nanoparticles used as nanofertilizer or nanopesticide in agriculture (Brar et al., 2010; Servin et al., 2015). Mattiello and Marchiol (2017) studied the effects of titanium oxide nanoparticles on seed germination percentage, mitotic index, root elongation of barley (*Hordeum vulgare*) and duration of the growth cycle, plant biomass and yield were also influenced by titanium oxide nanoparticles compared to control plants.

Silver nanoparticles have unique properties and can be used in various fields as antimicrobial, anticancer, larvicidal, catalytic, and wound-healing activities and also in agriculture as nanofertilizer (Firdhouse and Lalitha, 2015). In one case study, the effect of silver nanoparticles (AgNPs) on the growth of three different crop species, wheat (*T. aestivum*, var. UP2338), cowpea (*Vigna sinensis*, var. Pusa Komal), and brassica (*Brassica juncea*, var. Pusa Jai Kisan) were analyzed. The optimum growth promotion and increased root nodulation were observed in cowpea, while improved shoot parameters were recorded in *Brassica* after application of AgNPs (Pallavi et al., 2016).

The biocidal activity of silver nanocrystals depends on their size, shape, and surface coatings. Therefore, the development of AgNPs with well-controlled morphological and physicochemical features for physiological application in humans is necessary to expand their biomedical applications. The bacterial strains *Bacillus subtilis* (PTCC 1023), *Lactobacillus acidophilus* (PTCC 1608), *Klebsiella pneumoniae* (PTCC 1053), *Escherichia coli* (PTCC 1399), *Enterobacter cloacae* (PTCC 1238), and *Staphylococcus aureus* (PTCC 1112) effectively produced AgNPs for use as pest control in agriculture (Minaeian et al., 2008; Soni and Dhiman, 2017).

AgNPs are among the most attractive nanomaterials, and have been widely used in a range of biomedical applications, including diagnosis, treatment, drug delivery, medical device coatings, personal health care, and crop improvement. Recently, AgNPs have become of focus in biomedical applications because of their antibacterial, antifungal, antiviral, and antiinflammatory activity (El-Badawy et al., 2010; Zhong et al., 2010). Plants exposed to AuNPs exhibited increase in growth and crop yield of the plant at lower concentration. Some researchers have shown that the gold nanoparticle (AuNP) exposure has improved free radical scavenging activity, antioxidant enzymatic activities, and alter miRNA expression, which regulates different biological processes in plants. These modulations led to improved growth and yields of plants (Siddiqi and Husen, 2016; Ndeh et al., 2017).

Copper is as an essential trace element with distinct biological roles to play in all organisms including plants, animals, and microbes. In addition, copper-based compounds have been reported as one of the first fungicides used for disease management in plants. Copper-based chemicals are effectively used as antimicrobials in agriculture (Johnson, 1935; Banik and Pérez-de-Luque, 2017). Copper nanoparticles (CuNPs) were synthesized using bacterial strain *Pseudomonas fluorescens* with was 49 nm and spherical and hexagonal shapes. The application of the CuNPs induced the production of fruits with greater firmness. Vitamin C, lycopene, and the 2,2'-azino-bis(3-ethylbenzothiazoline-6-sulphonic acid (ABTS) antioxidant capacity increased in tomato fruits compared to the control (Shantkriti and Rani, 2014; Lopez-Vargas et al., 2018).

## 12.2 Synthesis of nanoparticles by fungal microbes

Nanotechnology has gained momentum in modulating metals into nanosize, shapes, and controlled disparity due to their potential use for human benefits. An endophytic fungus, *Pencillium* sp., isolated from healthy leaves of *Curcuma longa* (turmeric) was subjected to extracellular biosynthesis of AgNPs. Antibacterial activity against *Pseudomonas aeruginosa* and *K. pneumoniae* showed maximum zone of inhibition of 21 and 15 mm (Singh et al., 2013).

AgNPs synthesized from the supernatant of endophytic fungus *Alternaria* sp. isolates of *Raphanus sativus* leaves, had spherical size with range of 4-30 nm, which inhibit the growth of human pathogenic bacteria, suggesting the possibility of using AgNPs as efficient antibacterial agents (Singh et al., 2017). Three endophytic fungi, *Aspergillus tamarii* PFL2, *Aspergillus niger* PFR6, and *Penicllium ochrochloron* PFR8, isolated from an ethno-medicinal plant *Potentilla fulgens* L. were used for the biosynthesis of AgNPs (Devi and Joshi, 2015). Silver and AuNPs were synthesized using the filamentous fungus *Neurospora crassa* (Longoria et al., 2011). AgNPs were also synthesized by a nonpathogenic and agriculturally important fungus *Trichoderma asperellum* and *Candida albicans* (PTCC 5011) (Mukherjee et al., 2008; Minaeian et al., 2008). *Aspergillus terrus* strain CZR1 and *Aspergillus terrus* TFR2, *Aspergillus flavus* strain CZR2 and TFR1, *Aspergillus tubengensis* strain TFR3, and *Aspergillus japonicas* strain AJP01 synthesized the metallic nanoparticles of gold, silver, titanium, zinc, magnesium, and iron (Tarafdar, 2015).

## 12.3 Synthesis of nanoparticles by bacteria and actinobacteria

Nanotechnology is an emerging and rapidly growing field in science and agriculture. Among microorganisms, prokaryotes have received the most attention for biosynthesis of nanometals (Mandal et al., 2006). It has been shown that bacteria and actinomycetes are the best candidates for nanoparticle synthesis. Synthesis of nanoparticles using bacterial origin has emerged as a novel area of research in pharmaceutical engineering and agriculture (Sunkar and Nachiyar, 2012a). For instance, some bacterial species have the ability to use specific defense mechanisms to quell stresses like toxicity of heavy metal ions. It has been shown that some species can survive and grow even at high metal ion concentrations (e.g., *Pseudomonas stutzeri* and *P. aeruginosa*) (Bridges et al., 1979; Haefeli et al., 1984). *Bacillus* sp. isolated from the medicinal plants *Adhatoda beddomei* (malabar nut) and *Garcinia xanthochymus* (egg tree) synthesized AgNPs through the reduction of silver nitrate ($AgNO_3$) (Pissuwan et al., 2006; Kitov et al., 2008; Sunkar and Nachiyar, 2012b). Bacterial strain *E. cloacae* Ism26 (KP988024) isolated from soil contaminated with industrial waste was able to synthesize AgNPs (El-Baghdady et al., 2018). Extracellular synthesis of AgNPs by the culture supernatant of *E. coli* (DH5a) strain have been reported and were used as fertilizer to enhance the carbohydrate content in *Helianthus annuus L.* (Ghorbani, 2013; Yaseen et al., 2016). Synthesis of palladium nanoparticles within the periplasmic space or on the outer membrane of sulfate-reducing bacteria *Desulfovibrio desulfuricans* and on the S-layer protein of *Bacillus sphaericus* strain has been shown. The palladium nanoparticles were small in size and largely monodispersed, between 0.2 and 8 nm and occasionally from 9 to 12 nm with occasional larger nanoparticles (Omajali et al., 2015).

A number of researchers have suggested that actinobacteria are capable of producing metal and metal oxide nanoparticles that can be exploited in the green synthesis of nanomaterials and used in biological systems including agriculture. Green synthesis of AuNPs by actinobacteria *Streptomyces fulvissimus* strain, isolated from rice fields of Guilan Province, Iran had been reported by Meysam et al., 2015; Subramanian et al., 2016.

## 12.4 Synthesis of nanoparticles by cynobacteria

Agricultural practices are strongly dependent on the application of chemical fertilizers and pesticides, intensive tillage, and overirrigation, which have undoubtedly helped many countries meet the food requirements of their populations but have caused environmental and health problems, such as deterioration of soil fertility and microbial flora, overuse of land and water resources, pollution, and increased cost of crop production (Singh et al., 2016). Cynobacteria are important microbes in agriculture and food. They are also used as biofertilizer or organic fertilizer that contains living organisms and harnesses naturally occurring inputs like solar energy, macronutrients, and water to ensure soil fertility and plant growth. Small-scale farmers using biofertilizers have the potential to provide larger and more sustainable yields and healthier soils for themselves and their communities (Sharma et al., 2012). Recent research has shown that cynobacteria are excellent sources for the synthesis of nanoparticles. Three species of cynobacteria—*Limnothrix* sp. 37-2-1, *Anabaena* sp. 66-2, and *Synechocystis* sp. 48-3 (are part of the domestic Florida International University algae culture collection)—were able to synthesize AgNPs and cyanobacterium *Gloeocapsa* sp. were also able to synthesize AgNPs (Patel et al., 2014; AL-Katib et al., 2015). Husain et al. (2015) reported that *Microchaete* sp. NCCU-342 and *Cylindrospermum stagnale* NCCU-104 cynobacterial strains synthesized AgNPs with size ranging from 38 to 88 nm. In terms of size, *Cylindrospermum stagnale* NCCU-104 was the best organism with 38 and 40 nm. But in terms of time, *Microcheate* sp. NCCU-342 was the best (it took 30 hours for AgNP synthesis). The AgNPs were synthesized using *Plectonema boryanum* UTEX 485, a filamentous cyanobacterium. Cyanobacteria involved metabolic processes from the utilization of nitrate at 25°C and also organics released from the dead cyanobacteria at 25−100°C (Lengke et al., 2007). A cynobacterial strain, *Lyngbya majuscule*, was isolated from the Aloqair area (Al-Ahsa Government, Saudi Arabia) and intracellularly synthesized AuNPs. AuNPs alone or in combination with *Lyngbya majuscula* extract produced an inhibitory effect on isoproterenol-induced changes in serum cardiac injury markers, ECG, arterial pressure indices, and antioxidant capabilities of the heart of mice (Bakir et al., 2018).

## 12.5 Green synthesis of nanometals and their translocation in plants

Nanoparticles have been synthesized for better utilization of nanotechnology in plant systems, as they are preferred due to their potential nontoxic behavior. Biological methods of nanoparticle synthesis using plant or plant extracts offer numerous benefits (Kouvaris et al., 2012; Jae and Beom, 2009). Ankamwar et al. (2005) synthesized stable AuNPs and AgNPs using *Emblica officinalis*. The antioxidant and hepatoprotective activities of AgNp prepared from *E. officinalis* were studied (Rosarin et al., 2013; Bhuvaneswari et al., 2014). Few reports are available regarding biosynthesis of Ag, ZnO, and Cu nanoparticles from amla and its antimicrobial properties (Pinto and Nazareth, 2016; Anbukkarasi et al., 2015; Caroling et al., 2013; Ramesh et al., 2014; Saini et al., 2008). Bankar et al. (2010) used banana peel extract for the synthesis of AgNPs. Some researchers have synthesized metallic nanoparticles using Papaya callus and plant extracts (Mude et al., 2009; Sankar et al., 2014). Antioxidative as well as scavenging activity of these nanoparticles have also been show by researchers as the antimicrobial and 2,2-diphenyl-1-picrylhydrazyl (DPPH) activity have been studied recently on AgNPs (Banala et al., 2015; Kokila et al., 2016). Free radical scavenging activities have also been studied on PtNP (Watanabe et al., 2009). Many enzymatic biosensors have been designed (Hutter and Maysinger, 2013; Shi et al., 2014; Dimcheva and Horozova, 2013; Rathee et al., 2016).

The application of chemically synthesize nanomaterials is considered as toxic to nature, so nanomaterial synthesis from plant systems is preferred (Prasad et al., 2014). Green nanotechnology is safe process, energy efficient, and reduces waste and greenhouse gas emissions with minimal influence on the environment (Prasad et al., 2014, 2016).

The entry, translocation, and accumulation of metallic nanoparticles depend on the plant species and the size, kind, chemical composition, and stability of the nanometals. Metallic nanoparticles are adsorbed to plant surfaces including root, shoot,

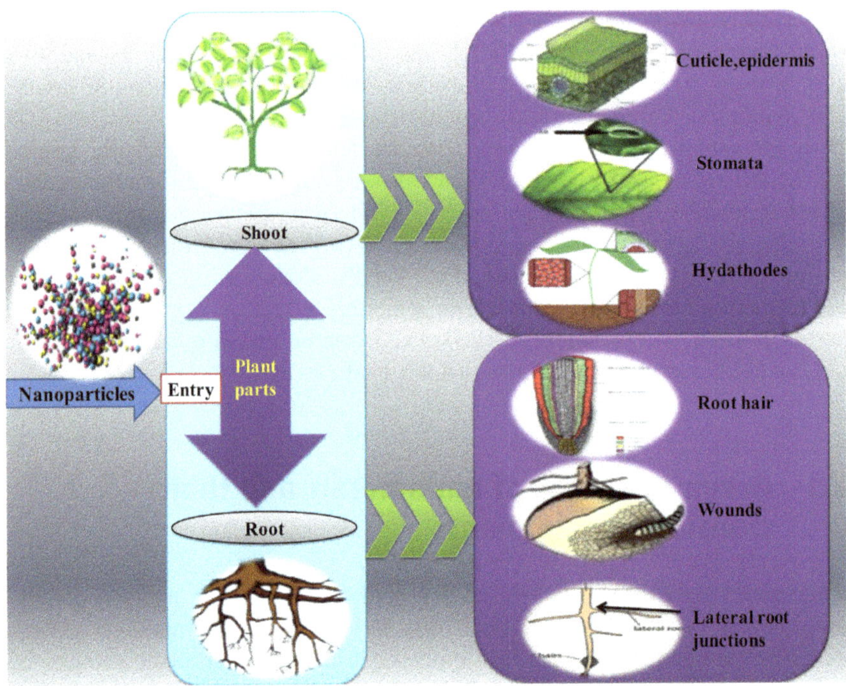


**Figure 12.1** Route of entry of nanoparticles and translocation in plants.

and taken up through natural nano- or micrometer-scale plant openings. Several routes have been found for nanoparticle association and uptake in plants (Singh et al., 2015; Tarafdar, 2015) (Fig. 12.1).

## 12.6 Synthesis of nanoparticles by microbes

The nanoparticles term used for the synthesized particles of the range in between 1-100 nm and having at least one of three possible dimensions. Nanoparticles can be made of materials diverse in chemical nature, with the most common being metals, metal oxides, silicates, nonoxide ceramics, polymers, organics, carbon, and biomolecules. Nanoparticles exist synthesized in the shape of spheres, cylinders, platelets, tubular, cubic, etc. Microbial synthesis of metal nanoparticles can take place either intracellularly or extracellularly (Ahmad et al., 2003a,b; Jain et al., 2011; Kalishwaralal et al., 2010; Saifuddin et al., 2009). Microbes are efficient candidates for the production of metallic nanoparticles both intracellularly and extracellularly (Table 12.1). Nanoparticles have important physicochemical properties like shape, size, and distribution of particles and roughness, topography, purity, stability, dispersion, reactivity, and hydrophobocity (Kumar and Dixit, 2017).

**Table 12.1** Some of the nanoparticles synthesized by microbes.

| Metallic nanoparticles | Strain | Microbe | Source | References |
|---|---|---|---|---|
| Silver | *Bacillus* strain CS 11 | Bacteria | Industrialized area | Das et al. (2014) |
| Silver Selenium | *Bacillus* sp. JAPSK2 | Bacteria | Coal mine | Singh et al. (2014) |
| Titanium | *Lactobacillus* sp. | Bacteria | Yoghurt | Prasad et al. (2007) |
| Selenium | *Stenotrophomonas maltophilia* SeITE02 a | Bacteria | Polluted site | Zonaro et al. (2015) |
| Tellurium | *Ochrobactrum* sp. MPV1 | Bacteria | Polluted site | Zonaro et al. (2015) |
| Silver | *Aspergillus flavus* NJP08 | Fungi | Metal-rich region | Jain et al. (2011) |
| Tellurium | *Aspergillus welwitschiae* KY766958 | Fungi | Kitchen drainage tub | Abo Elsoud et al. (2018) |
| Silver | *Halococcus salifodinae* BK3 | Archea | Water | Srivastava et al. (2013) |
| Gold Silver Palladium Platinum | *Calothrix pulvinata*, strain ALCP 745 A | Cynobacteria | Black soil (Switzerland) | Brayner et al. (2007) |
| Gold Silver Palladium Platinum | *Anabaena flos-aquae*, strain ALCP B24 | Cynobacteria | Black soil (Switzerland) | Brayner et al. (2007) |
| Gold Silver Palladium Platinum | *Leptolyngbya foveolarum*, strain ALCP 671B | Cynobacteria | Black soil (France) | Brayner et al. (2007) |
| Silver | *Nostoc* sp. strain HKAR-2 | Cynobacteria | Rajgiri (Gujrat) | Sonker et al. (2017) |
| Silver | *Actinomycetes* sp. | Actinomycetes | Metal polluted soil samples | Abdeen et al. (2014) |
| Silver | *Thermoactinomyces* sp. | Actinomycetes | Marine sediment | Deepa et al. (2013) |
| Gold | *Streptomyces griseoruber* | Actinomycetes | Soil | Ranjitha and Rai (2017) |
| Gold | *Streptomyces viridogens* strain HM10 | Actinomycetes | Himalayan mountain | Balagurunathan et al. (2011) |
| Gold | *Rhodococcus* sp. | Actinomycetes | Tree (*Ficus carica*) | Ahmad et al. (2003a,b) |

Different chemical methods are used for synthesis of metallic nanoparticles, which use toxic chemicals. Various biological entities like bacteria, cynobacteria, fungi, higher plants, actinomycetes, and viruses have been used as alternatives for the synthesis of nanoparticles, because of less toxic effect and high efficient synthesis. Different biological sources have been used for the synthesis of nanoparticles and are being used in agriculture for precision farming such as silver nanoparticles, titanium dioxide nanoparticles, and zinc oxide nanoparticles.

Many techniques are used to characterize nanoparticles including microscopy and spectroscopy. Microscopic techniques such as transmission electron microscopy, scanning electron microscopy, atomic force microscopy, high-resolution microscopes, etc., were invented to observe nanosize particles (Joshi et al., 2008). UV-Vis spectrophotometer consists of light source, reference and sample beams, a monochromator, and a detector. The UV spectrum for a compound is obtained by exposing a sample of the compound to ultraviolet light source as xenon lamp. The optical absorption spectra of metal nanoparticles shifts to longer wavelengths with increasing particle size (Brause et al., 2002). The position and shape of the particle, plasmon absorption of metal nanoparticles are strongly dependent on the size of nanoparticles, electric medium, and surface absorbed species (Mulvaney, 1996). Fourier transform infrared spectroscopy is one of the most widely used tools for the detection of functional groups in pure compounds and mixtures. It is an effective tool in detecting the shape of nanometer-sized materials. This spectroscopy is widely used to study the nature of surface adsorbents in nanoparticles (Dablemont et al., 2008). X-ray diffraction is also one of the most important characterization techniques used to observe the structural properties of nanoparticles. It gives enough information and phase of nanoparticles (Khan et al., 2017). The zeta potential is a parameter commonly used to characterize metal nanoparticles. It is the apparent surface potential that is related to the surface charge. The zeta potential is an important parameter to assess when studying nanoparticles in suspension as it affects particle agglomeration, sedimentation, interaction, and complexation with other media constituents (Hunter, 1993; Skoglund et al., 2017) (Fig. 12.2).

## 12.7 Applications of nanoparticles

### *12.7.1 Nanotechnology in agri sector*

Recently from last few decades, nanotechnology has been applied in the different levels of agriculture and also gaining attraction at the research and industrial level for further exploration. Some of the main goals in agricultural nanotechnology are the development of novel nanocomposites, nanopesticides, nanofertilizers, etc., for designed targeted delivery systems inside cells.

Nanotechnology in agriculture in general includes:

1. Nanomaterials for quality management of preharvest and postharvest agri products.
2. Research and development areas such as photocatalysis, nanotubes, bioremediation of resistant pesticides, disinfectants, nanoscale carriers, nanolignocellulosic materials, nanobarcode technology, quantum dots for staining bacteria, and nanobiosensors.

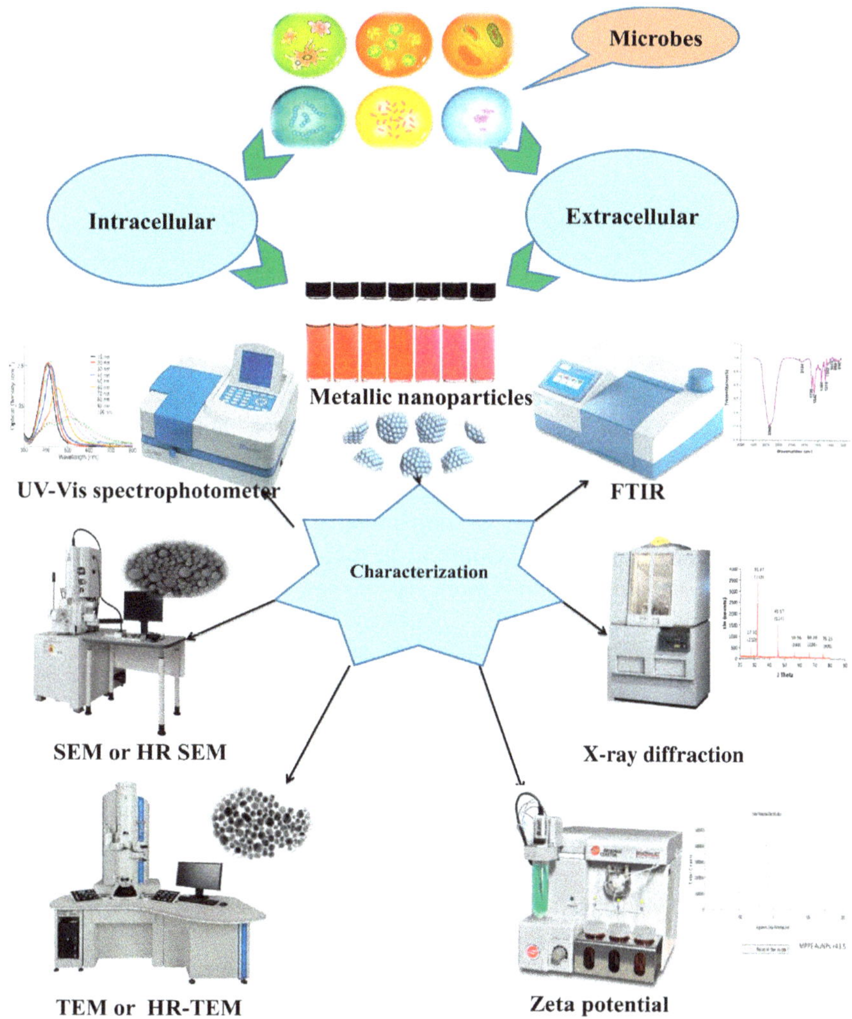


**Figure 12.2** Schematic representation of synthesis and general characterization of nanoparticles.

3. On-site applications for agriculture include desalination, photocatalysis, removal of heavy metals, and nanosensors.

### 12.7.2 Nanotechnology in the food sector

Nanotechnology is used in the food sector for nanosensors, targeted delivery of required components, food safety, new product developments, packaging, etc. (McClements et al., 2009; Huang et al., 2010; Yu and Huang, 2013). Agricultural nanotechnology emerged in the late 1990s and was quickly adopted in many fields.

However, there is still potential to improve agricultural production. Agro nanotechnology can be used to develop healthy seeds that can improve plant germination, growth, yield, and quality. This technology also has the potential to increase the storage period for vegetables and fruits. Organic pesticides and fertilizers can also be developed by the proper use of agro nanotechnology.

### 12.7.3 Nanotechnology in sustainable agriculture

Nanomaterials have gained importance in the agri and food sector, notably from targeted inside cell delivery to preservation and packaging due to their small size and unique physicochemical features. The biogenic synthesis of nanoparticles using plant extracts is also reliable, ecofriendly, and cost effective. The aqueous extract from plants acts as both reducing and stabilizing agent for nanoparticle synthesis. Nanoparticles have been applied as sensors for a large number of applications where normal sensors do not worked including the detection of analytes at normal as well as very low concentrations (Fig. 12.3).

The growth of nanotechnology in agriculture requires more scientific calculations for its successful implementation at the larger commercial and local level. Nanoparticles must help make soils more capable and fertile with greater productivity and better environmental security. Application of nano particles in the soil maintain their ionic balance. The availability of nutrients to plants depends on this equilibrium and thus the use of nanotechnology must not interrupt it.

Metallic nanoparticles can enhance crop production and quality. For example, in bitter melon (*Momordica charantia*) plant nanoparticles have been shown to increase plant biomass, fruit yield, and phytomedicine content (Kole et al., 2013). The positive effects of nanoparticles on plant growth and physiological and genetic factors have been investigated mostly at seedling stages (Nair et al., 2010; Rico et al., 2011).

An innovative breakthrough in the field of biopesticides is the employment of engineered nanomaterials or bionanotechnology. Nanopesticides play a key role in the control of host pathogens due to enhanced solubility, specificity, and stability (Khot et al., 2012, Bhattacharyya et al., 2016). This is a promising way to use nanopesticide delivery systems to increase agriculture production and to help reduce destructive environmental impacts (Jampilek and Kralova, 2015; Kookana et al., 2014). Among the various types of nanoparticles, metal nanoparticles exhibit the best antibacterial and antifungal activities because of electrostatic interaction of the nanoparticles with bacterial cell membrane and their greater and lasting accumulation in cytoplasm (Abdel-Aziz et al., 2016; Duhan et al., 2017). Microorganisms play an important role in soil health and crop productivity in agriculture (Srivastava et al., 2013; 17). The relevant applications of nanotechnology at small scale was achieved with the growth and vigor of the seeds of legumes (cowpea), cabbage (*brassica*), and cucumber when treated by nano-863. For example, Anjali et al. (2012) synthesized nanocapsules from *Azadirachta indica* for disease and pest control in plants and Milani et al. (2012) worked on delivery of nutrients to specific sites by coating different macronutrients fertilizers with zinc oxide to improve

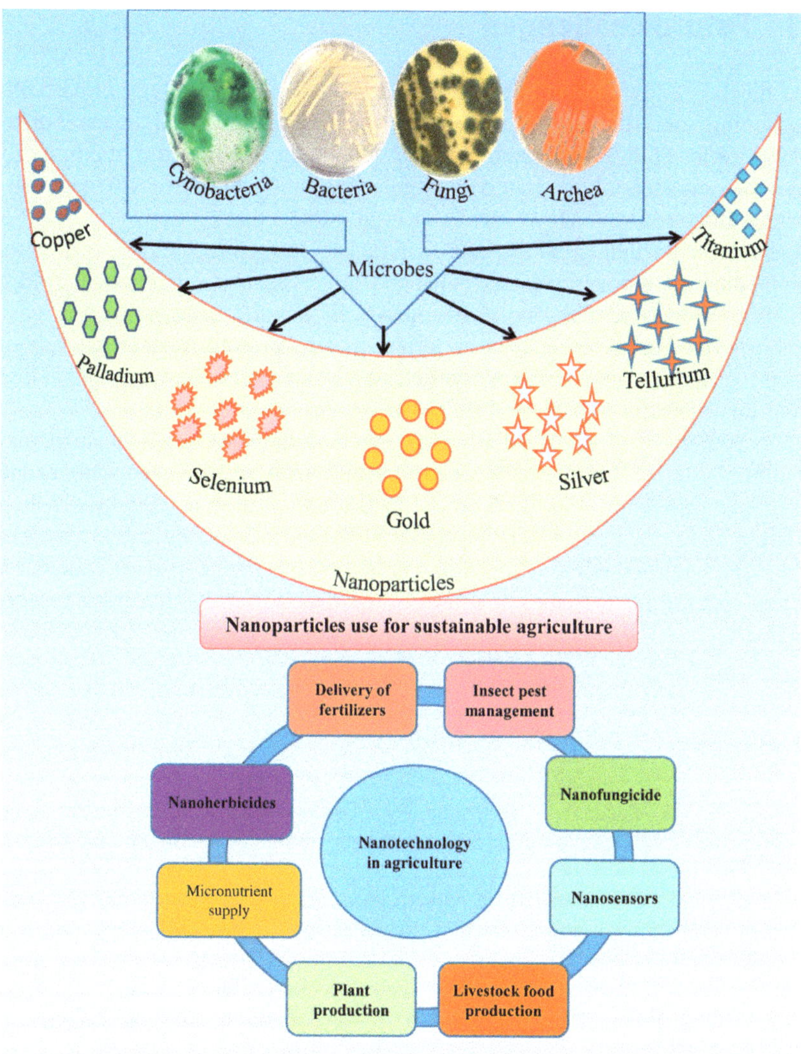


**Figure 12.3** Schematic representation of applications of different nanoparticles in agriculture.

nutrient absorption by plants. $TiO_2$ nanoparticles were used by McMurray et al. (2006) for photocatalytic degradation of agrochemicals in contaminated waters, which helped remove pesticides from the water. Vamvakaki and Chaniotakis (2007) used liposome-based nano-biosensor, a highly sensitive biochemical sensor, to monitor environmental conditions for plant health and growth. Torney et al. (2007) produced mesoporus silica nanoparticles that by carrying DNA or RNA can be directly delivered to plant cells for genetic transformation and to trigger defense responses when activated by pathogens.

## 12.8 Future challenges

While much is known about the positive aspects of nanotechnology in particular in the agriculture and food sectors, little is known about the risk to humans. For example, Nurkiewicz et al. (2008) found that the inhalation of nanosized titanium dioxide engineered nanotechnology materials engineered nanomaterial (ENM) reached systemic circulation in rats. However, Tsuji et al. (2005) and Nurkiewicz et al. (2008) demonstrated that inhalation exposure of rats to low concentrations of nanosized titanium dioxide exposure in rats enhanced microvascular dysfunction. Radomski et al. (2005) observed that due to nanoparticles, platelet aggregation and vascular thrombosis occurred. Hence, it is important to obtain more reliable data about the influence of nanomaterials or nanoparticles on animal and plant systems as well to prevent their harmful impact on living beings.

Nanotechnology is a new technology that has the potential to change current agriculture systems. But there are challenges that must be addressed such as debates surrounding genetically modified crops. Future exploration and improved technology larger scales is also needed. Nanoparticles have often been called "magic bullets" loaded with herbicides, fungicides, nutrients, fertilizers, or direct nucleic acids targeting specific plant tissues to achieve their desired results, but more research is needed.

## References

Abdeen, S., Geo, S., Sukanya, S., Praseetha, P.K., Dhanya, R.P., 2014. Biosynthesis of silver nanoparticles from Actinomycetes for therapeutic applications. Int. J. Nano Dimens. 5 (2), 155–162.

Abdel-Aziz, H.M., Hasaneen, M.N., Omer, A.M., 2016. Nano chitosan-NPK fertilizer enhances the growth and productivity of wheat plants grown in sandy soil. Span. J. Agric. Res. 14 (1), 0902.

Abo Elsoud, M.M., Al-Hagarb, O.E.A., Abdelkhalek, E.S., Sidkey, N.M., 2018. Synthesis and investigations on tellurium myconanoparticles. Biotechnol. Rep. 18, e00247.

Ahmad, A., Mukherjee, P., Senapati, S., Mandal, D., Khan, M.I., Kumar, R., et al., 2003a. Extracellular biosynthesis of silver nanoparticles using the fungus *Fusarium oxysporum*. Coll. Surf. B 28, 313–318.

Ahmad, A., Senapati, S., Khan, M.I., Kumar, R., Ramani, R., Srinivas, V., et al., 2003b. Intracellular synthesis of gold nanoparticles by a novel alkalotolerant actinomycete, *Rhodococcus* species. Nanotechnology 14, 824–828.

AL-Katib, M., AL-Shahri, Y., AL-Niemi, A., 2015. Biosynthesis of silver nanoparticles by cyanobacterium *Gloeocapsa* sp. Int. J. Enhanc. Res. Sci. Technol. Eng. 4 (9), 60–73.

Anbukkarasi, V., Srinivasan, R., Elangovan, N., 2015. Antimicrobial activity of green synthesized zinc oxide nanoparticles from *Emblica officinalis*. Int. J. Pharm. Sci. Rev. Res. 33 (2), 110–115.

Anjali, C.H., Sharma, Y., Mukherjee, A., Chandrasekaran, N., 2012. Neem oil (*Azadirachta indica*) nanoemulsion—a potent larvicidal agent against *Culex quinquefasciatus*. Pest Manage. Sci. 68 (2), 158–163.

Ankamwar, B., Damle, C., Ahmad, A., Sastry, M., 2005. Biosynthesis of gold and silver nanoparticles using *Emblica officinalis* fruit extract, their phase transfer and transmetallation in an organic solution. J. Nanosci. Nanotech 5 (10), 1665–1671.

Auffan, M., Rose, J., Bottero, J.Y., Lowry, G.V., Jolivet, J.P., Wiesner, M.R., 2009. Towards a definition of inorganic nanoparticles from an environmental, health and safety perspective. Nat. Nanotech 4 (10), 634.

Avendano, R., Chaves, N., Fuentes, P., Sánchez, E., Jiménez, J.I., Chavarría, M., 2016. Production of selenium nanoparticles in *Pseudomonas putida* KT2440. Sci. Rep. 6, 37155.

Bakir, E.M., Younis, N.S., Mohamed, M.E., El Semary, S.A., 2018. Cyanobacteria as nanogold factories: chemical and anti-myocardial infarction properties of gold nanoparticles synthesized by *Lyngbya majuscule*. Mar. Drugs 16, 217.

Balagurunathan, R., Radhakrishnan, M., Rajendran, R.B., Velmurugan, D., 2011. Biosynthesis of gold nanoparticles by actinomycete *Streptomyces viridogens* strain HM10. Indian J. Biochem. Biophys. 48, 331–335.

Banala, R.R., Nagati, V.B., Karnati, P.R., 2015. Green synthesis and characterization of Carica papaya leaf extract coated silver nanoparticles through X-ray diffraction, electron microscopy and evaluation of bactericidal properties. Saudi J. Biol. Sci. 22 (5), 637–644.

Banik, S., Pérez-de-Luque, A., 2017. In vitro effects of copper nanoparticles on plant pathogens, beneficial microbes and crop plants. Span. J. Agric. Res. 15 (2), e1005.

Bankar, A., Joshi, B., Kumar, A.R., Zinjarde, S., 2010. Banana peel extract mediated novel route for the synthesis of palladium nanoparticles. Mater. Lett. 64 (18), 1951–1953.

Bhattacharyya, A., Duraisamy, P., Govindarajan, M., Buhroo, A.A., Prasad, R., 2016. Nanobiofungicides: emerging trend in insect pest control. Advances and Applications Through Fungal Nanobiotechnology. Springer, Cham, pp. 307–319.

Bhuvaneswari, R., Chidambaranathan, N., Jegatheesan, K., 2014. Hepatoprotective effect of *Embilica officinalis* and its silver nanoparticles against ccl4 induced hepatotoxicity in wistar albino rats. Dig. J. Nanomater. Biostruct. (DJNB) 9, 223–235.

Brar, S.K., Verma, M., Tyagi, R.D., Surampalli, R.Y., 2010. Engineered nanoparticles in wastewater and wastewater sludge—evidence and impacts. Waste Manage 30, 504–520.

Brause, R., Moeltgen, H., Kleinermanns, K., 2002. Characterization of laser ablated and chemically reduced silver colloids in aqueous solution by UV/Vis spectroscopy and STM/SEM microscopy. Appl. Phys. B 75, 711–716.

Brayner, R., Barberousse, H., Hemadi, M., Djediat, S., Yepremian, C., Coradin, T., et al., 2007. Cyanobacteria as bioreactors for the synthesis of Au, Ag, Pd, and Pt nanoparticles via an enzyme-mediated route. J. Nanosci. Nanotechnol. 7, 1–13.

Bridges, K., Kidson, A., Lowbury, E.J.L., Wilkins, M.D., 1979. Gentamicin- and silver-resistant *Pseudomonas* in a burns unit. BMJ 1 (6161), 446–449.

Caroling, G., Tiwari, S.K., Ranjitham, A.M., Suja, R., 2013. Biosynthesis of silver nanoparticles using aqueous broccoli extract-characterization and study of antimicrobial cytotoxic effects. Asian J. Pharm. Clin. Res. 6, 165–172.

Dablemont, C., Lang, P., Mangeney, C., Piquemal, J.Y., Petkov, V., Herbst, F., et al., 2008. FTIR and XPS study of Pt nanoparticle functionalization and interaction with alumina. Langmuir 24, 5832–5841.

Das, V.L., Thomas, R., Varghese, R.T., Soniya, E.V., Mathew, J., Radhakrishnan, E.K., 2014. Extracellular synthesis of silver nanoparticles by the *Bacillus* strain CS 11 isolated from industrialized area. 3 Biotech 4, 121–126.

Deepa, S., Kanimozhi, K., Panneerselvam, A., 2013. Antimicrobial activity of extracellularly synthesized silver nanoparticles from marine derived actinomycetes. Int. J. Curr. Microbiol. Appl. Sci. 2 (9), 223–230.

Devi, L.S., Joshi, S.R., 2015. Ultrastructures of silver nanoparticles biosynthesized using endophytic fungi. J. Microsc. Ultrastruct. 3, 29–37.

Dimcheva, N., Horozova, E., 2013. Direct electrochemistry of *Penicillium chrysogenum* catalase adsorbed on spectroscopic graphite. Bioelectrochemistry 90, 1–7.

Duhan, J.S., Kumar, R., Kumar, N., Kaur, P., Nehra, K., Duhan, S., 2017. Nanotechnology: the new perspective in precision agriculture. Biotechnol. Rep. 15, 11–23.

El-Badawy, A., Feldhake, D., Venkatapathy, R., 2010. State of the Science Literature Review: Everything Nanosilver and More. US Environmental Protection Agency, Washington, DC.

El-Baghdady, K.Z., El-Shatoury, E.H., Abdullah, O.M., Khalil, M.M.H., 2018. Biogenic production of silver nanoparticles by *Enterobacter cloacae* Ism26. Turk. J. Biol. 42, 319–328.

Fernandez-Llamosas, H., Castro, L., Blázquez, M.L., Díaz, E., Carmona, M., 2016. Biosynthesis of selenium nanoparticles by *Azoarcus* sp. CIB. Microb. Cell Fact. 15 (109), 1–10.

Firdhouse, M.J., Lalitha, P., 2015. Biosynthesis of silver nanoparticles and its applications. J. Nanotechnol. 1–18. Article ID 829526.

Ghorbani, H.R., 2013. Biosynthesis of silver nanoparticles by *Escherichia coli.* Asian J. Chem. 25 (3), 1247–1249.

Golinska, P., Wypij, M., Ingle, A.P., Gupta, I., Dahm, H., Rai, M., 2014. Biogenic synthesis of metal nanoparticles from actinomycetes: biomedical applications and cytotoxicity. Appl. Microbiol. Biotechnol. 98 (19), 8083–8097.

Gurunathan, S., Kalishwaralal, K., Vaidyanathan, R., Venkataraman, D., Pandiana, S.R., Muniyandi, J., et al., 2009. Biosynthesis, purification and characterization of silver nanoparticles using *Escherichia coli.* Colloids Surf. B. Biointerf. 74, 328–335.

Haefeli, C., Franklin, C., Hardy, K., 1984. Plasmid-determined silver resistance in *Pseudomonas stutzeri* isolated from a silver mine. J. Bacteriol. 158 (1), 389–392.

Hu, T., Li, H., Li, J., Zhao, G., Wu, W., Liu, L., et al., 2018. ). Absorption and biotransformation of selenium nanoparticles by wheat seedlings (*Triticum aestivum* L.). Front. Plant Sci. 9, 597.

Huang, Q., Yu, H., Ru, Q., 2010. Bioavailability and delivery of nutraceuticals using nanotechnology. J. Food Sci. 75 (1), 50–57.

Hulkoti, N.I., Taranath, T.C., 2014. Biosynthesis of nanoparticles using microbes—a review. Colloids Surf. B: Biointerf. 121, 474–483.

Hunter, R.J., 1993. Introduction to Modern Colloid Science.. Oxford University Press, Oxford.

Husain, S., Sardar, M., Fatma, T., 2015. Screening of cyanobacterial extracts for synthesis of silver nanoparticles. World J. Microbiol. Biotechnol. 31, 1279–1283.

Hutter, E., Maysinger, D., 2013. Gold-nanoparticle-based biosensors for detection of enzyme activity. Trends Pharmacol. Sci. 34 (9), 497–507.

Jae, Y.S., Beom, S.K., 2009. Rapid biological synthesis of silver nanoparticles using plant leaf extracts. Bioproc. Biosyst. Eng. 32, 79–84.

Jain, N., Bhargava, A., Majumdar, S., Tarafdar, J.C., Panwar, J., 2011. Extracellular biosynthesis and characterization of silver nanoparticles using *Aspergillus flavus* NJP08: a mechanism perspective. Nanoscale 3, 635–641.

Jampilek, J., Kralova, K., 2015. Application of nanotechnology in agriculture and food industry, its prospects and risks. Ecol. Chem. Eng. S. 22 (3), 321–361.

Johnson, G.F., 1935. The early history of copper fungicides. Agric. Hist. 9, 67–79.

Joshi, M., Bhattachryya, A., Ali, W., 2008. Characterization techniques for nanotechnology applications in textiles. Indian J. Fibre Textil. Res. 33, 304–317.

Kalishwaralal, K., Deepak, V., Pandian, S.R.K., Kottaisamy, M., Barath Mani Kanth, S., Kartikeyan, B., et al., 2010. Biosynthesis of silver and gold nanoparticles using *Brevibacterium casei*. Coll. Surf. B 77, 257–262.

Khan, I., Saeed, K., Khan, I., 2017. Nanoparticles: properties, applications and toxicities. Arabian J. Chem. xxx, xxx–xxx.

Khot, L.R., Sankaran, S., Maja, J.M., Ehsani, R., Schuster, E.W., 2012. Applications of nanomaterials in agricultural production and crop protection: a review. Crop. Prot. 35, 64–70.

Kitov, P.I., Mulvey, G.L., Griener, T.P., Lipinski, T., Solomon, D., Paszkiewicz, E., et al., 2008. In vivo supra molecular templating enhances the activity of multivalent ligands: a potential therapeutic against the *Escherichia coli* O157 AB5 toxins. Proc. Natl. Acad. Sci. USA 105, 16837–16842.

Kokila, T., Ramesh, P.S., Geetha, D., 2016. Biosynthesis of AgNPs using Carica papaya peel extract and evaluation of its antioxidant and antimicrobial activities. Ecotoxicol. Environ. Saf. 134 (2), 467–473.

Kole, C., Kole, P., Randunu, K.M., Choudhary, P., Podila, R., Ke, P.C., et al., 2013. Nanobiotechnology can boost crop production and quality: first evidence from increased plant biomass, fruit yield and phytomedicine content in bitter melon (*Momordica charantia*). BMC Biotechnol. 13, 37.

Kookana, R.S., Boxall, A.B., Reeves, P.T., Ashauer, R., Beulke, S., Chaudhry, Q., et al., 2014. Nanopesticides: guiding principles for regulatory evaluation of environmental risks. J. Agric. Food Chem. 62 (19), 4227–4240.

Kouvaris, P., Delimitis, A., Zaspalis, V., Papadopoulos, D., Tsipas, S.A., Michailidis, N., 2012. Green synthesis and characterization of silver nanoparticles produced using *Arbutus unedo* leaf extract. Mater. Lett. 76, 18–20.

Kumar, A., Dixit, C.K., 2017. Methods for characterization of nanoparticles. In: Nimesh, S., Ramesh, C., Gupta, N. (Eds.), Advances in Nanomedicine for the Delivery of Therapeutic Nucleic Acids. Woodhead Publishing, Cambridge, pp. 43–58.

Lengke, M.F., Fleet, M.E., Southam, G., 2007. Biosynthesis of silver nanoparticles by filamentous cyanobacteria from a silver(I) nitrate complex. Langmuir 23, 2694–2699.

Longoria, E.C., Nestor, A.R.V., Borja, M.A., 2011. Biosynthesis of silver, gold and bimetallic nanoparticles using the filamentous fungus *Neurospora crassa*. Coll. Surf. B 83, 42–48.

Lopez-Vargas, E.R., Ortega-Ortiz, H., Cadenas-Pliego, G., Romenus, K.A., de la Fuente, M. C., Benavides-Mendoza, A., et al., 2018. Foliar application of copper nanoparticles increases the fruit quality and the content of bioactive compounds in tomatoes. Appl. Sci. 8 (1020), 1–15.

Lyu, S., Wei, X., Chen, J., Wang, C., Wang, X., Pan, D., 2017. Titanium as a beneficial element for crop production. Front. Plant Sci. 8, 597.

Mandal, D., Bolander, M.E., Mukhopadhyay, D., Sarkar, G., Mukherjee, P., 2006. The use of microorganisms for the formation of metal nanoparticles and their application. Appl. Microbiol. Biotechnol. 69, 485–492.

Mattiello, A., Marchiol, L., 2017. Application of nanotechnology in agriculture: assessment of TiO2 nanoparticle effects on barley. In: Janus Dr, Magdalena (Ed.), Application of Titanium Dioxide. InTech, pp. 23–39.

McClements, D.J., Decker, E.A., Park, Y., Weiss, J., 2009. Structural design principles for delivery of bioactive components in nutraceuticals and functional foods. Crit. Rev. Food Sci. Nutr. 49, 577–606.

McMurray, T.A., Dunlop, P.S.M., Byrne, J.A., 2006. The photocatalytic degradation of atrazine on nanoparticulate TiO2 films. J. Photochem. Photobiol. A: Chem. 182 (1), 43–51.

Meysam, S.N., Hosein, S.B.G., Naimeh, K., 2015. Biosynthesis of gold nanoparticles using *Streptomyces fulvissimus* isolate. Nanomed. J. 2 (2), 153–159.

Minaeian, S., Shahvedrdi, A.R., Nohi, A.S., Shahverdi, H.R., 2008. Extracellular biosynthesis of silver nanoparticles by some bacteria. J. Sci. Islam Azad Univ. 17–1-4.

Milani, N., McLaughlin, M.J., Stacey, S.P., Kirby, J.K., Hettiarachchi, G.M., Beak, D.G., et al., 2012. Dissolution kinetics of macronutrient fertilizers coated with manufactured zinc oxide nanoparticles. J. Agric. Food Chem. 60 (16), 3991–3998.

Mude, N., Ingle, A., Gade, A., Rai, M., 2009. Synthesis of silver nanoparticles using callus extract of *Carica papaya*—a first report. J. Plant Biochem. Biotechnol. 18 (1), 83–86.

Mukherjee, P., Roy, M., Mandal, B.P., Dey, G.K., Mukherjee, P.K., Ghatak, J., et al., 2008. Green synthesis of highly stabilized nanocrystalline silver particles by a non-pathogenic and agriculturally important fungus *T. asperellum*. Nanotechnology 19, 103–110.

Mulvaney, P., 1996. Surface plasmon spectroscopy of nanosized metal particles. Langmuir 1996 (12), 788–800.

Nair, R., Varghese, S.H., Nair, B.G., Maekawa, T., Yoshida, Y., Kumar, D.S., 2010. Nanoparticulate material delivery to plants. Plant Sci. 179, 154–163.

Ndeh, N.T., Maensiri, S., Maensiri, D., 2017. The effect of green synthesized gold nanoparticles on rice germination and roots. Adv. Nat. Sci. Nanosci. Nanotechnol. 8 (035008), 10pp.

Nurkiewicz, T.R., Porter, D.W., Hubbs, A.F., Cumpston, J.L., Chen, B.T., Frazer, D.G., et al., 2008. Nanoparticle inhalation augments particle-dependent systemic microvascular dysfunction. Part. Fibre Toxicol. 5 (1), 1.

Omajali, J.B., Mikheenko, I.P., Merroun, M.L., Woo, J., Macaskie, L.E., 2015. Characterization of intracellular palladium nanoparticles synthesized by *Desulfovibrio desulfuricans* and *Bacillus benzeovorans*. J. Nanopart. Res. 17, 264.

Pallavi, A., Mehta, C.M., Srivastava, R., Arora, S., Sharma, A.K., 2016. Impact assessment of silver nanoparticles on plant growth and soil bacterial diversity. 3 Biotech 6, 254.

Patel, V., Berthold, D., Puranik, P., Gantar, M., 2014. Screening of cyanobacteria and microalgae for their ability to synthesize silver nanoparticles with antibacterial activity. Biotechnol. Rep. 5, 112–119.

Pinto, G.M., Nazareth, R., 2016. Green synthesis and characterization of zinc oxide nanoparticles. J. Chem. Pharm. Res. 6, 427–432.

Pissuwan, D., Valenzuela, S.M., Cortie, M.B., 2006. Therapeutic possibilities of plasmonically heated gold nanoparticles. Trends Biotechnol. 24, 62–67.

Prasad, K., Jha, A.K., Kulkarni, A.R., 2007. Lactobacillus assisted synthesis of titanium nanoparticles. Nanosc. Res. Lett. 2, 248–250.

Prasad, R., Kumar, V., Prasad, K.S., 2014. Nanotechnology in sustainable agriculture: present concerns and future aspects. Afr. J. Biotechnol. 13 (6), 705–713.

Prasad, R., Pandey, R., Barman, I., 2016. Engineering tailored nanoparticles with microbes: quo vadis. WIREs Nanomed. Nanobiotechnol. 8, 316–330.

Radomski, A., Jurasz, P., Alonso-Escolano, D., Drews, M., Morandi, M., Malinski, T., et al., 2005. Nanoparticle-induced platelet aggregation and vascular thrombosis. Br. J. Pharmacol. 146 (6), 882–893.

Ramesh, P., Rajendran, A., Meenakshisundaram, M., 2014. Green syntheis of zinc oxide nanoparticles using flower extract *Cassia auriculata*. J. NanoSci. NanoTech 2 (1), 41–45.

Ranjitha, V.R., Rai, V.R., 2017. Actinomycetes mediated synthesis of gold nanoparticles from the culture supernatant of *Streptomyces griseoruber* with special reference to catalytic activity. 3 Biotech 7, 299.

Rathee, K., Dhull, V., Dhull, R., Singh, S., 2016. Biosensors based on electrochemical lactate detection: a comprehensive review. Biochem. Biophys. Rep. 5, 35–54.

Rico, C.M., Majumdar, S., Duarte-Gardea, M., Peralta-Videa, J.R., Gardea-Torresdey, J.L., 2011. Interaction of nanoparticles with edible plants and their possible implications in the food chain. J. Agric. Food Chem. 59, 3485–3498.

Rosarin, F.S., Arulmozhi, V., Nagarajan, S., Mirunalini, S., 2013. Antiproliferative effect of silver nanoparticles synthesized using amla on Hep2 cell line. Asian Pac. J. Trop. Med. 6 (1), 1–10.

Sahoo, P.K., Kalyan Kamal, S.S., Kumar, T.J., Sreedhar, B., Singh, A.K., Srivastava, S.K., 2009. Synthesis of silver nanoparticles using facile wet chemical route. Def. Sci. J 59 (4), 447–455.

Saifuddin, N., Wong, C.W., NurYasumira, A.A., 2009. Rapid biosynthesis of silver nanoparticles using culture supernatant of bacteria with microwave irradiation. J. Chem. 6, 61–70.

Saini, A., Sharma, S., Chhibber, S., 2008. Protective efficacy of *Emblica officinalis* against *Klebsiella pneumoniae* induced pneumonia in mice. Indian J. Med. Res. 128, 188–193.

Sankar, R., Manikandan, P., Malarvizhi, V., Fathima, T., Shivashangari, K.S., Ravikumar, V., 2014. Green synthesis of colloidal copper oxide nanoparticles using Carica papaya and its application in photocatalytic dye degradation. Spectrochim. Acta A: Mol. Biomol. Spectrosc. 121, 746–750.

Servin, A., Elmer, W., Mukherjee, A., Torre-Roche, R.D., Hamdi, H., White, J.C., et al., 2015. A review of the use of engineered nanomaterials to suppress plant disease and enhance crop yield. J. Nanopart. Res. 17, 1–21.

Shantkriti, S., Rani, P., 2014. Biological synthesis of copper nanoparticles using *Pseudomonas fluorescens*. Int. J. Curr. Microbiol. Appl. Sci. 3 (9), 374–383.

Sharma, R., Khokhar, M.K., Jat, R.L., Khandelwal, S.K., 2012. Role of algae and cyanobacteria in sustainable agriculture system. Wudpecker J. Agric. Res. 1 (9), 381–388.

Shi, Y., Pan, Y., Zhang, H., Zhang, Z., Li, M.J., Yi, C., et al., 2014. A dual-mode nanosensor based on carbon quantum dots and gold nanoparticles for discriminative detection of glutathione in human plasma. Biosens. Bioelectron. 56, 39–45.

Siddiqi, K.S., Husen, A., 2016. Engineered gold nanoparticles and plant adaptation potential. Nanosc. Res. Lett. 11 (400), 1–10. Available from: https://doi.org/10.1186/s11671-016-1607-2.

Singh, D., Rathod, V., Ninganagouda, S., Herimath, J., Kulkarni, P., 2013. Biosynthesis of silver nanoparticle by endophytic fungi *Pencillium* sp. isolated from *Curcuma longa* (turmeric) and its antibacterial activity against pathogenic gram negative bacteria. J. Pharm. Res. 7, 448–453.

Singh, N., Saha, P., Rajkumar, K., Jayanthi, A., 2014. Biosynthesis of silver and selenium nanoparticles by *Bacillus* sp. JAPSK2 and evaluation of antimicrobial activity. Der. Pharm. Lett. 6 (1), 175–181.

Singh, A., Singh, N.B., Hussain, I., Singh, H., Singh, S.C., 2015. Plant-nanoparticle interaction: an approach to improve agricultural practices and plant productivity. Int. J. Pharm. Sci. 4 (8), 25–40.

Singh, J.S., Kumar, A., Rai, A.N., Singh, D.P., 2016. Cyanobacteria: a precious bioresource in agriculture, ecosystem, and environmental sustainability. Front. Microbiol. 7, 529.

Singh, T., Jyoti, K., Patnaik, A., Singh, A., Chauhan, R., Chandel, S.S., 2017. Biosynthesis, characterization and antibacterial activity of silver nanoparticles using an endophytic fungal supernatant of *Raphanus sativus*. J. Gen. Eng. Biotechnol. 15, 31–39.

Skoglund, S., Hedberg, J., Yunda, E., Godymchuk, A., Blomberg, E., Odnevall Wallinder, I., 2017. Difficulties and flaws in performing accurate determinations of zeta potentials of metal nanoparticles in complex solutions—four case studies. PLoS One 12 (7), e0181735.

Soni, N., Dhiman, R.C., 2017. Silver nanoparticles for pest control: a bionanopesticide approach. Indo Global J. Pharm. Sci. 7 (1), 8.

Sonker, A.S., Richa, R., Pathak, J., Rajneesh, R., Kannaujiya, V.K., Sinha, R.P., 2017. Characterization and in vitro antitumor, antibacterial and antifungal activities of green synthesized silver nanoparticles using cell extract of *Nostoc* sp. strain HKAR-2. Can. J. Biotech. 1 (1), 26–37.

Srivastava, P., Braganca, J., Ramanan, S.R., Kowshik, M., 2013. Synthesis of silver nanoparticles using haloarchaeal isolate *Halococcus salifodinae* BK3. Extremophiles 17 (5), 821–831.

Subramanian, K.S., Muniraj, I., Uthandi, S., 2016. Role of actinomycete-mediated nanosystem in agriculture. In: Subramanian, G., et al., (Eds.), Plant Growth Promoting Actinobacteria. Springer, Singapore, pp. 237–247.

Sunkar, S., Nachiyar, C.V., 2012a. Microbial synthesis and characterization of antibacterial silver nanoparticles using the endophytic bacterium *Bacillus cereus*: a novel source in the benign synthesis. Global J. Med. Res. 12, 43–49.

Sunkar, S., Nachiyar, C.V., 2012b. Biogenesis of antibacterial silver nanoparticles using the endophytic bacterium *Bacillus cereus* isolated from *Garcinia xanthochymus*. Asian Pac. J. Trop. Biomed. 2, 953–959.

Tarafdar, J.C. (2015). Nanoparticle production, characterization and its application to horticultural crops. Conference Paper, Winter School on "Utilization of Degraded Land and Soil through Horticultural Crops for Agricultural Productivity and Environmental Quality," December 3–23, 2015 at ICAR-National Research Centre on Seed Spices, Tabiji, Ajmer.

Torney, F., Trewyn, B.G., Lin, V.S.Y., Wang, K., 2007. Mesoporous silica nanoparticles deliver DNA and chemicals into plants. Nat. Nanotechnol. 2 (5), 295.

Tsuji, J.S., Maynard, A.D., Howard, P.C., James, J.T., Lam, C.W., Warheit, D.B., et al., 2005. Research strategies for safety evaluation of nanomaterials, part IV: risk assessment of nanoparticles. Toxicol. Sci. 89 (1), 42–50.

Vamvakaki, V., Chaniotakis, N.A., 2007. Pesticide detection with a liposome-based nano-biosensor. Biosens. Bioelectron. 22 (12), 2848–2853.

Watanabe, A., Kajita, M., Kim, J., Kanayama, A., Takahashi, K., Mashino, T., et al., 2009. In vitro free radical scavenging activity of platinum nanoparticles. Nanotechnology 20 (45), 455105.

Yaseen, A.A.M., Merzah, S., Khalidah, M.S., Basim, S., Wasan, M.H., 2016. Effect of foliar spray of nano silver and organic fertilizer (Algastar) and salicylic acid on some morphological characteristics and carbohydrate content in (*Helianthus annuus* L.). J. Agric. Ecol. Res. Int. 9 (3), 1–7.

Yu, H., Huang, Q., 2013. Bioavailability and delivery of nutraceuticals and functional foods using nanotechnology. Bio-Nanotechnology: A Revolution in Food, Biomedical and Health Sciences. Wiley, pp. 593–604.

Zhong, W., Xing, M.M., Maibach, H.I., 2010. Nanofibrous materials for wound care. Cutan. Ocul. Toxicol. 29 (3), 143–152.

Zonaro, E., Lampis, S., Turner, R.J., Qazi, S.J.S., Vallini, G., 2015. Biogenic selenium and tellurium nanoparticles synthesized by environmental microbial isolates efficaciously inhibit bacterial planktonic cultures and biofilms. Front. Microbiol. 6, 584.

# *Trichoderma*-mediated biocontrol and growth promotion in plants: an endophytic approach

*Jaisingh Patel[1], Basavaraj Teli[2], Raina Bajpai[2], Jhumishree Meher[2], Md. Mahtab Rashid[2], Arpan Mukherjee[3] and Sudheer Kumar Yadav[4]*

[1]Department of Plant food and Environmental Sciences, Dalhousie University, Nova Scotia, Canada, [2]Department of Mycology and Plant Pathology, Institute of Agricultural Sciences, Banaras Hindu University, Varanasi, India, [3]Institute of Environment and Sustainable Development, Banaras Hindu University, Varanasi, India, [4]Department of Botany, Institute of Science, Banaras Hindu University, Varanasi, India

## 13.1 Introduction

Most of the plants on Earth are potential reservoirs for endophytes, which are indigenous microbes residing in plant tissue without producing any observable external symptoms. These indigenous microbes help in assimilation and processing of nutrients, defense response induction, and secondary metabolites synthesis (Pandey et al., 2017). During the course of evolution, certain microbes have been able to infiltrate the tissues of plants using enzymes that degrade the plant cell wall, such as different types of cellulase and pectinase. After infiltration, these microbes established themselves inside of the plant and coevolved (Pandey et al., 2016). The coevolvement may have led to the microbes' adaptation to the internal environment of the plant host through communication with the host. These endophytic microbes can settle inside the plant tissues in an obligate or facultative mode and, as described earlier, can adjudicate in plant without producing any instant external or negative symptoms. Instead, there have been reports showing their beneficial result of microbes on the host plant, which opens up an array of opportunities for the discovery of products and processes may be relevant and applicable to the field of agriculture (Pandey et al., 2012). A significant role has been known to be played by the endophytic bacteria and fungi in the growth of plants, the fertility of soil, and biocontrol activities. For example, phytopathogens are controlled by the stable microbic endophytes that possess an arsenal of defense activities including antibiotics and antifungal metabolite production, de novo synthesis of structural compounds, induction and expression of molecular-based immunity, competition against other organism for occupation of the niche, and/or induced systemic resistance (ISR) (Pandey et al., 2018). The host plants also receive benefits from the endophytic microorganisms through the plant growth stimulation, phytopathogens

**Role of Plant Growth Promoting Microorganisms in Sustainable Agriculture and Nanotechnology.**
**DOI: https://doi.org/10.1016/B978-0-12-817004-5.00013-0**

plant growth and seedling emergence comparatively (Ahmad and Baker, 1988). The *Trichoderma fusant* strain T-22 is produced through the protoplast fusion of auxotrophic mutants from *T. harzianum* strains T-12 and T-95, and has been found to colonize the rhizosphere of the maize plant better than other strains. In a greenhouse experiment, the application of isolate T-22 showed significant increase, an average of 66%, in the shoot and root growth of sweet corn when compared to the untreated control (Sivan and Harman, 1991; Bjorkman et al., 1998). In another study, it was found that the T-203 strain of *T. harzianum* was able to colonize and live inside the plant roots just as mycorrhizal fungi (Kleifeld and Chet, 1992). Additional research revealed that the application of *Trichoderma velutinum* GI/8 as a spore suspension in the seedlings of lettuce plants significantly increased the length of root, the area of leaf, and overall weight in comparison to the untreated control. Apart from the endophytic growth of the fungus within the roots, extensive colonization of hyphae was observed on the root surface and epidermis, which grew parallel to the direction of root cells (Zachow et al., 2010). A well-reported endophytic strain of *T. gamsii*, YIM PH30019 had no pathogenicity to its plant host (*P. notoginseng*) and even conferred supportive biocontrol efficacy when observed in vitro and in field conditions with continuous cropping soil (Chen et al., 2016). Apart from the species of *Trichoderma*, various other classes of endophytic microbes can modify the plant responses to abiotic stresses. For example, in *Theobroma cacao*, plant growth is promoted and in *Trichoderma hamatum* the onset of drought is delayed by isolate DIS 219b (Bae et al., 2009).

## 13.3 Growth promotion mechanisms

As it is evident from research, various *Trichoderma* strains are able to initiate growth promotion in plants through one or more mechanisms. For example, *T. harzianum* T-22-mediated growth promotion was demonstrated in glasshouses and fields in a broad range of crops. Recently, many other mechanisms have been discovered that explains the *Trichoderma's* growth promotion. These mechanisms include phytohormone synthesis by plants and/or microbes, enhancement in soil nutrient solubilization, increased nutrient uptake and translocation, enhancement in the development of roots, and increase in photosynthesis rate, carbohydrate metabolism, and defense mechanisms in plants (Altomare et al., 1999; Harman, 2004b, 2006; Mastouri et al., 2010).

### 13.3.1 Nutrient uptake and solubilization

The nutrient content of the soil is a major factor affecting the bioactivity and proliferation of fungi to promote plant growth. The promotion of plant growth is exceptional in soil with low nutrient and/or mineral content. In many instances, nitrogen was determined to be the limiting factor in the crop yield (Stewart and Hill, 2014). The mechanism behind this promotion of plant growth was presumed to be the improved uptake of minerals and nitrogen (in the form of ammonia) by the roots. A report has shown that the maize plant roots colonized by *T. harzianum* T-22 demands 40% less nitrogen fertilizer as compared to the uncolonized ones. In addition to this, T-22 has also shown to solubilize a wide extent of soil nutrients which are otherwise unavailable to the plants in solubilized form. These soil nutrients include rock phosphate, $Fe^{3+}$, $Cu^{2+}$, $Mn^{4+}$ and ZnO in certain soils (Altomare et al., 1999; Harman et al., 2004a). Likewise, *T. harzianum* 1295-27 have shown their ability to solubilize phosphate and micronutrients, thus making them available for uptake by the plants (Altomare et al., 1999; Whipps, 2001). Apart from the ability of *Trichoderma* species to reduce insoluble $Cu^{2+}$ to soluble $Cu^{1+}$, $Fe^{3+}$ to $Fe^{2+}$, and $Mn^{4+}$ to $Mn^{2+}$, there have also been reports of the species producing many organic acids like citric, fumaric, or gluconic, thus lowering the pH of soil (Benítez et al., 2004) and hence permitting phosphate solubilization (Stewart and Hill, 2014).

### 13.3.2 Secondary metabolites

In an experiment conducted by Windham et al. (1986), the application of *T. harzianum* and *T. koningii* to autoclaved soil improved the rate of seedling emergence when compared to control in tobacco and tomato plants. The dry weights of roots and shoots also increased by 313% and 318% 8 weeks after emergence, respectively, and the germination time was shortened significantly by 1–3 days in maize, tomato, and tobacco plants. The secondary metabolites produced by *Trichoderma* species have been reported to play a role in the promotion of plant growth. The isolation, identification, and bioactivity of Koninginin A (a secondary metabolite from *T. koningii*) and 6-pentylalpha-pyrone (a secondary metabolite from *T. harzianum*), which acted as regulators of plant growth, were described by Cutler et al. (1989). There are separate reports proving the production of secondary metabolites, such as *T. harzianum* T-22, T-39, A6, and *T. atroviride* P1, by various *Trichoderma* biocontrol agents in a liquid media. The spraying of 6 ppm of $10^{-6}$ M solutions onto the seedlings of tomato plants led to increase in plant height and leaf area (Vinale et al., 2008). The auxin-like secondary metabolites of *Trichoderma* species have optimal activity at lower concentrations ($10^{-5}$, $10^{-6}$ M); while at higher concentrations it shows an inhibitory effect (Brenner, 1981). One of the major advantages of the formation of secondary metabolites is that they have a much longer shelf life, along with greater efficiency against the soil-borne phytopathogens, especially bacteria. The formulation consortia can be made of different antimicrobial metabolites for individual pathogens, and regardless of the geographical location it can be utilized where a particular disease has a higher incidence (Keswani et al., 2014).

### 13.3.3 Plant hormones production

Research on PGPR (plant growth-promoting rhizobacteria) has proved that these bacteria utilize tryptophan to synthesize IAA (indole acetic acid), which in turn leads to the growth of plant roots through increased cell elongation and division.

1-aminocyclopropane-1-carboxylate (ACC) is an immediate precursor of phytohormone ethylene to give α-ketobutyrate and ammonia. The bacterial production of ACC deaminase, an enzyme that catalyzes cleaving in ACC, also stimulates the growth of plants.(Todorovic and Glick, 2008). Ethylene production under abiotic and biotic stresses causes plant growth inhibition, thus making it an important molecule for plant signaling (Pierik et al., 2006). Reports show that the inoculation of plants with bacteria that produce ACC deaminase leads to lowers levels of ethylene production in those plants, resulting in longer plant roots and reduced plant growth inhibition (Glick et al., 1998). IAA is also known to induce the ACC deaminase activity thus it not only promotes the growth of plants through direct induction of elongation and cell division, but also indirectly by inhibiting the synthesis of ethylene. Such mechanisms have been lately reported for the *Trichoderma* species. For example, *T. atroviride* promoted tomato plants growth in plants grown under hydroponics resulting in an increase in the yield of marketable fruits. This may be due to slowed in vitro production of IAA by the fungus. Another factor of this growth was the l-tryptophan, tryptamine, and tryptophol that were added in the culture medium. The growth of the tomato seedlings was enhanced by the fungus due to increasing l-tryptophan concentration which suggests that IAA can be synthesized by the fungus through a tryptophan-dependent pathway (Gravel et al., 2007). The root length of tomato seedlings reduced when IAA was applied exogenously to the untreated control, but when IAA was applied to the *T. atroviride*-inoculated seedlings, the length of roots increased significantly. This implies that the fungus is able to partially degrade IAA in vitro as well as produce ACC deaminase enzyme in vitro. On the basis of the aforementioned observations, two mechanisms were postulated by the authors of this chapter through which the fungus would be able to reduce the detrimental effects of IAA on elongation of root *viz.* These mechanisms are (1) degradation of IAA to a level that does not inhibit root elongation and (2) regulation of the ethylene concentration through ACC deaminase action leading to the reduction of its precursor molecule ACC. Similar mechanisms of action have been demonstrated by several other *Trichoderma* species. For instance, when grown in a medium with ACC as the only source of nitrogen, *Trichoderma asperellum* produces higher ACC deaminase levels (Stewart and Hill, 2014). The promotion of plant growth by the *Trichoderma* species is likely due to a synergistic result of regulation of IAA concentration within the plant rhizosphere, and ethylene concentration within the plant roots (Gravel et al., 2007).

The *Trichoderma* species is the ubiquitous fungus commonly found as saprobes in root and soil ecosystems (Vinale et al., 2008; Kodsueb et al., 2008). They can be isolated easily from the soil, decaying wood, and various other organic material (Zeilinger and Omann, 2007). Various reports support the use of *Trichoderma* species as biocontrol agents against the phytopathogens (Harman et al., 2004a; Zeilinger and Omann, 2007). Several phytopathogenic fungi such as *Botrytis cinerea*, *Fusarium* spp., *Pythium* spp., *Phytophthora palmivora*, *P. parasitica*, and *Rhizoctonia* spp. are controlled by the application of *Trichoderma* species (Benítez et al., 2004; Zeilinger and Omann, 2007). The mechanism of biocontrol utilized by the *Trichoderma* species is a consolidation of various mechanisms (Howell, 2003;

Zeilinger and Omann, 2007). The chiefly employed mechanisms among these are antibiosis and mycoparasitism (Howell, 2003; Vinale et al., 2008). The mycoparasitic action of *Trichoderma* species depends on the recognition, binding and enzymatic disruption of cell wall of the host fungi (Woo and Lorito, 2007). *Trichoderma* species have a very useful application as mycofungicides because of the following reasons: fast growing nature, high capacity of reproduction, diversity in control mechanisms, excellent rhizosphere competence, tolerance to soil fungicides, capability to modify the rhizosphere, survival ability under unfavorable conditions, efficient utilization of soil nutrients, strong aggressiveness against phytopathogenic fungi, promotion of plant growth, and inhibition of a broad range of fungal diseases (Vinale et al., 2006). The ability of the *Trichoderma* species to colonize and grow plant roots is described as rhizosphere competence. From a taxonomical point of view, the *Trichoderma* species is very complex; there have recently been many taxonomic studies done on them (Woo et al., 2006; Samuels, 2006). A high level of genetic diversity is found among these species (Harman, 2006). Therefore, we can conclude that only a few of the *Trichoderma* species, which are available have been put into utilization as mycofungicides however, *Trichoderma* species are still most the commonly used biocontrol fungal agent, and have been formulated as biofertilizers, biofungicides, and soil amendments for commercial purposes (Vinale et al., 2006).

## 13.4 Effect of *Trichoderma* on pathogens

*Hypocrea/Trichoderma* spp. often behave as necrotroph, hyperparasite, or mycoparasite against other parasitic phytopathogenic fungi (Harman, 2011). The concept of mycoparasitism is demonstrated through a survey of more than 1100 *Trichoderma* isolates obtained from 75 molecularly characterized species, each of the species with mycoparasitic activity against several disease-causing fungi such as *Botrytis cinereal*, *Sclerotinia sclerotiorum*, and *Alternaria alternate* (Druzhinina et al., 2011). However, *Trichoderma* is also classified as a saprophyte, feeding on dead material. In this way, the genus could be defined as a mycotroph instead of mycoparasitic fungus, which more accurately refers to the nature of *Trichoderma* (Druzhinina ct al., 2011). *Trichoderma* belongs to the family of asexual fungi (i.e., deuteromycetes) that exist in the soil of all climatic zones. It is an opportunistic fungus, growing and producing its spores very fast, which is the source of cellulases and antibiotics (Vinale et al., 2008).

In the different *Trichoderma* species, several genes, coding proteases, and oligopeptides transporters have been found to be involved in mycoparasitic action (Seidl et al., 2009; Suárez et al., 2007). *Trichoderma* can produce enzymes (such as chitinase, cellulase, glucanases) and antibiotics against other fungi. It has the ability to secrete several secondary metabolite compounds such as non-ribosomal peptide synthases (NRPs) in response to its mycoparasitic activity (Kubicek et al., 2011). Glucanases are also produced by *Trichoderma* for biocontrol activity. Glucanase can lose the activity of gene modification when the genes that code for enzymes (that reduce the mycoparasitic activity of the *Trichoderma*) are mutated. This can lead to deletion of the gene that codes for the b-1,6-glucanase, which reduces the biocontrol and mycoparasitic activity of the *Trichoderma virens* against the phytopathogen (Djonovic et al., 2006). However, overexpression of the b-glucanases-encoding genes increases the mycoparasitic activity of the biocontrol against *P. ultimum*, *R. solani*, and *Rhizopus oryzae*. The mycoparasitism of the *Trichoderma* can be induced by chitinases, proteases, and glucanases encoded by Prb1/Sp1 (Viterbo and Horwotz, 2010). One study (Catalano et al., 2011) suggests that the role of laccase for the colonization of the *T. virens* in the rhizosphere of the plant is significant. The oligopeptides are released by the activity of such proteases on the other fungi, which then binds to receptors for the nitrogen starvation in the *H. atroviridis* (Seidl et al., 2009). The study by Seidl et al. (2009) suggests the use of class IV of the G-protein-coupled receptors (GCPRs) in *H. atroviridis* to detect oligopeptides (Kubicek et al., 2011). Two paralogues of the class IV GPCRs are reported in the *H. virens*, *H. Atroviridis*, and *H. jecorina* (Kubicek et al., 2011). There may be several other types of GPCRs in the *Trichoderma*, used for defense against pathogens, such as Gpr1 protein in the *T. atroviride* v2.0 genome (with the identification number 160995). Another study suggests the presence of a cyclic adenosine monophosphate (AMP) receptor similar to GPCRs in the *H. atroviridis*, and that a signaling cascade is performed by the three heterotrimeric G-proteins (G$\alpha$, G$\beta$, and G$\gamma$) in response to the GPCRs. Mutation in the G$\alpha$ subunit Tga1 causes complete loss of the mycoparasitic activity, reduction in the chitinase activity and less production of antimicrobial compounds such as 6-pentyl pyrone in the *H. atroviridis* (Rocha-Ramírez et al., 2002; Reithner et al., 2005). However, mutation of the tgaA resulted in the little loss of the mycoparasitic activity against the pathogen *Athelia rolfsii* (anamorph *Sclerotium rolfsii*) in the *H. virens* (Mukherjee et al., 2004).

A well-known system of pathway signaling in fungi is mitogen-activated protein kinase (MAPK) (Schmoll, 2008). The genome of *Trichoderma* holds genes for the three MAPKs known as pathogenicity MAPKs: (1) TmKA (also called as Tvk1 and Tmk1), (2) the cell integrity kinase (TmkB), and (3) osmoregulatory MAPK (Hog1) (Schmoll, 2008). Mutation of the gene TmkA in the *T. virens* strain "P" affects the pathogenicity against *S. rolfsii* in the form of loss, however, no effect on pathogenicity against *R. solani* has been noted (Mukherjee et al., 2003; Viterbo et al., 2005). Deletion of the same gene TmkA in the *T. virens* strain "Q" showed positive effects on the biocontrol efficiency of the *Trichoderma* against both *Pythium ultimum* and *R. solani* (Mendoza-Mendoza et al., 2007). The metabolic profiling of both "P" and "Q" strains supply information about the results of the mutation study of TmKA. Mutation in a homologue of Tmk1 in the *Hypocrea atroviridis* resulted in reduced mycoparasitic activity against *R. solani*, and also supported increased production of chitinase and other antifungal compounds (Reithner et al., 2007).

However, the function of the other two MAPKs (TmkB and Hog1) is not well understood due to the poor growth of the mutants. *T. virens* consisting of a mutant of TmKB is defective toward the mycoparasitic against *S. rolfsii* (Kumar et al., 2010). However, the mutant of Hog1 protein is related to the osmotic tolerance of the *Trichoderma*, and has no mycoparasitic activity (Delgado-Jarana et al., 2006). Another signaling cascade related with cyclic adenosine monophosphate (cAMP) is an especially important pathway for the sexual development, differentiation, virulence, stress, nutritional status, and cell cycle control for the common fungi (Kronstad et al., 1998; Sarma et al., 2014). This signaling cascade is the membrane-associated enzyme adenylate cyclase regulated by the cAMP signaling pathway. The synthesis product of adenylate cyclase is the intracellular messenger cAMP, which is regulated by the alpha subunit of heterotrimeric G-proteins in several fungi. The function of cAMP in the eukaryotic organism is activated by the cAMP-dependent protein kinases with two cytosolic subunits (Dickman and Yarden, 1999). The function of cAMP in the *Trichoderma ressi* is produced by the endoglucanase effectors, giving it its ability to antagonize *P. ultimum* (the causal organism of zucchini *P. ultimum* blight) (Seidl et al., 2006).

The exogenous cAMP manages supercoiling around the nylon membrane in the *Trichoderma harzianum* system as well as the substances (such as caffeine, dinitrophenol, and aluminum tetrafluoride) responsible for increasing the intracellular level of cAMP and repressing the *N*-acetyl-α D-glucosaminidase synthesis (Silva et al., 2004). cAMP plays a significant role during stress signaling in *T. atroviride* and *T. viride* (Casas-Flores et al., 2006). Production of conidia by *Trichoderma* is the major mechanism for the survival of this organism, and this production induced by several environmental factors such as nutrient stress and blue light detection. Photoinduction in *T. viride* increases the level of intracellular level of the cAMP (Gresi and Kolarova, 1988) and the formation of the conidia seen by exogenous application of cAMP in both dark and light condition (Nemčovič and Farkaš, 1998). The light responses in the *Trichoderma* have been reported to be regulated by the protein kinases (Casas-Flores et al., 2006). Mutation in the gene encoding for adenylate cyclase decreases the level of the cAMP below the limit of detection and the mutants showed only 5%–6% of the wild-type growth. However, sporulation was not observed in the mutants in dark conditions and spores failed to germinate in the water. This caused a loss in the mycoparasitic activity against phytopathogens including *R. solani*, *Pythium* sp., and *S. rolfsii* due to a reduction in the secondary metabolite production. Mukherjee et al. (2007) reported that the cAMP signaling cascade played an integral role in the increase of secondary metabolite production by *T. virens* and Tac1. This cAMP signaling cascade is used for regulation of mycoparasitism, secondary metabolism, growth, and germination of spores.

## 13.5 Effect of *Trichoderma* on plants

The growth of *Trichoderma* spp. is mostly observed in the rhizospheric region due to its capacity to penetrate and colonize the plant roots internally (Harman et al., 2004a). The colonization of plant roots occurs both externally and internally via *Trichoderma* spp. With the help of electron microscopy, the physical interaction between *Trichoderma* and the plant was observed, which was only limited to the first few cell layers of plant epidermis and root outer cortex (Yedidia et al., 1999). The relationship between plant and *Trichoderma* has evolved into a symbiotic relationship rather than a parasitic relationship as the fungus' growing nutritional niche acts as guard by protecting the plant from disease (Vinale et al., 2008). The attraction of *Trichoderma* to plant roots is the consequences of exchanging and interacting chemical signals between the two, similar to any other kind of biological interactions (Mukherjee et al., 2012a,b). There are multiple positive outcomes of this "endophytic" colonization of the roots by *Trichoderma*; for example, enhanced growth response and higher uptake of nutrients are benefits of this colonization. Another example of the beneficial effects of this "intimate" association between plants and *Trichoderma* is the induction of systemic or localized resistance against both fungi and bacteria attacking root and foliar parts. When *Trichoderma* intercellularly colonizes the root epidermis and outer cortical layer the colonization leads to the deposition of callose to create a wall around the *Trichoderma* thallus. Changes in the plant's transcriptome and the proteome are associated with this phenomenon. Here, root colonization is also reported to be regulated genetically by chemical triggers initiated by the attachment and penetration of organisms resembling any mycoparasitism (Mukherjee, 2012). Colonization of roots by *Trichoderma* strains are well-known, and recently, site-specific application of *Trichoderma* conidia to different parts of plant-like fruit, flowers and foliage to control plant diseases is also being observed (Harman, 2000; Elad, 1994). More than a decade has passed since the first report of a plant's internal colonization was discovered, however the genetic studies regarding the nutritional relationship were accounted for only recently. In the process of acquiring sucrose by *T. virens* from plant cells, Vargas et al. (2009, 2011) have characterized enzyme invertase and a sucrose transporter playing a significant part in it. This leads us to conclude that throughout the symbiosis, the sink activity of roots is influenced by the fungal cell's sucrolytic actions, ultimately directing carbon partitioning to the roots and raising the speed of photosynthesis in leaves. Additionally, for regulation of the elicitor protein Sm1, hydrolysis of sucrose was vital. *T. virens* also has a highly specific sucrose/$H^{(+)}$ symporter very similar to plants that are triggered in the early phases of root colonization (Vargas et al., 2011). Through this commonality, the participation of this sucrose transporter gene in active sucrose transference from plant to fungal cells has been detected via gene deletion technique. Mukherjee et al. (2012a, b) reported that the sucrose-dependent network in the fungal cells maintains this symbiotic relationship between plants and *Trichoderma*.

Among all the components, the most interesting condition is that in spite of root colonization by *Trichoderma* spp., they don't behave as plant pathogens except in the cases of apples, maize and alfalfa, in which they pathogenically causes diseases on crops. Also, few strains secrete metabolites that are extremely phytotoxic. *Trichoderma* is known to be a prominent producer of certain enzymes, all of which are necessary for plant cell wall degradation. Therefore, a remarkable phenomenon has been studied in *Trichoderma* spp. and also in other nonpathogenic microorganisms such as *Fusarium*, *Rhizoctonia* strains, mycorrhizae, and other fungi that also infect the roots and have the inherent ability to behave as plant pathogens, but whose infection is restricted to only superficial cells in plant roots (Harman et al., 2004a).

### 13.5.1 Defense induction

On invasion by *Trichoderma*, plants respond immediately with rapid ion fluxes and an oxidative burst, followed by deposition of callose and synthesis of polyphenols. This results in the entire plant acquiring varying degrees of tolerance to pathogen invasion with the involvement of salicylate (SA) and jasmonate/ethylene (JA/ET) signaling (Shoresh et al., 2010). This response resembles the response triggered by PGPR and has most frequently been described as JA/ET-mediated ISR. Recent findings suggested that SA-mediated systemic acquired resistance (SAR) response (meaning a response invoked by necrotrophic pathogens) can be triggered by higher inoculum doses of *Trichoderma* (Segarra et al., 2007; Contreras-Cornejo et al., 2011; Salas-Marina et al., 2011; Yoshioka et al., 2012). This idea was introduced from the molecular cross-talk between plant and *Trichoderma:* the communication between MAPK from cucumber and a MAPK from *T. virens*, which presumably triggered the downstream defense responses (Viterbo et al., 2005; Shoresh et al., 2006). To elicit an immune response in plants, *Trichoderma* spp. produce Xylanase and peptaibols (peptaibiotics with high content of alpha-amino isobutyric acid) like alamethicin and trichovirin II (Leitgeb et al., 2007; Viterbo et al., 2007; Luo et al., 2010; Druzhinina et al., 2011). Recently, in maize plants a hybrid enzyme polyketide synthases/non-ribosomal peptide synthases (PKS/NRPS) was identified which is involved in defense responses (Mukherjee et al., 2012a,b). Sm1/Epl1 is an elicitor that is abundantly secreted as a small cysteine-rich hydrophobin-like protein of the cerato-platanin (CP) family, produced by *Trichoderma* spp. (Djonovic et al., 2006; Seidl et al., 2006). In maize plants, deletion of this *Trichoderma* gene impairs elicitation of ISR (Djonovic et al., 2007). The monomeric form of Sm1 is in a glycosylated state which is essential for elicitation properties. While in the nonglycosylated state, it is susceptible to oxidative-driven dimerization in plants rendering Sm1 inactive as an inducer of ISR (Vargas et al., 2008). Recently, *Ceratocystis platani* cerato-platanin's 3D structure has been resolved and the carbohydrate residue (an oligomer of *N*-acetyl glucosamine) that binds to it has been identified (de Oliveira et al., 2011). It was predicted that its structure and carbohydrate-binding properties may suggest a mechanism for the elicitation properties of Sm1 since the CP protein family is highly conserved.

The induction of plant defense responses has been well documented in antagonistic fungi (De Meyer et al., 1998; Yedidia et al., 1999; Hanson and Howell, 2004; Harman et al., 2004b). Various plants, both mono- and dicotyledonous species, when pretreated with *Trichoderma* showed increased resistance to pathogen attack (Harman et al., 2004a). At the site of inoculation, *Trichoderma* spp. colonizes a plant and reduces disease caused by one or more different pathogens (via induced localized acquired resistance). *Trichoderma* can also reduce disease when the biocontrol fungus was inoculated at different times or sites than those of the pathogen (via ISR). *Trichoderma* species enhance the defense system but is not involved in the production of pathogenesis-related proteins (PR proteins), which is similar to that occurring in plant resistance elicited by rhizobacteria (van Loon et al., 1998; Harman et al., 2004a).

In a recent study using a high-density oligonucleotide microarray approach, Alfano and co-workers (2007) investigated the plant genes involved in *T. hamatum* 382 resistance inductions at a molecular level. Interestingly, *Trichoderma*-induced genes were associated with biotic or abiotic stresses, as well as RNA, DNA, and protein metabolism. In particular, BCA was found to induce genes that codify for extensin and extensin-like proteins but not those genes coding for proteins belonging to the PR-5 family (thaumatin-like proteins), which are considered the main molecular markers of SAR. Different classes of metabolites may act as elicitors or resistance inducers during the interaction of *Trichoderma* with the plant (Harman et al., 2004b; Woo et al., 2006; Woo and Lorito, 2007). These molecules include: (1) proteins with enzymatic activity, such as xylanase (Lotan and Fluhr, 1990), (2) avirulence-like gene products able to induce defense reactions in plants (Woo et al., 2004), and (3) low-molecular-weight compounds released from fungal or plant cell walls by the activity of *Trichoderma* enzymes (Harman et al., 2004a; Woo et al., 2006; Woo and Lorito, 2007). Short oligosaccharides comprised of two types of monomers, with and without an amino acid residue were purified and characterized from the low-molecular-weight degradation products released from fungal cell walls (Woo et al., 2006; Woo and Lorito, 2007). These compounds, when applied to leaves or when injected into root or leaf tissues, elicited a reaction in the plant. By activating the mycoparasitic gene expression cascade, it also stimulated the biocontrol ability of *Trichoderma*. A small protein (Sm1) elicitor secreted by *T. virens*, which is involved in the activation of plant defense mechanisms as well as the induction of systemic resistance, was identified and demonstrated recently by Djonovic et al. (2006). This elicitor may also stimulate the biological activity of resident antagonistic microbial populations or introduced *Trichoderma* strains, and promote an ISR effect in the plant in addition to their innate antimicrobial effect. Against pathogens, other secondary metabolites, like peptaibols, may act as elicitors of plant defense mechanisms. In fact, in tobacco plants, the application of peptaibols activated an observed defense response (Benítez et al., 2004). Wiest et al. (2002) and purified a peptaibol synthetase from *T. virens*, which facilitated an understanding of the role of this class of compounds in plant defense response. Subsequently, cloning of the corresponding gene was done. Several studies have shown that increased levels of defense-related plant enzymes, including various peroxidases, chitinases, β-1,3-glucanases, and the lipoxygenase-pathway hydroperoxide lyase (Howell et al., 2000; Yedidia et al., 1999; Harman et al., 2004b) resulted in root colonization by *Trichoderma* strains. A transient increase in the production of phenylalanine ammonia lyase in both shoots and roots was observed in cucumber plants with the addition of *T. asperellum* T-203, however, within 2 days this effect decreased to background levels in both plant organs. However, the expression of many defense-related genes increased several times over if leaves were subsequently inoculated with the bacterial pathogen *Pseudomonas syringae pv.*

*lachrymans*. Subsequent accumulation of antimicrobial compounds may result due to changes in plant metabolism.

In cotton, the ability of *T. virens* to induce phytoalexin production and localized resistance has previously been established. An increase in phenolic glucoside levels in leaves of cucumber plants was observed with root colonization by strain T-203; their aglycones (which are phenolic glucosides with the carbohydrate moieties removed) are strongly inhibitory to a range of bacteria and fungi. When the roots are inoculated first with T-203, this, in turn, augments the levels of inhibitory compounds and whose leaves are then challenged with *P. syringae pv. lachrymans*. When either organism was applied alone this effect was not seen. However, this response is not restricted to interactions with T-203. Strains P1 or T-22 were found to induce more antimicrobial compounds than in nontreated controls groups consisting of leaf extracts from other inoculated plants (Woo et al., 2004). *Trichoderma* strains also strongly stimulate plants to produce their own antimicrobial compounds, not only produce antibiotic substances directly. Therefore, significant changes are observed in the metabolic machinery of plants with root colonization by these fungi (Harman et al., 2004a). Proteomes from 5-day-old maize seedlings (grown from seeds that were either treated or not treated with T-22) were fractionated by 2D gel electrophoresis and approximately 40% of the proteins that were seen in the presence of T-22 were not visible in gels that contained proteins from untreated plants. Together, the data indicates that modification in plant metabolism by these fungi results most often in benefits for the plant.

### *13.5.2 Plant resistance due to* Trichoderma

A variety of strains of *T. virens*, *T. asperellum*, *T. atroviride*, and *T. harzianum* produce a signaling response during symbiotic interaction that leads to induction or alteration of diverse metabolites in plants. This causes an increase in the plant's resistance to a wide range of pathogenic microorganisms and viruses. Moreover, this type of response is highly associated with many plants' species. When the fungal structures are inoculated in the rhizospheric zone, the spores or other propagative bodies communicate with the root surface for proper germination, even infecting some of the outer root cells. During the interaction, various *Trichoderma* spp. produces at least three classes of substance that elicit plant defense responses towards the inciting plant pathogens in order to prevent further infection in the roots zone. In some cases, the resistance response will be basal or through elicitation of induced systemic resistance/ systemic acquired resistance (ISR/SAR) defense-related genes. This is based on the site of pathogen infection in the plant (i.e., localized) but in most cases, the systemic resistance is more common in *Trichoderma* systems. These results indicate that the initial reactions, which include the production of PR proteins and phytoalexins, have features in common with SAR. As seen in the interaction between *T. asperellum* and cucumber plant, a longer-term response shows low levels of expression of PR proteins before the pathogen infection. This response therefore has elements in common with ISR. *Trichoderma* spp. directly inhibits the activity of other fungi through the production of antibiotics, and sometimes will indirectly increase competition for nutrients accumulated through these relationships with other fungi.

### *13.5.3 Effect on rhizosphere by* Trichoderma

Organic nutrients present in the soil not only act upon plants but also affects the activities of BCAs in the soil (Hoitink and Boehm, 1999). Similarly, the biotic and abiotic factors along with the accumulation of organic matter correlate the activity of *Trichoderma* in the rhizospheric zone (Simon and Sivasithamparam, 1989; Wakelin et al., 1999). Generally, composts serve as a flawless substrate for the BCAs, which incite the establishment of various BCAs into the surrounding soil environment (Hoitink and Boehm, 1999; Leandro et al., 2007). The *Trichoderma* not only interplay with the host plant, but use various mechanisms (i.e., antagonism and production/activation of ISR that is activated by the presence/activity of pathogens in the surrounding areas). Furthermore, these mechanisms may exhibit based on the availability and concentration of nutrients within the soil habitat (Hoitink et al., 2006). Krause et al. (2001) demonstrated that the soil inoculation of *T. hamatum*, which has the capacity to carry high microbial load significantly inhibiting the severity of radish plant dampening, or crown and root rot of poinsettia plants caused by *Rhizoctonia* species. However, *T. hamatum* treated with the compost-amended medium mixed with soil reduced the severity of Phytophthora leaf blight by inducing SAR in cucumber plants (Khan et al., 2004). The in vitro culturing of various *Trichoderma* strains on this medium containing high organic matter helps in our understanding of the quick responses of *Trichoderma*, which will lead to accurate selection of appropriate strains for large acreage. This ultimately will boost the organic farming potential worldwide.

## Acknowledgment

Sudheer K. Yadav is grateful to the Indian Council of Medical Research, New Delhi, India, for financial assistance [Grant 3/1/3/JRF-2012/HRD-66(80689)].

## References

Ahmad, J.S., Baker, R., 1988. Rhizosphere competence of benomyl-tolerant mutants of *Trichoderma* spp. Can. J. Microbiol. 34 (5), 694–696.

Alfano, G., Lewis Ivey, M.L., Cakir, C., Bos, J.I.B., Miller, S.A., Madden, L.V., et al., 2007. Systemic modulation of gene expression in tomato by *Trichoderma hamatum* 382. Phytopathology 97, 429–437.

Altomare, C., Norvell, W.A., Bjorkman, T., Harman, G.E., 1999. Solubilization of phosphates and micronutrients by the plant-growth-promoting and biocontrol fungus *Trichoderma harzianum* Rifai 1295-22. Appl. Environ. Microbiol. 65 (7), 2926–2933.

Bae, H., Sicher, R.C., Kim, M.S., Kim, S.H., Strem, M.D., Melnick, R.L., et al., 2009. The beneficial endophyte *Trichoderma hamatum* isolate DIS 219b promotes growth and delays the onset of the drought response in *Theobroma cacao*. J. Exp. Bot. 60 (11), 3279–3295.

Benítez, T., Rincón, M.A., Limón, M.C., Codón, C.A., 2004. Biocontrol mechanisms of *Trichoderma* strains. Int. Microbiol. 7, 249–260.

Bjorkman, T., Blanchard, L.M., Harman, G.E., 1998. Growth enhancement of shrunken-2 (sh2) sweet corn by *Trichoderma harzianum* 1295-22: effect of environmental stress. J. Am. Soc. Hortic. Science 123 (1), 35–40.

Brenner, M.L., 1981. Modern methods for plant-growth substance analysis. Annu. Rev. Plant Physiol. 32, 511–538.

Casas-Flores, S., Rios-Momberg, M., Rosales-Saavedra, T., Martínez-Hernández, P., Olmedo-Monfil, V., Herrera-Estrella, A., 2006. Cross talk between a fungal blue-light perception system and the cyclic AMP signaling pathway. Eukaryot. Cell 5 (3), 499–506.

Catalano, V., Vergara, M., Hauzenberger, J.R., Seiboth, B., Sarrocco, S., Vannacci, G., et al., 2011. Use of a nonhomologous end-joining-deficient strain (delta-ku70) of the biocontrol fungus *Trichoderma virens* to investigate the function of the laccase gene lcc1 in sclerotia degradation. Curr. Genet. 57, 13–23.

Chaverri, P., Samuels, G.J., 2013. Evolution of habitat preference and nutrition mode in a cosmopolitan fungal genus with evidence of inter kingdom host jumps and major shifts in ecology. Evolution 67, 2823–2837.

Chaverri, P., Gazis, R., Samuels, G.J., 2011. *Trichoderma amazonicum*, a new endophytic species on *Hevea brasiliensis* and *H. guianensis* from the Amazon basin. Mycologia 103, 139–151.

Chen, J.L., Sun, S.Z., Miao, C.P., Wu, K., Chen, Y.W., Xu, L.H., et al., 2016. Endophytic *Trichoderma gamsii* YIM PH30019: a promising biocontrol agent with hyperosmolar, mycoparasitism, and antagonistic activities of induced volatile organic compounds on root-rot pathogenic fungi of *Panax notoginseng*. J. Ginseng Res. 40 (4), 315–324.

Contreras-Cornejo, H.A., Macías-Rodríguez, L., Cortés-Penagos, C., López-Bucio, J., 2009. *Trichoderma virens*, a plant beneficial fungus, enhances biomass production and promotes lateral root growth through an auxin-dependent mechanism in *Arabidopsis*. Plant Physiol. 149, 1579–1592.

Contreras-Cornejo, H.A., Macias-Rodriguez, L., Beltran-Pena, E., Herrera-Estrella, A., Lopez-Bucio, J., 2011. *Trichoderma*-induced plant immunity likely involves both hormonal and camalexin dependent mechanisms in *Arabidopsis thaliana* and confers resistance against necrotrophic fungi *Botrytis cinerea*. Plant Signal. Behav. 6, 1554–1563.

Contreras-Cornejo, H.A., Macías-Rodríguez, L., Alfaro-Cuevas, R., López-Bucio, J., 2014. *Trichoderma* spp. improve growth of *Arabidopsis* seedlings under salt stress through enhanced root development, osmolite production, and Na(+) elimination through root exudates. Mol. Plant Microb. Interact. 27, 503–514.

Cutler, H.G., Himmelsbach, D.S., Arrendale, R.F., Cole, P.D., Cox, R.H., 1989. Koninginin-a: a novel plant-growth regulator from *Trichoderma koningii*. Agric. Biol. Chem. 53 (10), 2605–2611.

De Meyer, G., Bigirimana, J., Elad, Y., Hofte, M., 1998. Induced systemic resistance in *Trichoderma harzianum* T39 biocontrol of *Botrytis cinerea*. Eur. J. Plant Pathol. 104, 279–286.

de Oliveira, A.L., Gallo, M., Pazzagli, L., Benedetti, C.E., Cappugi, G., Scala, A., et al., 2011. The structure of the elicitor cerato-platanin (CP), the first member of the CP fungal protein family, reveals a double wb-barrel fold and carbohydrate binding. J. Biol. Chem. 286, 17560–17568.

Delgado-Jarana, J., Sousa, S., Gonzalez, F., Rey, M., Llobell, A., 2006. ThHog1 controls the hyperosmotic stress response in *Trichoderma harzianum*. Microbiology 152 (6), 1687–1700.

Dickman, M.B., Yarden, O., 1999. Serine/threonine protein kinases and phosphatases in filamentous fungi. Fungal Genet. Biol. 26 (2), 99–117.

Djonovic, S., Pozo, M.J., Dangott, L.J., Howell, C.R., Kenerley, C.M., 2006. Sm1, a proteinaceous elicitor secreted by the biocontrol fungus *Trichoderma virens* induces plant defense responses and systemic resistance. Mol. Plant Microb. Interact. 19, 838–853.

Djonovic, S., Vargas, W.A., Kolomiets, M.V., Horndeski, M., Wiest, A., Kenerley, C.M., 2007. A proteinaceous elicitor Sm1 from the beneficial fungus *Trichoderma virens* is required for induced systemic resistance in maize. Plant Physiol. 145, 875–889.

Druzhinina, I.S., Seidl-Seiboth, V., Herrera-Estrella, A., Horwitz, B.A., Kenerley, C.M., Monte, E., et al., 2011. *Trichoderma*: the genomics of opportunistic success. Nat. Rev. Microbiol. 16, 749–759.

Elad, Y., 1994. Biological control of grape grey mould by *Trichoderma harzianum*. Crop Protect. 13, 35–38.

Glick, B.R., Penrose, D.M., Li, J.P., 1998. A model for the lowering of plant ethylene concentrations by plant growth-promoting bacteria. J. Theoret. Biol. 190 (1), 63–68.

Gravel, V., Martinez, C., Antoun, H., Tweddell, R.J., 2006. Control of greenhouse tomato root rot [*Pythium ultimum*] in hydroponic systems, using plant-growth promoting microorganisms. Can. J. Plant Pathol. 28, 475–483.

Gravel, V., Antoun, H., Tweddell, R.J., 2007. Growth stimulation and fruit yield improvement of greenhouse tomato plants by inoculation with *Pseudomonas putida* or *Trichoderma atroviride*: possible role of indole acetic acid (IAA). Soil Biol. Biochem. 39, 1968–1977.

Gresi, M., Kolarova, N., 1988. Membrane potential, ATP, and cyclic AMP changes induced by light in *Trichoderma viride*. Exp. Mycol. 12 (4), 295–301.

Hanson, L.E., Howell, C.R., 2004. Elicitors of plant defense responses from biocontrol strains of *Trichoderma virens*. Phytopathology 94, 171–176.

Harman, G.E., 2000. Myths and dogmas of biocontrol: changes in perceptions derived from research on *Trichoderma harzianum* T-22. Plant Dis. 84, 377–393.

Harman, G.E., 2006. Overview of mechanisms and uses of *Trichoderma* spp. Phytopathology 96 (2), 190–194.

Harman, G.E., 2011. Multifunctional fungal plant symbionts: new tools to enhance plant growth and productivity. New Phytol. 189, 647–649.

Harman, G.E., Howell, C.R., Viterbo, A., Chet, I., Lorito, M., 2004a. *Trichoderma* species-opportunistic, avirulent plant symbionts. Nat. Rev. Microbiol. 2, 43–56.

Harman, G.E., Petzoldt, R., Comis, A., Chen, J., 2004b. Interactions between *Trichoderma harzianum* strain T22 and maize inbred line Mo17 and effects of these interactions on diseases caused by *Pythium ultimum* and *Colletotrichum graminicola*. Phytopathology 94 (2), 147–153.

Harman, G.E., Herrera-Estrella, A.H., Benjamin, A., Matteo, L., 2012. Special issue: *Trichoderma*—from basic biology to biotechnology. Microbiology 58, 1–2.

Hermosa, R., Viterbo, A., Chet, I., Monte, E., 2012. Plant-beneficial effects of *Trichoderma* and of its genes. Microbiology 158 (1), 17–25.

Hoitink, H.A.J., Boehm, M.J., 1999. Biocontrol within the context of soil microbial communities: a substrate-dependent phenomenon. Annu. Rev. Phytopathol. 37, 427–446.

Hoitink, H.A.J., Madden, L.V., Dorrance, A.E., 2006. Systemic resistance induced by *Trichoderma* spp.: interactions between the host, the pathogen, the biocontrol agent, and soil organic matter quality. Phytopathology 96, 186–189.

Howell, R.C., 2003. Mechanisms employed by *Trichoderma* species in the biological control of plant diseases: the history and evolution of current concepts. Plant Dis. 87, 4–10.

Howell, C.R., Hanson, L.E., Stipanovic, R.D., Puckhaber, L.S., 2000. Induction of terpenoid synthesis in cotton roots and control of *Rhizoctonia solani* by seed treatment with *Trichoderma virens*. Phytopathology 90, 248–252.

Jablonka, E., 2013. Epigenetic inheritance and plasticity: the responsive germline. Progr. Biophys. Mol. Biol. 111, 99–107.

Kaewchai, S., Soytong, K., Hyde, K.D., 2009. Mycofungicides and fungal biofertilizers. Fungal Divers. 38, 25–50.

Keswani, C., Mishra, S., Sarma, B.K., Singh, S.P., Singh, H.B., 2014. Unraveling the efficient applications of secondary metabolites of various *Trichoderma* spp. Appl. Microbiol. Biotechnol. 98 (2), 533–544.

Khan, J., Ooka, J.J., Miller, S.A., Madden, L.V., Hoitink, H.A.J., 2004. Systemic resistance induced by *Trichoderma hamatum* 382 in cucumber against Phytophthora crown rot and leaf blight. Plant Dis. 88, 280–286.

Kleifeld, O., Chet, I., 1992. *Trichoderma harzianum*—interaction with plants and effect on growth response. Plant Soil 144, 267–272.

Kodsueb, R., McKenzie, E.H.C., Lumyong, S., Hyde, K.D., 2008. Diversity of saprobic fungi on Magnoliaceae. Fungal Divers. 30, 37–53.

Krause, M.S., Madden, L.V., Hoitink, H.A.J., 2001. Effect of potting mix microbial carrying capacity on biological control of Rhizoctonia damping-off of radish and Rhizoctonia crown and root rot of poinsettia. Phytopathology 91, 1116–1123.

Kronstad, J., De Maria, A., Funnell, D., Laidlaw, R.D., Lee, N., De Sá, M.M., et al., 1998. Signaling via cAMP in fungi: interconnections with mitogen-activated protein kinase pathways. Arch. Microbiol. 170 (6), 395–404.

Kubicek, C.P., Herrera-Estrella, A., Seidl-Seiboth, V., Martinez, D.A., Druzhinina, I.S., Thon, M., et al., 2011. Comparative genome sequence analysis underscores mycoparasitism as the ancestral life style of *Trichoderma*. Genom. Biol. 12, R40.

Kumar, A., Scher, K., Mukherjee, M., Pardovitz-Kedmi, E., Sible, G.V., Singh, U.S., et al., 2010. Overlapping and distinct functions of two *Trichoderma virens* MAP kinases in cell-wall integrity, antagonistic properties and repression of conidiation. Biochem. Biophys. Res. Commun. 398 (4), 765–770.

Leandro, L.F.S., Guzman, T., Ferguson, L.M., Fernandez, G.E., Louws, F.J., 2007. Population dynamics of *Trichoderma* in fumigated and compost amended soil and on strawberry roots. Appl. Soil Ecol. 35, 237–246.

Leitgeb, B., Szekeres, A., Manczinger, L., Vagvolgyl, C., Kredics, L., 2007. The history of alamethicin: a review of most extensively studied peptaibol. Chem. Biodivers. 4, 1027–1051.

Lotan, T., Fluhr, R., 1990. Xylanase, a novel elicitor of pathogenesis related proteins in tobacco, uses a nonethylene pathway for induction. Plant Physiol. 93, 811–817.

Luo, Y., Zhang, D.D., Dong, X.W., Zhao, P.B., Chen, L.L., Song, X.Y., et al., 2010. Antimicrobial peptaibols induce defense responses and systemic resistance in tobacco against tobacco mosaic virus. FEMS Microbiol. Lett. 313, 120–126.

Mastouri, F., Björkman, T., Harman, G.E., 2010. Seed treatment with *Trichoderma harzianum* alleviates biotic, abiotic, and physiological stresses in germinating seeds and seedlings. Phytopathology 100 (11), 1213–1221.

Mendoza-Mendoza, A., Rosales-Saavedra, T., Cortes, C., Castellanos-Juarez, V., Martinez, P., Herrera-Estrella, A., 2007. The MAP kinase TVK1 regulates conidiation, hydrophobicity and the expression of genes encoding cell wall proteins in the fungus *Trichoderma virens*. Microbiology 153 (7), 2137–2147.

Mukherjee, M., 2012. *Trichoderma* genes involved in interactions with fungi and plants. In: Gupta, V.K., Ayyachamy, M. (Eds.), Biotechnology of Fungal Genes. Taylor and Francis, pp. 153–171.

Mukherjee, P.K., Latha, J., Hadar, R., Horwitz, B.A., 2003. TmkA, a mitogen-activated protein kinase of *Trichoderma virens*, is involved in biocontrol properties and repression of conidiation in the dark. Eukaryot. Cell 2 (3), 446–455.

Mukherjee, P.K., Latha, J., Hadar, R., Horwitz, B.A., 2004. Role of two G-protein alpha subunits, TgaA and TgaB, in the antagonism of plant pathogens by *Trichoderma virens*. Appl. Environ. Microbiol. 70 (1), 542–549.

Mukherjee, M., Mukherjee, P.K., Kale, S.P., 2007. cAMP signaling is involved in growth, germination, mycoparasitism and secondary metabolism in *Trichoderma virens*. Microbiology 153 (6), 1734–1742.

Mukherjee, P.K., Buensanteai, N., Moran-Diez, M.E., Druzhinina, I.S., Kenerley, C.M., 2012a. Functional analysis of non-ribosomal peptide synthetases (NRPSs) in *Trichoderma virens* reveals a polyketide synthase (PKS)/NRPS hybrid enzyme involved in induced systemic resistance response in maize. Microbiology 158, 155–165.

Mukherjee, M., Mukherjee, P.K., Horwitz, B.A., Zachow, C., Berg, G., Zeilinger, S., 2012b. *Trichoderma*–plant–pathogen interactions: advances in genetics of biological control. Indian J. Microbiol. 52 (4), 522–529.

Mukherjee, A.K., Kumar, A.S., Kranthi, S., Mukherjee, P.K., 2014. Biocontrol potential of three novel *Trichoderma* strains: isolation, evaluation and formulation. 3 Biotech 4 (3), 275–281.

Naseby, D.C., Pascual, J.A., Lynch, J.M., 2000. Effect of biocontrol strains of *Trichoderma* on plant growth, *Pythium ultimum* populations, soil microbial communities and soil enzyme activities. J. Appl. Microbiol. 88, 161–169.

Nemčovič, M., Farkaš, V., 1998. Stimulation of conidiation by derivatives of cAMP in *Trichoderma viride*. Folia Microbiol. 43 (4), 399–402.

Nieto-Jacobo, M.F., Steyaert, J.M., Salazar-Badillo, F.B., Nguyen, D.V., Rostás, M., Braithwaite, M., et al., 2017. Environmental growth conditions of *Trichoderma* spp. affects indole acetic acid derivatives, volatile organic compounds, and plant growth promotion. Front. Plant Sci. 8, 102.

Pandey, P.K., Yadav, S.K., Singh, A., Sarma, B.K., Mishra, A., Singh, H.B., 2012. Cross-species alleviation of biotic and abiotic stresses by the endophyte *Pseudomonas aeruginosa* PW09. J. Phytopathol. 160, 532–539.

Pandey, P.K., Singh, S., Singh, A.K., Samanta, R., Yadav, R.N.S., Singh, M.C., 2016. Inside the plant: bacterial endophytes and abiotic stress alleviation. J. Appl. Nat. Sci. 8 (4), 1899–1904.

Pandey, P.K., Singh, S., Singh, M.C., Singh, A.K., Pandey, P., Pandey, A.K., et al., 2017. Inside the plants: bacterial endophytes and their natural products. Int. J. Curr. Microbiol. Appl. Sci. 6 (6), 33–41.

Pandey, P.K., Singh, S., Singh, M.C., Singh, A.K., Yadav, S.K., Pandey, A.K., et al., 2018. Diversity, ecology, and conservation of fungal and bacterial endophytes. In: Sharma, S., Varma, A. (Eds.), Microbial Resource Conservation. Springer, Cham, pp. 393–430.

Pierik, R., Tholen, D., Poorter, H., Visser, E.J.W., Voesenek, L.A.C.J., 2006. The Janus face of ethylene: growth inhibition and stimulation. Trends Plant Sci. 11 (4), 176–183.

Piotrowski, M., Volmer, J.J., 2006. Cyanide metabolism in higher plants: cyanoalanine hydratase is a NIT4 homolog. Plant Mol. Biol. 61, 111–122.

Reithner, B., Brunner, K., Schuhmacher, R., Peissl, I., Seidl, V., Krska, R., et al., 2005. The G protein α subunit Tga1 of *Trichoderma atroviride* is involved in chitinase formation and differential production of antifungal metabolites. Fungal Genet. Biol. 42 (9), 749–760.

Reithner, B., Schuhmacher, R., Stoppacher, N., Pucher, M., Brunner, K., Zeilinger, S., 2007. Signaling via the *Trichoderma atroviride* mitogen-activated protein kinase Tmk1 differentially affects mycoparasitism and plant protection. Fungal Genet. Biol. 44 (11), 1123–1133.

Richards, C.L., Bossdorf, O., Pigliucci, M., 2010. What role does heritable Epigenetic variation play in phenotypic evolution? Bioscience 60, 232–237.

Rocha-Ramírez, V., Omero, C., Chet, I., Horwitz, B.A., Herrera-Estrella, A., 2002. *Trichoderma atroviride* G-protein α-subunit gene tga1 is involved in mycoparasitic coiling and conidiation. Eukaryot. Cell 1 (4), 594–605.

Salas-Marina, M.A., Silva-Flores, M.A., Uresti-Rivera, E.E., Castro-Longoria, E., Herrera-Estrella, A., Casas-Flores, S., 2011. Colonization of Arabidopsis roots by *Trichoderma atroviride* promotes growth and enhances systemic disease resistance through jasmonic acid/ethylene and salicylic acid pathways. Eur. J. Plant Pathol. 131, 15–26.

Samuels, G.J., 2006. *Trichoderma*: systematics, the sexual state, and ecology. Phytopathology 96, 195–206.

Sarma, B.K., Yadav, S.K., Patel, J.S., Singh, H.B., 2014. Molecular mechanisms of interactions of *Trichoderma* with other fungal species. Open Mycol. J. 8, 140–147.

Sarma, B.K., Yadav, S.K., Singh, S., Singh, H.B., 2015. Microbial consortium-mediated plant defense against phytopathogens: readdressing for enhancing efficacy. Soil Biol. Biochem. 87, 25–33.

Schmoll, M., 2008. The information highways of a biotechnological workhorse–signal transduction in *Hypocrea jecorina*. BMC Genom. 9 (1), 430.

Segarra, G., Casanova, E., Bellido, D., Odena, M.A., Oliveira, E., Trillas, I., 2007. Proteome, salicylic acid, and jasmonic acid changes in cucumber plants inoculated with *Trichoderma asperellum* strain T34. Proteomics 7, 3943–3952.

Seidl, V., Schmoll, M., Scherm, B., Balmas, V., Seiboth, B., Migheli, Q., et al., 2006. Antagonism of Pythium blight of zucchini by *Hypocrea jecorina* does not require cellulase gene expression but is improved by carbon catabolite derepression. FEMS Microbiol. Lett. 257 (1), 145–151.

Seidl, V., Song, L., Lindquist, E., Gruber, S., Koptchinskiy, A., Zeilinger, S., et al., 2009. Transcriptomic response of the mycoparasitic fungus *Trichoderma atroviride* to the presence of a fungal prey. BMC Genom. 10 (1), 567.

Shoresh, M., Gal-On, A., Leibman, D., Chet, I., 2006. Characterization of a mitogen-activated protein kinase gene from cucumber required for *Trichoderma*-conferred plant resistance. Plant Physiol. 142, 1169–1179.

Shoresh, M., Harman, G.E., Mastouri, F., 2010. Induced systemic resistance and plant responses to fungal biocontrol agents. Annu. Rev. Phytopathol. 48, 21–43.

Silva, R.N., da Silva, S.P., Brandão, R.L., Ulhoa, C.J., 2004. Regulation of N-acetyl-β-D-glucosaminidase produced by *Trichoderma harzianum*: evidence that cAMP controls its expression. Res. Microbiol. 155 (8), 667–671.

Simon, A., Sivasithamparam, K., 1989. Pathogen suppression: a case study in biological suppression of *Gaeumannomyces graminis* var, *tritici* in soil. Soil Biol. Biochem. 21, 331–337.

Singh, H.B., Singh, B.N., Singh, S.P., Sarma, B.K., 2013. Exploring different avenues of *Trichoderma* as a potent biofungicidal and plant growth promoting candidate—an overview. Annu. Rev. Plant Pathol. 5, 315–426.

Sivan, A., Harman, G.E., 1991. Improved rhizosphere competence in a protoplast fusion progeny of *Trichoderma harzianum*. J. Gen. Microbiol. 137, 23–29.

Stewart, A., Hill, R., 2014. Applications of *Trichoderma* in plant growth promotion. In: Gupta, V.K., Schmoll, M., Herrera-estrella, A., Upadhyay, R.S., Druzhinina, I., Tuohy, M.G. (Eds.), Biotechnology and Biology of Trichoderma. Elsevier publications, pp. 415–428.

Suárez, M.B., Vizcaíno, J.A., Llobell, A., Monte, E., 2007. Characterization of genes encoding novel peptidases in the biocontrol fungus *Trichoderma harzianum* CECT 2413 using the TrichoEST functional genomics approach. Curr. Genet. 51 (5), 331–342.

Todorovic, B., Glick, B.R., 2008. The interconversion of ACC deaminase and D-cysteine desulfhydrase by directed mutagenesis. Planta 229 (1), 193–205.

van Loon, L.C., Bakker, P.A.H.M., Pieterse, C.M.J., 1998. Systemic resistance induced by rhizobacteria. Annu. Rev. Phytopathol. 36, 453–483.

Vargas, W.A., Djonović, S., Sukno, S.A., Kenerley, C.M., 2008. Dimerization controls the activity of fungal elicitors that trigger systemic resistance in plants. J. Biol. Chem. 283, 19804–19815.

Vargas, W.A., Mandawe, J.C., Kenerley, C.M., 2009. Plant-derived sucrose is a key element in the symbiotic association between *Trichoderma virens* and maize plants. Plant Physiol. 151, 792–808.

Vargas, W.A., Crutcher, F.K., Kenerley, C.M., 2011. Functional characterization of a plant-like sucrose transporter from the beneficial fungus *Trichoderma virens*. Regulation of the symbiotic association with plants by sucrose metabolism inside the fungal cells. New Phytol. 189, 777–789.

Vásquez, M.M., César, S., Azcón, R., Barea, J.M., 2000. Interactions between arbuscular mycorrhizal fungi and other microbial inoculants (*Azospirillum*, *Pseudomonas*, *Trichoderma*) and effects on microbial population and enzyme activities in the rhizosphere of maize plants. Appl. Soil Ecol. 15, 261–272.

Vinale, F., Marra, R., Scala, F., Ghisalbert, E.L., Lorito, M., Sivasithamparam, K., 2006. Major secondary metabolotes produced by two commercial *Trichoderma* strains active against different phytopathogens. Lett. Appl. Microbiol. 43, 143–148.

Vinale, F., Sivasithamparam, K., Ghisalberti, E.L., Marra, R., Woo, S.L., Lorito, M., 2008. *Trichoderma*–plant–pathogen interactions. Soil Biol. Biochem. 40 (1), 1–10.

Viterbo, A., Horwotz, B.A., 2010. Mycoparasitism. In: Borkovich, K.A., Ebbole, D.J. (Eds.), Cellular and Molecular Biology of Filamentous Fungi. ASM Press, Herndon, pp. 676–694.

Viterbo, A., Harel, M., Horwitz, B.A., Chet, I., Mukherjee, P.K., 2005. *Trichoderma* mitogen-activated protein kinase signaling is involved in induction of plant systemic resistance. Appl. Environ. Microbiol. 71 (10), 6241–6246.

Viterbo, A., Wiest, A., Brotman, Y., Chet, I., Kenerley, C.M., 2007. The 18mer peptaibols from *Trichoderma virens* elicit plant defense responses. Mol. Plant Pathol. 8, 737–746.

Waghunde, R.R., Shelake, R.M., Sabalpara, A.N., 2016. *Trichoderma*: a significant fungus for agriculture and environment. Afr. J. Agric. Res. 11 (22), 1952–1965.

Wakelin, S.A., Sivasithamparam, K., Cole, A.L.J., Skipp, R.A., 1999. Saprophytic growth in soil of a strain of *Trichoderma koningii*. New Zeal. J. Agric. Res. 42, 337–345.

Whipps, J.M., 2001. Microbial interactions and biocontrol in the rhizosphere. J. Exp. Bot. 52, 487–511.

Wiest, A., Grzegorski, D., Xu, B., Goulard, C., Le Doan, S., Ebbole, D.J., et al., 2002. Identification of peptaibols from *Trichoderma virens* and cloning of a peptaibol synthetase. J. Biol. Chem. 277, 20862–20868.

Windham, M.T., Elad, Y., Baker, R., 1986. A mechanism for increased plant-growth induced by *Trichoderma* spp. Phytopathology 76 (5), 518–521.

Woo, L.S., Lorito, M., 2007. Exploiting the interactions between fungal antagonists, pathogens and the plant for biocontrol. In: Vurro, M., Gressel, J. (Eds.), Novel Biotechnologies for Biocontrol Agent Enhancement and Management. Springer, Dordrecht, pp. 107–130.

Woo, S.L., Formisano, E., Fogliano, V., Cosenza, C., Mauro, A., Turrà, D., et al., 2004. Factors that contribute to the mycoparasitism stimulus in *Trichoderma atroviride* strain P1. J. Zhejiang Univ. Sci. 30, 421.

Woo, S.L., Scala, F., Ruocco, M., Lorito, M., 2006. The molecular biology of the interactions between *Trichoderma* spp., phytopathogenic fungi, and plants. Phytopathology 96, 181–185.

Yadav, S.K., Dave, A., Sarkar, A., Singh, H.B., Sarma, B.K., 2013. Co-inoculated biopriming with *Trichoderma*, *Pseudomonas* and *Rhizobium* improves crop growth in *Cicer arietinum* and *Phaseolus vulgaris*. Int. J. Agric. Environ. Biotechnol. 6 (2), 255–259.

Yadav, S.K., Singh, S., Singh, H.B., Sarma, B.K., 2017. Compatible rhizosphere-competent microbial consortium adds value to the nutritional quality in edible parts of chickpea. J. Agric. Food Chem. 65 (30), 6122–6130.

Yedidia, I., Benhamou, N., Chet, I., 1999. Induction of defense responses in cucumber plants (*Cucumis sativus* L.) by the biocontrol agent *Trichoderma harzianum*. Appl. Environ. Microbiol. 65, 1061–1070.

Yedidia, I., Srivastva, A.K., Kapulnik, Y., Chet, I., 2001. Effect of *Trichoderma harzianum* on microelement concentrations and increased growth of cucumber plants. Plant Soil 235, 235–242.

Yoshioka, Y., Ichikawa, H., Naznin, H.A., Kogure, A., Hyakumachi, M., 2012. Systemic resistance induced in *Arabidopsis thaliana* by *Trichoderma asperellum* SKT-1, a microbial pesticide of seedborne diseases of rice. Pest Manage. Sci. 68, 60–66.

Zachow, C., Fatehi, J., Cardinale, M., Tilcher, R., Berg, G., 2010. Strain specific colonization pattern of *Rhizoctonia* antagonists in the root system of sugar beet. FEMS Microbiol. Ecol. 74 (1), 124–135.

Zeilinger, S., Omann, M., 2007. *Trichoderma* biocontrol: signal transduction pathways involved in host sensing and mycoparasitism. Gene Regul. Syst. Biol. 1, 227–234.

# Fungal endophytes: potential biocontrol agents in agriculture

*Ajay Kumar Gautam*[1] *and Shubhi Avasthi*[2]
[1]Faculty of Science, School of Agriculture, Abhilashi University, Mandi, Himachal Pradesh, India, [2]School of Studies in Botany, Jiwaji University Gwalior, Madhya Pradesh, India

## 14.1 Introduction

Endophytic microbes are an intriguing group of organisms that associate with the internal tissue of plant stems, petioles, roots, and leaves without causing apparent symptoms of disease. These microbes reside within the various tissues and organs of terrestrial and some aquatic plants for all or part of their own life cycle; these infections are inconspicuous and the infected host tissues are at least transiently symptomless (Stone et al., 2000; Faeth and Fagan, 2002; Mandyam and Jumpponen, 2005; Yuan et al., 2007; Ilyas et al., 2009; Rodriguez et al., 2009; Shankar and Krishnamurthy 2010; Ding et al., 2009). In 1866 De Barry introduced the term endophyte for organisms that live inside plant tissues (Taylor and Taylo, 2000). He included all organisms that live asymptomatically inside plant tissues at some stage of their own life cycle (Petrini, 1991). The term endophyte is applied broadly in accord with its literal definition and the wide spectrum of potential hosts and inhabitants. The interactions of mycorrhizal fungi with the roots of their hosts are designated as endophytic by some authors; however, Brundrett (2006) distinguishes mycorrhizal from endophytic interactions. He distinguished mycorrhizal interactions possessing synchronized plant–fungus development and nutrient transfer at specialized interfaces.

An endophyte is a bacterium or fungus that has established equilibrium with their host over the course of evolution. This relationship with the plant host may vary from symbiotic to bordering on pathogenic. The types of symbiosis adapted by endophytes include facultative saprobic, parasitic, exploitative, and mutualistic (Clay and Schardl, 2002). However, in most cases, this relationship is asymptomatic and may even benefit the plant hosts. Some endophytes may exhibit a mutualistic interaction with one plant species, but not with another (Hardoim et al., 2015), while some species associate with a wide range of hosts. In addition, association of endophytes with specific host tissues has also been observed. Endophytes are ubiquitous and have been found in all species of plants studied to date; however, relationships between most endophytes and their plant hosts are still not well understood.

Endophytic microbes, especially endophytic fungi, have been recognized as potential sources of a diverse array of bioactive secondary metabolites (Tan and

**Role of Plant Growth Promoting Microorganisms in Sustainable Agriculture and Nanotechnology.**
**DOI: https://doi.org/10.1016/B978-0-12-817004-5.00014-2**

Zou, 2001; Gangadevi and Muthumary, 2008; Sonaimuthu et al., 2010; Pradeep et al., 2010; Bhardwaj and Agrawal, 2014). Although these fungal endophytes usually associate with their hosts asymptomatically, in many cases they may produce beneficial or pathogenic effects (Photita et al., 2001; Neubert et al., 2006; Wei et al., 2007). Endophytic fungi have been found to create interesting associations with plants by secreting selected secondary metabolites that promote growth (Dai et al., 2008), improve resistance to stress (Lewis, 2004; Malinowski et al., 2004), and protect from diseases and insects (Wilkinson et al., 2000; Tanaka et al., 2005; Vega et al., 2008). The secondary metabolites produced by endophytic fungi have also been identified as sources of anticancer, insecticidal, antidiabetic, immunosuppressive, and biocontrol compounds. These novel compounds with unusual and valuable chemistry and biology are helpful remedies for health problems of humans, animals, and plants. Because of the production of a wide spectrum of useful chemical compounds, the endophyte has been proposed as "the chemical synthesizer inside the plant" (Owen and Hundley 2004).

Endophytic fungi contribute significantly to fungal diversity on earth. As they affect the structure and community of plants, placed them in unique group of fungi. (Sanders, 2004; Hyde and Soytong, 2008). They have massive biological diversity and distribution, especially in temperate and tropical rainforests. This fungal group is studied extensively for detection of novel compounds useful in the industrial, pharmaceutical, medical, and agricultural sectors. The use of these chemical compounds in agriculture is a noteworthy discovery that underscores their potential use as biocontrol agents (Mane and Vedamurthy, 2018).

Fungal endophytes are well known to contribute to plant fitness by enabling adaptation of the plant host to biotic and abiotic stresses. These microbes have established a symbiotic relationship with their respective hosts and confer pest and disease resistance to host plants by enhancing their growth, increasing their fitness to resist various stresses, and promoting accumulation of secondary metabolites. These critical changes are consequences of increased production of bioactive components in their hosts which in turn support defense mechanisms of plants against pests. Currently, endophytic fungi are the subject of rigorous study focused on the isolation and identification of bioactive compounds and the application of these compounds in management of agricultural crop pathogens.

In recent decades endophytic fungi have attracted the attention of mycologists all over the world because of the benefits they provide to host plants in development and defense. These organisms are a source of secondary metabolites of potential interest due to their important roles in the regulation of plant communities and associated herbivores. Plant–endophyte interactions and their potential use in pest control are receiving increased attention; however, research is ongoing and much remains to be discovered. The present chapter is an effort to compile the studies of endophytic fungi with reference to their biological characteristics, species diversity, modes of infection and transmission, production of bioactive compounds, role in management of plant diseases, and their advantages over chemical methods of crop protection. We hope this chapter will provide readers with information useful for understanding the complexity of endophyte–plant relationships, the possibility of enhancing plant defense, and the role of endophytes in the control of crop pests and diseases.

## 14.2 Biological characteristics of fungal endophytes

The word "endophyte" comes from the Greek "endon," meaning inside or within and "phyton" meaning plant. De Bary (1866) defined endophytes for the first time as "any organisms that grow within plant tissues" (i.e., spend all or part of their life cycle colonizing the healthy tissues of the host plant, either inter- or intracellularly). However, this definition continues to change as research on endophytes advances (Bacon and White, 2000). The most suitable definition was provided by Petrini (1991), "any organism that at some part of its life cycle colonizes the internal plant tissues without causing any type of harm to the host plant." Fungal endophytes, their range of symbiotic relationships, and their vast applications have been highlighted by the extensive studies of numerous researchers. Generally fungal endophytes may be obligate or facultative endophytes and are associated with all types of plants. Endophytic fungi exhibit numerous biological characteristics that are affected by various factors, such as host species, host developmental stage, inoculum density, and environmental conditions (Patle et al., 2018). A diagrammatic preview of the biological characteristics of endophytic fungi is presented in Figs. 14.1 and 14.2.

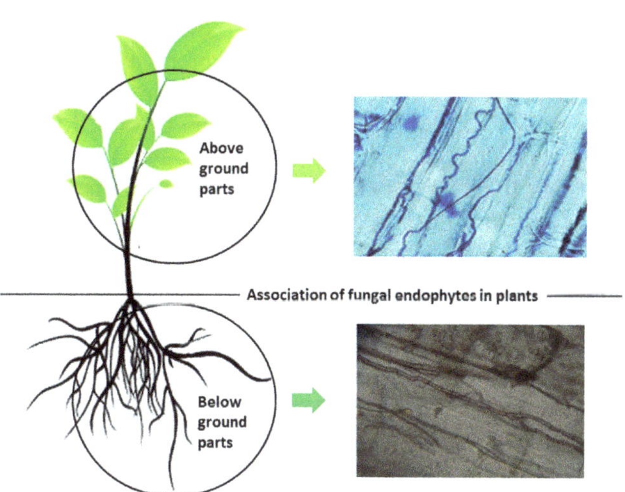


**Figure 14.1** General presentation on association of endophytic fungi with plant tissues.

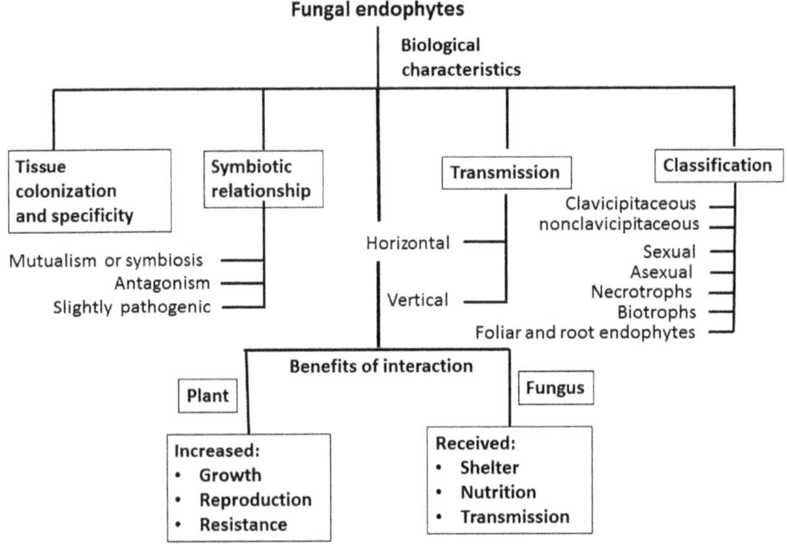


**Figure 14.2** Biological characteristics of endophytic fungi.

## 14.2.1 Tissue colonization and specificity

Endophytic fungi have been recovered from almost all classes of vascular plants and grasses examined to date from environments including hot deserts, arctic tundras, mangroves, temperate and tropical forests, grasslands, savannas, and croplands (Zhang et al., 2006). Detailed literature review of last 100 years revealed that nearly all plants in natural ecosystems are symbiotic with mycorrhizal fungi and/or fungal endophytes (Khiralla et al., 2017). The first endophytic fungus described was *Sphaeria typhena* by Person in 1772 which is now known as *Epichloe typhina* (Pers.) Tul. However, the paleontological studies of fossilized plants revealed the association of endophytes fungi even with 400-million-year-old plant fossils (Khan, 2007). This was proven by Krings et al. (2007) who described three fungal endophytes in petrographic thin sections of *Rhynie chert* and *Nothia aphylla*. Besides vascular plants, endophytes are also known to associate with mosses and other nonvascular plants, ferns and other seedless plants, conifers, and flowering plants. Endophytes colonize internal plant tissues under the epidermal cell layers and live within the intercellular spaces of the tissues without causing any obvious harm or symptomatic infection to the plant host. It appears that these fungi may penetrate and interact with living cells asymptomatically (Strobel, 2003; Selim et al., 2012). Endophytic fungi associate with nearly 300,000 land plant species, with each plant hosting one or more fungal species (Arnold, 2008). The assemblages of fungi that colonize plant roots are diverse in contrast to endophytic growth in the aboveground plant parts (Vandenkoornhuyse et al., 2002).

Research on endophytes has revealed that these microorganisms may inhabit specific tissues or areas of a plant host, while other species may systematically reside in different tissues of the same plant hosts. Similarly, assemblages of fungal endophytes may be found in specific plant parts and tissues, demonstrating the characteristic specificity of tissue colonization by these organisms. Differences in the assemblages of endophytic fungi in various plant parts have been documented. Similarly, variation in assemblages of these fungi in the outer bark and inner tissue, such as xylem, has also been investigated by many researchers. Several studies have been conducted on colonization by the most common endophytic fungi in different hosts growing distantly or adjacent to one another. Generally, the host–endophyte relationship can be explained in terms of host specificity, host recurrence, host selectivity, or host preference (Cohen, 2006; Selim et al., 2012; Sharma and Gautam, 2018). Schulz and Boyle (2005) proposed a hypothesis that a balance between endophytic virulence and plant defense response results in asymptomatic colonization. If this balance of host–pathogen interaction becomes disturbed, either disease of the plant results or the fungus is killed.

It has been found that both habitat and host cause variations in the assemblages of colonizing endophytes, with some endophytes adapting to very specialized habitats. The diversity and density of colonization are believed to increase during the course of the vegetation period. Particularly, asexual sporulation increases in autumn at the end of the vegetation period (Smalla et al., 2001). A study carried out by Agostinelli et al. (2018) revealed that xylem is a more selective substrate for endophytes than bark. The endophyte assemblages in xylem are correlated with the degree of host vitality. It has been noted that high vitality of host trees provides reduced quality habitats to wood associated endophytes.

### 14.2.2 Symbiotic relationship with host plants

The interaction between endophytic fungi and their host plants plays a fundamental role in determining their relationship. Their relationship can be described in terms of host specificity, host recurrence, host selectivity, or host preference (Zhou and Hyde, 2001; Cohen, 2006). Symbiosis is the major relationship established by endophytes with their host. In a symbiotic relationship the endophyte infects the tissues of healthy plants for all or nearly all of their life cycle without producing any disease symptoms. This interaction remains asymptomatic for many years and only becomes parasitic when the host is stressed (Firáková et al., 2007; Limsuwan et al., 2009). In such interactions plants strictly limit the growth of endophytes, and these endophytes use many mechanisms to gradually adapt to their living environments. In order to maintain stable symbiosis, endophytes produce several compounds that promote growth of plants and help them adapt better to the environment (Lee et al., 2004; Das and Varma, 2009; Dudeja et al., 2012). A variety of relationships ranging from mutualism or symbiosis to antagonism or slightly pathogenic can coexist between endophytes and their host plants (Wilson, 1995; Schulz and Boyle, 2005; Arnold, 2007; Hyde and Soytong, 2008). The fossil record indicates the association of fungal symbionts with plants can be traced back to approximately 400 million years ago, the Ordovician Period when plants are believed to have first become established on land (Redecker et al., 2000; Remy et al., 1994). Symbiotic metabolism between two partners can entail interaction on many levels including induction of metabolic pathways in the partner organism (i.e., endophyte induces host metabolism and host induces endophyte metabolism), which leads to sharing of parts of a specific pathway (Ludwig-Müller, 2015). Investigation of the genetic basis of symbiotic lifestyle of fungi revealed that fungal plant pathogens could also exhibit nonpathogenic lifestyles, potentially dependent on genetic mutations. In symbiotic relationships endophytic bioactive compounds are less toxic to host cells. Symbiotic lifestyle switching adds a new dimension to fungal taxonomy and fungal ecology (Alvin et al., 2014).

### 14.2.3 Transmission and acquisition of endophytes

Endophytes can colonize nearly all plant parts and endophyte populations are highly diverse. Variability in the endophyte population depends on various factors such as host species, host developmental stage, inoculum density, and environmental conditions (Dudeja and Giri, 2014). Transmission of these highly diverse fungi is an important aspect that also determines their variability in plant species. Endophytic fungi can be transmitted to new host plants either horizontally or vertically. The means of propagation and the transmission routes of many endophytic fungi are still under investigation. Broadly, horizontal transmission includes the transfer of endophyte inoculum to another plant, while vertical transmission is production of infected seed by an infected plant. Fungal endophytes reportedly use either of these routes for their transmission. Endophytes can be transmitted via both routes i.e. horizontally or vertically; however, preservation of their ability to produce ascospores for longer duration affects their transmission (Saikkonen et al., 2002).

The horizontal route of transmission is the predominant mechanism among endophytic fungi. The reproductive structures of endophytic fungi involved in horizontal transmission can be transmitted via the soil, through air movement, or by vectors. Production of inoculums in an asymptomatic host is still a matter of investigation. In light of the nature of fungal endophytes to live inside living plant tissues, it is surprising that some saprophytes have been shown the ability to infect new hosts. In a study carried out by Sánchez Márquez et al. (2007), saprophytic fungi were incubated with dead plant tissue and healthy tissue. Interestingly, these saprophytes were able to produce fruitifications on both substrates. These results have opened a new field to study leaf litter as a source endophytic inoculum (Bills and Polishook, 1994). Fungal spores formed by endophytic fungi are transported via wind or rain dispersal, or by a vector from plant to plant and colonize new host plants following an infectious process. Infection likely occurs via formation of infectious structures, such as appresoria, or directly via the hyphal penetration of plant tissues. Horizontal transmission via ascospores has been well documented in sexual species of *Epichloë* (Schardl et al., 2004; Gao and Mendgen, 2006). It has been observed that fungal species with weak pathogenicity against insect herbivores are primarily transmitted horizontally and primarily found in woody and herbaceous plants (Higgins et al., 2007; Sieber, 2007).

In contrast, vertically transmitted endophytes are transferred directly from the host plants (parents) to their progenies (Saikkonen et al., 2002). Vertical transmission involves dispersion through seeds. These fungi cannot produce reproductive structures on their hosts, but develop and gather their mycelium in the aleurone layer near the embryo of a developing seed. Upon seed germination, young seedlings are colonized by hyphae and then give rise to asymptomatic infected plants. Therefore, these endophytic species are vertically transmitted in a fashion similar to maternally inherited characteristics (Schardl et al., 2004). Accordingly, the incidence of these endophytes is very high in natural populations of their hosts (Arroyo García et al., 2002). True endophytes (such as species of the genus *Neotyphodium*) are mostly vertically transmitted through seeds from one plant to another (Hartley and Gange, 2009). Such species vertically transmitted through seeds are generally referred as seed-transmitted endophytes (Dongyi and Kelemu, 2004; Bennett et al., 2008; Schardl et al., 2013; Quesada-Moraga et al., 2014).

### 14.2.4 Classification

Fungal endophytes can be classified broadly into different ecological categories based upon host plant and the tissues they inhabit, modes of reproduction and transmission, source of nutrition, as well as expression of infection. Based on diversity, endophytic fungi mainly consist of members of the phylum Ascomycota; however, some taxa of Basidiomycota, Zygomycota and Oomycota are also endophytes (Zheng and Jiang, 1995; Sinclair and Cerkauskas, 1996; Stone et al., 2004; Rajamanikyam et al., 2017). There are various methods of grouping fungal endophytes, such as transmission mode and host type. Broadly, fungal endophytes can be classified into two ecological categories, clavicipitaceous and nonclavicipitaceous. While clavicipitaceous fungal endophytes most commonly inhabit grasses, the nonclavicipitaceous are found to be associated chiefly with vascular and nonvascular plant species (Rodriguez et al., 2009; Bamisile et al., 2018).

A number of criteria have been suggested by researchers to classify fungal endophytes further. The criteria suggested to classify fungal endophytes into subclasses of clavicipitaceous and nonclavicipitaceous are: host range, mode of reproduction, part of plant colonized, mode of transmission, source of nutrition, and ability to express symptoms in the host plant (Brem and Leuchtmann, 2001; Saikkonen et al., 2002; Rodriguez et al., 2009; Purahong and Hyde, 2011; Mane and Vedamurthy, 2018). Detailed classifications of endophytic fungi based on various criteria are presented in Table 14.1.

## 14.3 Species diversity of fungal endophytes

Fungal endophytes are ubiquitous organisms, as they have been isolated from almost every plant species examined (Stone et al., 2000). The relationship between these fungi and plant hosts varies from symbiotic to pathogenic. The fungi stimulate

**Table 14.1** Classification of endophytic fungi based on various criteria.

| Criteria | Categories | Description/examples | References |
|---|---|---|---|
| Host plants | Clavicipitaceous | Infects some grasses (e.g., *Penicillium* sp., *Phoma* sp., *Fusarium* sp.) | Rodriguez et al. (2009), González-Teuber et al. (2017) |
| | Nonclavicipitaceous | Common in vascular and nonvascular plant species (e.g., *Curvularia protuberate*, *Fusarium culmorum*, *agricultural fields Colletotrichum* spp.) | Redman et al. (2002), Márquez et al. (2007) |
| Mode of reproduction | Sexual | *Epichloë* sp. | Brem and Leuchtmann (2001), Moon et al. (1999), Leuchtmann et al. (2000), Schardl and Craven (2003) |
| | Asexual | *Neotyphodium* sp. (formally *Acremonium*) | |
| Mode of transmission | Vertically transmitted | Mostly transmitted through seeds. (e.g., *Neotyphodium*, *Epichloë*, etc.) | Saikkonen et al. (2002), Schardl et al. (2013), Quesada-Moraga et al. (2014), Hartley and Gange (2009), Higgins et al. (2007), Sieber (2007) |
| | Horizontally transmitted | Generally exhibit weak pathogenicity against insect herbivores (e.g., *Hormonema dematioides*, *Rhodotorula minuta*) | Pirttila et al. (2002) |
| Nutrition source | Necrotrophs | Mortify host cells in order to grow on the dead tissues (e.g., *Leptosphaeria maculans*) | Promputtha et al. (2007), Purahong and Hyde (2011), Delaye et al. (2013), Junker et al. (2012) |
| | Biotrophs | Develop and obtain nutrients within the tissue of a living host (e.g., *L. maculans* found in *Arabidopsis thaliana* plants became a necrotrophic pathogen when the plant was stressed) | Junker et al. (2012) |
| Expression of infection | Symptomatic | In rare cases normally asymptomatic endophytic fungi express symptoms (i.e., latent pathogens; e.g., *Fusarium* spp.) | Pinto et al. (2000), Schulz and Boyle (2005), Hyde and Soytong (2008), Porras-Alfaro and Bayman (2011) |
| | Asymptomatic | Nearly types of fungi | Selim et al. (2012) |
| Host plant part affected | Root endophytes | *Fusarium* spp., *Metarhizium* spp., *Piriformospora indica*, *Glomus* spp. | Gautam et al. (2013), Gautam (2014), Impullitti and Malvick (2013), Yan et al. (2015), Behie et al. (2015), Wilberforce et al. (2003), Wyrebek et al. (2011) |
| | Foliar endophytes | *Aspergillus* spp., *Penicillium* spp., *Alternaria* spp., *Colletotrichum gloeosporioide*, *Colletotrichum truncatum*, *Fusarium oxysporum*, *Phoma* spp. | Gonzalez-Teuber et al. (2014), González-Teuber et al. (2017) |

plant growth, increase disease resistance, improve the plant's ability to withstand environmental stresses, and recycle nutrients (Sturz and Nowak, 2000). These fungi colonize specific host plants, parts, or tissues, or may inhabit a wide range of hosts. Like the range of hosts they inhabit, endophytic fungi are also highly diverse in terms of the fungal species capable of an endophytic lifestyle. Almost all groups or divisions of fungi can act as endophytes and infect plants of all ages as well as stage. Fungal endophytes can be classified broadly into different ecological categories based upon their species or their functional roles. Based on diversity, endophytic fungi mainly consist of members of Ascomycota; however, some taxa of Basidiomycota, Zygomycota, and Oomycota are also endophytic (Zheng and Jiang, 1995; Sinclair and Cerkauskas, 1996; Stone et al., 2004; Rajamanikyam et al., 2017). A list of fungal endophytes isolated from host plants during the 21st century is provided in Table 14.2.

## 14.4 Fungal endophytes: chemical synthesizers inside plants

Fungal endophytes utilize plants as a shelter where they live asymptomatically within living tissues in a symbiotic and likely mutualistic relationship. They grow and establish themselves in host plants and simultaneously synthesize bioactive compounds. These bioactive compounds can be used by the plant during environmental stress and for defense against phytopathogens. Many of these compounds are also very useful in numerous industrial applications. Some of these compounds have potential applications in medicine, agriculture, and the food industry. A large number of secondary metabolites have been extracted, isolated and characterized from endophytic fungi throughout the world. These metabolites include alkaloids, steroids, terpenoids, peptides, polyketides, flavonoids, quinols, phenols, as well as some halogenated compounds (Tan and Zou, 2001; Strobel et al., 2004; Ratnaweera et al., 2015; Dissanayake et al., 2016a,b). These bioactive compounds have been reported to have cytotoxic, antimicrobial, antidiabetic, insecticidal, and anticancer activities (Zhang et al., 2006; Desale and Bodhankar, 2013). Because they synthesize a wide range of useful bioactive compounds, fungal endophytes are designated "chemical synthesizers inside plants" (Owen and Hundley, 2004).

### *14.4.1 Mode of infection by endophytic fungi*

Fungal endophytes are diverse and widespread symbionts that occur asymptomatically in the living tissues of all lineages of plants. They spend at least one phase of their life cycle by colonizing plant tissues intracellularly or intracellularly. This colonization remains asymptomatic and is carried out by fungi from several groups including clavicipitaceous endophytes (colonize above ground tissues of grasses), nonclavicipitaceous endophytes (colonize vascular and nonvascular plant species), and so-called dark septate endophytes (colonize roots). Several aspects (e.g., plant

**Table 14.2** Endophytic fungi isolated from host plants during the 21st century.

| Endophytic fungi | Host plants | References |
|---|---|---|
| *Alternaria alternata, A. tenuissima, Aspergillus flavus, A. fumigatus, A. niger, A. oryzae, A. parasiticus, Cladosporium cladosporioides, C. herbarum, Curvularia siddiquii, C. vercuculosa, Drechslera* sp., *Epicoccum nigrum, Fusarium moniliforme, F. solani, Helminthosporium* sp., *Humicola grisea, Penicillium citrinum, P. notatum, Rhizopus nigricans* | *Withania somnifera* | Alwadi and Baka (2001), Gautam (2014) |
| *Cladosporium* sp., *Penicillium* sp., *Trichoderma* sp. | *Populus trichocarpa* | Huang et al. (2018) |
| *Pythium ultimum, Sclerotium oryzae, Rhizoctonia solani, Pyricularia oryzae* | *Zea mays, Oryza sativa* | Potshangbam et al. (2017) |
| *Aspergillusis niger, A. flavus, A. nidulans, Penicillium chrysogenum* | *Cannabis sativa* | Gautam et al. (2013), Meenatchi et al. (2016) |
| *P. citrinum, Phoma* sp., *Rhizopus* sp., *Colletotrichum* sp., *Cladosporium* sp., *Curvularia* sp. | | |
| *Gloeosporium musae, Myxosporium* spp., *Deightoniella torulosa, Alternaria tenuis, Sphaceloma* spp., *Aureobasidium* spp., *Melida* spp., *Uncinula* spp., *Penicillium* spp., *Aspergillus* spp., *Sarcinella* spp., *Cladosporium* sp., *Cephalosporium* sp. | *Musa acuminata* | Cao et al. (2002) |
| *Rhizopus stolonifer, Drechslera, Cladosporium, Curvularia lunata, Chaetomium, Penicillim* spp., *Fusarium, Ulocladium consortiale, Mucor hiemalis, Scytalidium thermophilum, Phoma solani, Taeniolella exilis, Botryodiplodia theobromae* | *Boswellia sacra* | El-Nagerabi et al. (2014) |
| *Alternaria* sp., *Colletotrichum* sp., *Nigrospora* sp., *Phomopsis* sp., *Fusarium* sp., *Penicillium* sp., *Schizophyllum commune* | *Tectona grandis, Samanea saman* | Chareprasert et al. (2006) |
| *Balansia* sp., *Pestalotiopsis versicolor, Aspergillus aculeatus, A. carbonarius, A. flavus, A. japonicas, A. niger, A. pulvurulentus, F. moniliforme, Gilmaniella* sp., *Nigrospora* sp., *Penicilliumcitrinum, P. herquei, P. janthinellum, P. rubrum, P. rugulosum, P. simplicissimum, P. implicatum, Trichoderma koningii, T. nivale* | *Melia azedarach* | Geris dos Santos et al. (2003) |
| *Cladosporium* sp., *Acremonium* sp., *Trichoderma* sp., *Monilia* sp., *Fusarium* sp., *Spicaria* sp., *Humicola* sp., *Rhizoctonia* sp., *Cephalosporium* sp., *Botrytis* sp., *Penicillium* sp., *Chalaropsis* sp.,*Geotrichum* sp. | *Cephalotaxus mannii* | Saithong et al. (2010) |
| *Alternaria alternata (Fr.) Keissl., Aspergillus flavus, A. niger, Chaetomium globosum, Chaetomium* sp., *Chloridium* sp., *Cochlonema* sp., *Colletotrichum* sp., *Curvularia* sp., *Drechslera* sp., *Fusarium* spp., *Penicillium* spp., *Gliomastix* sp., *Humicola* sp., *Nigrospora* sp., *Pestalotiopsis* spp., *Phoma eupyrena, Phoma* sp., *Phomopsis* sp., *Phyllosticta* sp., *Scytalidium* sp., *Trichoderma* sp., *Trichoderma* spp., *Verticillium* sp. | *Azadirachta indica* | Chutulo and Chalannavar (2018) |
| *Fusarium oxysporum, Fusarium solani, Emericella nidulans* | *Ipomea batatas, Taxus baccata* | Hipol (2012), Tayung et al. (2011), Mirjalili et al. (2012) |
| *Alternaria* sp., *Colletotrichum* sp., *Phomopsis* sp., *Xylaria* sp. | *Artemisia capillaris, Azadirachta indica, A. lactiflora* | Huang et al. (2009) |

| | | |
|---|---|---|
| *Glomerella* spp., *Diaporthae/Phomopsis* sp., *Alternaria* spp., *Cochliobolus* sp., *Cladosporium* sp., *Emericella* sp. | *Aegle marmelos, Coccinia indica, Moringa oleifera* | Gokul Raj et al. (2014) |
| *Alternaria* sp., *Cladosporium* sp., *Chaetomium* sp., *Curvularia* sp., *Drechslera* sp., *Scopulariopsis* sp., *Acremonium* sp., *Aspergillus* sp., *Colletotrichum* sp., *Fusarium* sp., *Paecilomyces* sp., *Penicillium* sp. | *Glycine max* | Pimentel et al. (2006) |
| *Aspergillus fumigatus, Colletotrichum gloeosporioides, Diaporthe discoidispora, Diaporthe pseudomangiferae, Nodulisporium* sp., *Penicillium* sp., *Pestalotiopsis* sp., *Phyllosticta capitalensis, Xylaria* sp. | *Mangroves* | Rajamani et al. (2018) |
| *Muscodor albus* | *Cinnamomum zeylanicum* | Ezra et al. (2004) |
| *Sporidiobolus* sp., *Rhodotorula* sp., *Pilidium concavum, Corynespora cassiicola, Neodeightonia subglobosa, Aspergillus awamori, Aspergillus* sp. | *Fragaria x ananassa* | Ezra et al. (2004) |
| *Alternaria* sp., *Cladosporium* sp., *Curvularia* sp., *Fusarium* sp., *Phaeoacremonium* sp., *Trichoderma* sp. | *Aquilaria malaccensis* | Premalatha and Kalra (2013) |
| *Aspergillus* sp., *Penicillium* sp., *Eurotiomycetes* sp., *Acremonium* sp., *Colletotrichum* sp., *Fusarium* sp., *Nodulisporium* sp., *Pestalotiopsis* sp. | *Marchantiapolymorpha* | Hipol et al. (2015) |
| *Acremonium* sp., *Colletotrichum* sp., *Cochliobolus* sp., *Fusarium* sp., *Hypocrea* sp., *Nemania* sp. | *Lycium chinense* | Paul et al. (2014) |
| *Clonostachys* sp., *Colletotrichum* sp., *Trichoderma* sp. | *Hevea brasiliensis* | Vaz et al. (2018) |
| *Aspergillus niger, Bipolaris maydis, Meyerozyma guilliermondii, Fusarium verticillioides* | *Ocimum sanctum* | Chowdhary and Kaushik (2015) |
| *Fusarium proliferatum, Fusarium* sp., *F. solani, C. lunata, Trichoderma atroviride, Calonectria gracilis, Rhizoctonia solani, Bionectria ochroleuca* | *Musa acuminata* | Zakaria et al. (2016) |
| *Penicillium chrysogenum, P. chrysogenum, Fusarium oxysporum, F. nygamai* | *Tamarix nilotica, Cressa cretica* | Gashgari et al. (2016) |
| *Fusarium* sp., *Phaeoacremonium* sp., *Acremonium* sp., *Cladosporium* sp., *C. gloeosporioides Penz., Phomopsis archeri, A. flavus, Nigrospora sphaerica* | *Sesbania grandiflora* | Powthong et al. (2013) |
| *Phomopsis* sp., *Alternaria raphani, M. hiemalis, Monodictys paradoxa, Aspergillus fumigates, A. japonicas, A. niger, Fusarium semitectum* | *Vitex negundo* | Monali and Bodhankar (2013) |
| *Glomerella acutata, Epicoccum nigrum, Diaporthe* spp., *Penicillium chloroleucon, Diaporthe endophytica, Mucor circinelloides* | *Vitex negundo* | Sibanda et al. (2018) |
| *Funneliformis mosseae, Rhizophagus intraradices, Claroideoglomus etunicatum* | *Sesbania sesban* | Abd-Allah et al. (2015) |
| *Aspergillus flavus, Chaetomium globosum, Cochliobolus lunatus, Fusarium dimerum, F. oxysporum, P. chrysogenum* | *Calotropis procera* | Gherbawy and Gashgari (2013) |
| *Cladosporium omanense* | *Zygophyllum coccineum* | Halo et al. (2019) |

tissues colonized and systemic versus nonsystemic spread in the plant) are considered for grouping these fungi. It is generally believed that when all morphological, cultural, and microscopic characteristics of endophytic fungi are similar to their pathogenic form, their modes of penetration and colonization should be identical. However, studies have revealed that the penetration and colonization modes in endophytic fungi are different from pathogenic fungi. The endophytic fungi penetrated through the stomata along the anticlinal epidermal cells in contrast to the pathogenic fungi, which penetrated directly through the cell (Barbara et al., 2002; Jia et al., 2016). Similarly, differences in the mode of colonization were also observed between endophytic and pathogenic fungi. While plant pathogenic fungi showed localized and systemic colonization of their hosts inter and Intracellularly; the colonization of the host plants by endophytic fungi is limited, localized, and intercellular in nature. These differences may partially account for the asymptomatic nature of colonization by these fungi (Bacon and White, 2000; Barbara et al., 2002; Sieber, 2007; Li and Zhang, 2015; Jia et al., 2016; Mane and Vedamurthy, 2018).

### 14.4.2 Infection strategies of fungal endophytes in plants

Like pathogenic fungi, penetration and colonization are not easy tasks for endophytic fungi. The process of chemotaxis involves the production of specific chemicals by the host plant to protect against invaders including endophytes. Despite this defense, endophytic fungi obtain entry into the host tissue through stomata along the anticlinal epidermal cells. Production of secondary metabolites by host plants does not hinder this entry into the tissue; however, these secondary metabolites are obstacles for tissue colonization by the endophytic fungi. Endophytes usually produce the enzymes necessary for colonization of plant tissues; however, it has been reported that most fungal endophytes utilize the polysaccharides xylan and pectin from host plants. Fungal endophytes show lipolytic activity and produce nonspecific peroxidases, laccases, chitinase, and glucanase. These enzymes enhance the process of colonization by fungal endophytes (Krings et al., 2007; Selim et al., 2012; Ren and Dai, 2012). The fate of a fungus after entering a plant host to behave either as an endophyte or as a pathogen generally depends on its genetics. Genetic mutations are generally believed to be behind this shift. A mutation in a single locus can transform the fungi into a pathogen or a mutualistic endophyte (Freeman and Rodrigues, 1993). Some fungal isolates have been reported to behave like as a pathogen in cucurbits or as an endophyte in species of other plant families (Redman et al., 2001). Numerous secondary metabolites, such as alkaloids and essential oils, are reported to secreted by host plants during the colonization process of fungal endophytes (Agrios, 2005; Mehrotra and Aggarwal, 2003; Jia et al., 2016; Chen et al., 2016; Sibanda et al., 2018). The infection strategies adapted by fungal endophytes to overcome the obstacles generated by production of secondary metabolites include the secretion of matching detoxification enzymes. Endophytic fungi disrupt the defense mechanisms of host plants by producing enzymes, such as cellulases, lactase, xylanase, and protease, to decompose these secondary metabolites before or during penetration. After winning this entry war, the endophytic fungi colonizes the host tissue either for the whole lifetime (neutralism) or for an extended period of time (mutualism or antagonism) until environmental conditions becomes favorable (Sieber, 2007; Jia et al., 2016; Goyal et al., 2017; Sharma and Gautam, 2019).

### 14.4.3 Spectrum of bioactive compounds produced by fungal endophytes

Endophytic fungi are one of the most creative groups of microorganisms, producing a wide range of secondary metabolites. More than 20,000 bioactive metabolites are of microbial origin (Bérdy, 2005), and fungi are one of the most important and well known producers. Research on identifying new chemical compounds produced by endophytic fungi is ongoing; however, more exploration is still needed. Because these secondary metabolites play important biological roles for human life, research on fungal endophytes has gained a faster pace during recent decades. Endophytes isolated from medicinal plants possess strong fungicidal, bactericidal, and cytotoxic metabolites (Wang et al., 2007). Taxol is a highly functionalized diterpenoid, first isolated from the bark of the western yew, *Taxus brevifolia*, and is widely used as an anticancer drug (Wani et al., 1971). Similarly, bioactive compounds including enzymes and chemical metabolites are potential sources of novel natural agents (Kock et al., 2001; Donadio et al., 2002; Gunatilaka, 2006; Stadler and Keller, 2008; Rajamanikyam et al., 2017). These enzymes may have various applications such as degradation and biotransformation of organic compounds (Firáková et al., 2007; Pimentel et al., 2006). The metabolites may also have wider applications in the biotechnology sector (Tomita, 2003) and in health sciences. Because of their antimicrobial, anticancer, and antiviral activities, these bioactive compounds are used in pharmaceuticals (Chin et al., 2006; Selim et al., 2012). Their antimicrobial potential highlights their agriculture applications. Further, their potential exploitation in bioremediation is also currently being explored.

## 14.5 Fungal endophytes and plant pathogens

Use of fungal endophytes as biological control agents is a new, efficient, and widely used method for environmental remediation and control of insects or pathogens (Guo et al., 2008). There is now sufficient evidence that fungal endophytes play an important role in plant physiology and the ability to potentially support plant health and protect the host from disease. By colonizing internal plant tissue; they obtain nutrition, shelter, protection, and propagation opportunities from their hosts. The establishment of this symbiotic relationship, in return, benefits the host plant by potentially reducing environmental sensitivity and promoting overall health. Endophytes now offer the best alternative to traditional chemical disease control. Fungal endophytes utilize direct and indirect strategies to control plant diseases by increasing stress tolerance, improving fitness, and promoting the accumulation of bioactive compounds. A more thorough understanding of the mechanisms employed by fungal endophytes in their mutualistic association with plants is needed, both to optimize their efficacy and for registration as plant protection products. The possible mechanisms adopted by endophytes against pathogens may be direct inhibition, competition, antibiosis, mycoparasitism, indirect inhibition, induction, and improvement of resistance of host plants (Yu et al., 2010; Alvin et al., 2014; Nisa et al., 2015; Yao et al., 2017; Sibanda et al., 2018). The numerous aspects involved in disease control by fungal endophytes are discussed in this section.

## 14.5.1 Strategies of fungal endophytes to control plant disease

Endophytes are naturally occurring biocontrol agents with potential beneficial uses in the control of plant disease. They play a vital role in plant-pathogen interactions. Several studies on the plant–endophyte relationship have demonstrated that endophytes produce toxins that discourage insects and other grazing animals (Bultman and Murphy, 2000). The first report on the role of endophytic fungi in plant protection was given by Webber (1981), who reported the role of *Phomopsis oblonga* in protection of elm trees against the beetle *Physocnemum brevilineum*. Although studies on the plant–endophyte relationship are still scanty, researchers are continuously engaged in discovery of the diverse mechanisms and strategies adopted by endophytes to inhibit or stop the development of pathogens inside host plants. Two main strategies, namely direct and indirect inhibition, are reported to be involved in plant protection mechanisms of endophytes (Király et al., 2007; Gao et al., 2010; Zabalgogeazcoa, 2008; Dutta et al., 2014; Atugala and Deshappriya, 2015; Strobel, 2018). Diverse mechanisms by which fungal endophytes may counteract plant disease are presented in Fig. 14.3.

### 14.5.1.1 Direct mechanisms

Many recent studies found that endophytic fungi have the ability to protect hosts from diseases and limit the damage caused by pathogenic microorganisms. The bioactive compounds produced by fungal endophytes possess the potential to directly inhibit the growth of other organisms including plant pathogens. Possible mechanisms used by fungal endophytes or their secondary metabolites to suppress pathogens have been proposed by some researchers; however, our knowledge about the regulation between endophyte, pathogen, and plant is still limited. In direct inhibition, endophytes produce antibiotics, secrete lytic enzymes, and directly suppress pathogens through either competition, antibiosis, or mycoparasitism, etc. (Arnold et al., 2003; Giménez et al., 2007; Ganley et al., 2008; Mejía et al., 2008; Sansanwal et al., 2017; Strobel, 2018; Bamisile et al., 2018).

#### Competition

Competition is an important method used by fungal endophytes against infection and proliferation of plant pathogens. It involves competition between plant pathogen and endophytic fungi for space and other common resources (Mejía et al., 2008).

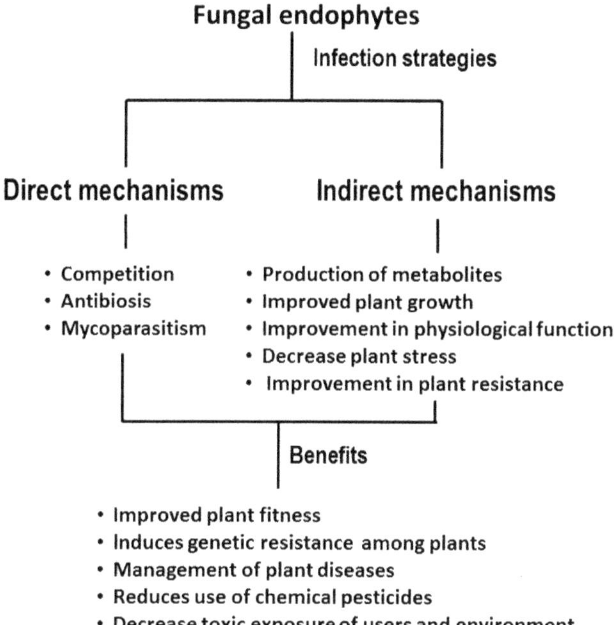


**Figure 14.3** Infection strategies of fungal endophytes against plant pathogens and benefits of this interaction.

Some endophytic organisms seem to be able to control pathogens with such mechanisms; they could be used in biocontrol of plant diseases. Atugala and Deshappriya (2015) evaluated the effect of the two most frequently isolated endophytic fungi on plant growth and blast disease incidence of two traditional rice varieties. They concluded that the tested endophytic fungi utilized a range of mechanisms of antagonistic activity including competition for substrate as well as antibiosis and mycoparasitism. Numerous such studies have been carried out across the globe (Richmond et al., 2004; Mejía et al., 2008; Atugala and Deshappriya, 2015; Bamisile et al., 2018), demonstrating that some endophytic fungi can be useful for developing management strategies for protection against numerous pathogens.

## Antibiosis

Fungal endophytes produce a wide spectrum of secondary metabolites with the ability to minimize attacks from all kinds of insect and pests. Many of these secondary metabolites have antifungal, antibacterial, and insecticidal properties that strongly inhibit the growth of other microorganisms including plant pathogens (Gunatilaka, 2006; Gautam et al., 2013). It has been reported that a single fungus can produce single or multiple kinds of antibiotics including terpenoids, alkaloids, aromatic compounds, and polypeptides, which have been proven effective against plant pathogens (Gao et al., 2010). Production and examination of antibiotic compounds produced by endophytic fungi have been carried out by many researchers

(Schulz and Boyle, 2005; Mejía et al., 2008). It has been proven experimentally that inoculation of plants with endophytes, as well as the application of endophytic culture filtrates, induces defense mechanisms in the host plant (Zabalgogeazcoa, 2008). After testing liquid broths of endophytic fungi against several plant pathogenic fungi, it was determined that endophyte cultures possess antibiotic properties (Liu et al., 2001; Park et al., 2005; Inácio et al., 2006; Kim et al., 2007; Pandya and Saraf, 2009; Wang and Chuan, 2011; Atugala and Deshappriya, 2015). As per the Li et al. (2004), it was found that the proteins secreted by endophytic fungal spp., are able to suppress plant pathogen activities directly by acting as pathogenicity-related (PR) proteins and may function by degrading pathogen cell walls (Li et al., 2004). A number of studies reported that fungal endophytes produce chemical compounds with antibiotics properties against pathogens. Dingle and McGee (2003) examined the interaction between *Puccinia recondita* f. sp. *tritici* (now widely referred to as *P. triticina*) and endophytic fungi and found that the presence of fungal endophytes suppressed leaf rust disease caused by this fungus. A chemical compound 3,11,12-trihydroxycadalene secreted by *Phomopis cassiae*, an endophytic fungus isolated from *Cassia spectabilis*, was effective against *Cladosporium sphaerospermum* and *Cladosporium cladsporioides* (Silva et al., 2005). Park et al. (2013) evaluated antifungal activities of 183 endophytic fungi isolated from 11 plant species from 11 locations in Korea against plant pathogenic fungi such as *Magnaporthe grisea* (causing rice blast disease), *Corticium sasaki* (causing rice sheath blight disease), *Botrytis cinerea* (causing tomato gray mold disease), *Phytophthora infestans* (causing tomato late blight disease), *P. recondita* (causing wheat leaf rust disease), and *Blumeria graminis* f. sp. *hordei* (causing barley powdery mildew disease). In another study, Arnold et al. (2003) examined the fungal endophytes associated with a woody angiosperm *Theobroma cacao* against the foliar pathogen *Phytophthora* sp. The study revealed that inoculation of leaf tissues by an assemblage of endophytes frequently isolated from naturally infected, asymptomatic hosts significantly limited damage by an important foliar pathogen.

## Mycoparasitism: hyperparasitism and predation

Distinct from competition and antibiosis, some endophytes exhibit hyperparasitism. These fungi act as mycoparasites, attacking plant pathogenic fungi to either suppress their growth or kill them. *Trichoderma* is one of the most popular examples of mycoparasitism. It detects a fungal pathogen before making any contact and produces low levels of an extracellular exochitinase, catalyzing discharge of cell-wall oligomers from the target fungus. This activity induces *Trichoderma* to release fungi toxicendochitinases, which also degrade the target fungal pathogen cell wall (Harmon et al., 2004). *Trichoderma* is also able to parasitize hyphae of plant pathogens, including *Rhizoctonia solani* (Grosch et al., 2006). *Acremonium strictum*, a noval endophytic fungi isolated from *Dactylis glomerata* L. and some grasses have been reported to have mycoparasitic activity against *Helminthosporium solani* Durieu and Mont. (Sánchez Márquez et al., 2007; Rivera Varas et al., 2007). The biocontrol potential of *Trichoderma* isolates from *Theobroma gileri*, a forest tree in

Ecuador, has been assessed against of cocoa diseases (Evans et al., 2003; Bailey et al., 2008; Mejía et al., 2008).

In mycoparasitism, binding of chemical compounds between mycoparasite and host fungus takes place initially (e.g., carbohydrates in the *Trichoderma* cell wall binds to lectins in the cell wall of the fungal host). After establishing a contact, hyphae of the mycoparasite inhabit the host fungus by various methods. In *Trichoderma* mycoparasitism, the hyphae coil and form appressoriae. Along with mechanical mechanisms, several lytic enzymes are involved in the degradation of cell walls of host fungi (e.g., chitinases, β-1, 3 glucanases, proteases, and lipases) (Cao et al., 2009; Ownley et al., 2010). These findings have shown that mycoparasitic activity of endophytic fungi can limit pest damage in agriculture crops and can be the best alternative to use of chemical pesticides.

### *14.5.1.2 Indirect mechanisms*

The association of fungal endophytes with plants has a direct impact on growth and development of the host plant. It affects hosts performance in terms of growth, morphology, biochemistry, and physiology. The production of secondary metabolites by these fungi along with establishment of a symbiotic relationship with plants is a potential reason for the overall improvement in health of the host. It is now believed that the fate of pathogen attack may be dependent on the association of endophytic mycobiota with a host plant (Király et al., 2007; Zabalgogeazcoa, 2008; Jia et al., 2016; Gao et al., 2010; Sansanwal et al., 2017). Such beneficial microbes can increase plant growth (root and shoot), improve nutrient uptake, increase efficiency of nitrogen fixation, decrease plant stress, decrease disease incidence, and improve resistance (Lu et al., 2000; Montesinos, 2003; Kuldau and Bacon, 2008; Jia et al., 2016). Due to these properties, fungal endophytes can play essential roles in agriculture and food safety and contribute to maintenance environmental equilibrium. Some of the important indirect mechanisms adapted by fungal endophytes in pest and disease control are given here.

#### Production of metabolites

Production of secondary metabolites during plant-fungal symbiotic association plays a major role in the adaptation of plants to biotic and abiotic components of the environment (Bourgaud et al., 2001). Association of fungal endophytes with plants is an important factor behind the production of a variety of metabolites. Use of endophytic fungi to control various diseases is a biological and environment friendly approach. Endophytes benefit host plants by preventing pathogenic organisms from colonizing (Arnold et al., 2003). Experimental studies using the dual culture technique revealed that fungal endophytes release certain chemical compounds that inhibit plant pathogenic fungi. Similarly, inhibitory studies of endophytic fungi against pathogenic bacteria also found them to be effective. It has been demonstrated that endophytes isolated from medicinal plants possess strong fungicidal, bactericidal, and cytotoxic metabolites (Zhang et al., 2006) that can reduce or kill pathogens by producing secondary metabolites in host tissue. Endophytic fungi have been reported to produce secondary metabolites, some of which are bioactive compounds that protect the host plant from pests and diseases (Gao et al., 2010; Sudha et al., 2016). These metabolites also serve as mediators for specific interactions and communications with the plant host. Secondary metabolite production and inhibition of pathogens by endophytes triggers fitness, competitiveness, and growth of host plants. These secondary metabolites are categorized into various functional groups: alkaloids, benzopyranones, chinones, flavonoids, phenolic acids, quinones, steroids, saponins, tannins, terpenoids, tetralones, xanthones, and many others. Studies on secondary metabolites of fungal endophytes and their antipathogenic effects have been conducted by various researchers and explored their applications in biocontrol of agriculture crop diseases. It was reported that endophytic fungi *Fusarium* spp. E4 and E5 could promote the growth of *Euphorbia pekinensis* and increased its terpenoids content (Yong et al., 2009). An endophytic fungus, *Muscodor albus*, isolated from small branches of *Cinnamomum zeylanicum* has been reported to produce 28 volatile compounds that effectively inhibit and kill other selected fungi and bacteria (Worapong et al., 2001; Strobel et al., 2001). *Muscodor crispansis*, another endophytic fungus of *Ananas ananassoides* from the Bolivian Amazon Basin, has been found to produce a mixture of antifungal and antibacterial volatile organic compounds with inhibitory activity against pathogenic fungi, namely *Pythium ultimum*, *Alternaria helianthi*, *Botrytis cinerea*, *Fusarium culmorum*, *F. oxysporum*, *Phytophtora cinnamomi*, *P. palmivora*, *Rhizoctonia solani*, *Sclerotinia sclerotiorum*, *Verticillium dahlia*, and bacteria including *Xanthomonas axonopodis* (Mitchell et al., 2010; Yuan et al., 2017). In ongoing investigations, researchers are searching for the production of secondary metabolites by endophytes with the capacity to inhibit insects across the globe (Demain, 2000; Johnson et al., 2013; Young et al., 2013; Lugtenberg et al., 2016).

### Plant growth (roots and shoots)

Endophytic fungi asymptomatically colonize their plant hosts and may be mutualistic organisms, which produce plant growth regulatory compounds and antimicrobial substances to enhance the growth, fitness, and competitiveness of the host in nature (Sudha et al., 2016). Colonization of host plants by fungal endophytes likely enhances root and shoot performance through a variety of mechanisms. One important mechanism is the symbiotic interaction between hosts and endophytes. Here, the plant provides food and shelter, and in return these fungi help the host plants to solubilize phosphate, enhance uptake of phosphorus, fix nitrogen, and produce siderophores as well as plant hormones such as auxin, abscisins, ethylene, gibberellins, and IAA, all of which are important for plant growth regulation (Malinowski and Belesky, 2000; Boddey et al., 2003; Loiret et al., 2004; Sandhiya et al., 2005; Firáková et al., 2007). The endophyte *Cladosporium sphaerospermum* isolated from *Glycine max* was examined by Hamayun et al. (2009) for production of bioactive molecules and their ability to prompt plant growth. The results revealed that *Cladosporium* produces GA3, GA4, and GA7 growth hormones, which were found to be effective in inducing plant growth in rice and soybean. Derivatives of indole acetic acid isolated from the endophytic fungi *Fusarium tricinctum* and *Alternaria alternata* were reported to enhance plant growth (Khan et al. 2013). In a similar study conducted by Li and Zhang (2015), an apestalotin analogue was isolated and characterized from *Pestalotiopsis microspora*. Further studies revealed that pestalotin induced a significant increase in the germination rate of *Distylium chinense* seeds. A study conducted by Johnson et al. (2014) on the root colonizing endophyte *Piriformospora indica* found that association of fungal endophytes with roots modulates phytohormones involved in growth and development of host plants and enhance nutrient uptake and translocation. Plant height, weight and other growth parameters such as fresh weight of shoots and roots were also reported to be influenced by colonization with endophytic fungi (Lopez and Sword, 2015; Jaber and Enkerli, 2016; Jaber and Enkerli, 2017). It is now known that endophytes can actively or passively promote plant growth through a variety of mechanisms; however, the molecular and biochemical mechanisms underlying how this habitat-adapted symbiotic interaction results in plant tolerance to high stress remain largely unknown.

## Improvement in physiological function

Fungal endophytes promote plant defenses against phytopathogenic organisms through enhancement of plant growth, which is achieved via a number of contributing mechanisms. An enhancement in plant growth in endophyte colonized host plants enables them to survive a variety of abiotic and biotic stresses. The promotion of physiological processes of host plants was also observed in many studies (Kuldau and Bacon, 2008; Gao et al., 2010). Many studies demonstrate that fungal endophytes produce phytohormones that may influence plant physiological functions. Endophytes have been found to enhance defense responses of host plants against pathogens. However, this enhanced defense requires more energy, reducing equivalents, and carbon skeletons, which are associated with accelerated primary metabolic pathways (Bolton, 2009). Colonization of a plant host by endophytic fungi is believed to improve nutrient uptake (phosphorus, nitrogen, and other essential nutrients), accelerate all metabolic activities including physiological functions, and enhance plant growth (Gao et al., 2010; Bamisile et al., 2018). Endophytic fungi are also known to produce compounds that interfere with plant cell division (Suryanarayanan and Rajagopal, 1998). *Colletotrichum* sp., an endophytic fungus in *Artemisia annua*, produces substances like IAA that can regulate plant processes (Lu et al., 2000). Dai et al. (2008) evaluated the effects of the endophytic fungus *Fusarium* sp. on growth of *Euphorbia pekinensis* and found that extract of this fungus functions as an auxin. The phytohormone effect may be involved in stimulating the host growth by endophytic fungi.

The endophytic fungi perform an important service for their hosts by fixing soil nitrogen (N). Yang et al. (2015) investigated the impact of *Phomopsis liquidambari* on N dynamics in rice. They found that under low N conditions the available nitrate and ammonium content found in the rhizosphere soil of endophyte-infected rice increased significantly. Fungal endophytes are also reported to improve the ability to suppress nitrification in soil. Cardoso et al. (2017) observed that the ability of *Brachiaria* species to suppress nitrification in soil by releasing an inhibitory compound called "brachialactone" from its roots is improved in plants growing with endophytic association.

## Decrease plant stress

In addition to promoting the growth, fitness, and competitiveness of host plants, endophytic microbes were found to enhance tolerance of plants to abiotic stresses (Marquez et al., 2007; Aly et al., 2011; Selim et al., 2012; White et al., 2016). The colonization of plants by endophytic fungi may help the plants to tolerate and withstand high stress environments such as drought, salt, and high temperatures (Malinowski and Belesky, 2000; Rodriguez et al., 2008; Rodriguez et al., 2009). The presence of endophytes may increase fitness in plants such as *Dichanthelium lanuginosum*, an herbal plant that survives in areas where soil can reach 57°C. These fungi induce tolerance to high soil temperature, water stress (Redman et al., 2002), and salt stress in barley (Waller et al., 2005). Hence, such symbioses are of great importance, as they may help plants to adapt to global climate change (Rodriguez et al., 2004; Selim et al., 2012; Trivedi et al., 2017). The beneficial effects of endophytes on plants have increased interest in the use of endophytic microbes to enhance agricultural production of crop plants. However, the molecular and biochemical mechanisms behind how these symbiotic interactions result in plant tolerance to high stress are still not fully known.

Colonization of plant tissues with fungal endophytes enables them to adapt to habitat-based stresses. It has been demonstrated that grasses with high tolerance to salinity and heat harbor symbiotic fungal endophytes (Rodriguez et al., 2008; Egamberdieva and Lugtenberg, 2014; Murphy et al., 2015). The grasses *Leymus mollis* (dunegrass) from several coastal beaches in the United States and *Dicanthium lanuginosum*, a grass found growing in geothermal soils in Yellowstone National Park, were shown to be symbiotic with the endophytes *Fusarium culmorum* and *Curvularia protuberata*, respectively, which confers salt and heat tolerance (Redman et al., 2002; Rodriguez et al., 2008).

## Improvement in plant resistance

Endophytic fungi are also capable of inducing resistance to disease. Many mechanisms have been proposed for this resistance, including improved root growth, nutrient uptake, and nitrogen fixation. Decreased plant stress directly influences the overall growth and development of the host plants, thus enabling them to resist phytopathogens (Montesinos, 2003; Bae et al., 2009; Sudha et al., 2016). Production of certain toxic compounds by endophyte colonized plants is also effective in repelling pests (Schardl et al., 2004; Akello et al., 2007).

Colonization by fungal endophytes may induce or improve the resistance of plant hosts against phytopathogens in three possible ways (i.e., competition, production of phytoalexins and/or biocidal compounds, and improving host resistance). The first mechanism involves competition between endophyte and pathogen for the same resources (Lockwood, 1992). In the second possible mechanism, fungal endophytes either stimulate the host to produce phytoalexins and/or biocidal compounds or itself produces secondary metabolites able to inhibit or control the growth of plant pathogens. Improvement in host resistance to pathogens by inducing host defense responses by endophytes is the third possible mechanism (Selim et al., 2012). Molecular studies on plants colonized with endophytic fungi showed important changes in plant metabolism, particularly production of biochemicals that may induce defense mechanisms and resistance against pathogens (Waller et al., 2005). Some proteins reported to be produced during endophytism suppress plant pathogen activities directly by acting as pathogenicity-related proteins (Li et al., 2004). Endophyte enhanced plant growth promotion is another conferred benefit that enables plants to develop resistance against pathogens (Johnson et al., 2014).

Many researchers have evaluated fungal endophytes for their abilities to induce resistance among host plants. Endophytic fungi *Cryptosporiopsis* cf. *quercina* and *Colletotrichum* sp. are effective against phytopathogens such as *Rhizoctonia cerealis*, *Phytophthora capsici*, *Pyricularia oryzae*, and *Gaeumannomyces graminis* (Li et al., 2000; Lu et al., 2000). The outcomes of studies conducted by Waller et al. (2005) revealed that barley plants inoculated with *Piriformospora indica* are resistant to *Fusarium culmorum* and *Blumeria graminis*, vascular and leaf pathogens, respectively, and exhibit an increase in yield and salt stress tolerance. Similar induced resistance was observed in *Festuca rubra* L. cultivars against dollar spot disease (caused by *Sclerotinia homoeocarpa*) when inoculated with endophyte *Epichloë festucae* (Clarke et al. 2006). Several other studies have also related endophytic fungi association with improved plant growth and induction of resistance against plant pathogens (Elena et al., 2011; Liao et al., 2014; Jaber and Enkerli, 2016; Jaber and Enkerli 2017; Bamisile et al., 2018).

### 14.5.2 Occupation of the ecological niche

Colonization of the inner space of plants with fungal endophytes generates a unique ecosystem with special types of interactions. This small world represents an ecological niche where some distinctive fungal endophytic species live. Both fungi and plant hosts establish a mutualistic approach and benefit from each other in terms of food and shelter. This mutual cooperation enables plants to cope with biotic and abiotic stresses. The process of colonization follows several steps including host recognition, spore germination, penetration of the epidermis, and tissue multiplication as in pathogens (Sharma and Gautam, 2019) but remains asymptomatic. Once these fungi colonize host tissue intercellularly or intracellularly, an endophytic niche is established. This unique association plays an important role in adaptation by plants to particular environments, while endophytes obtain a reliable source of nutrition from the plant fragments, exudates, and leachates and protect the host against other microorganisms (Rodriguez et al., 2004; Gao et al., 2010). Some endophytes possess the ability to infect a wide range of hosts and act as generalists, while others are specialists limited to one or a few hosts (Zabalgogeazcoa, 2008). There are many endophytes occupying unique biological niches (higher plants) growing in many unusual environments. Investigations to date on endophytic fungi have suggested that these organisms have very good potential to improve host plants and control their diseases.

### 14.5.3 Endophytic fungi: a source of potential biocontrol compounds

Currently endophytic fungi are viewed as an outstanding source of bioactive natural products, and numerous bioactive compounds are now available commercially for their multifarious uses. Research has shown that fungal endophytes regulate expression of infection or infestation after plant pathogen interactions. The mechanisms by which endophytes counteract pathogen development vary with the type of interactions. While some interactions may induce plant defense mechanisms or competition for resources, others produce antimicrobial compounds. Medicinal plants are reported to harbor endophytes (Strobel, 2003) that provide protection from infectious agents. Additionally, they are surprising producers of useful metabolites that can be used in management of plant pathogens, particularly fungi and nematodes (Petrini, 1991). Alteration in plant biochemistry is another possible mechanism that may induce defense mechanisms against plant pathogens. Pesatcin, ambuic acid, torreyanic acid, colutellin A, cryptocin, ergobalancine, chanoclavine aldehyde, fumiclavine, rubrofusarin, fonsecinone, asperpyrone, IAA, indole-3-ethanol (IEtOH), methylindole-3-carboxylate, indole-3-carboxaldehyde, diacetamide, cyclonerodiol, colletotric acid, are some bioactive compounds produced by endophytic fungi during interaction with their respective hosts (Gouda et al., 2016; Strobel, 2018). The potential bioactive compounds produced by endophytic fungi that can be used against plant pathogens are discussed in this section.

#### 14.5.3.1 Interaction with plant pathogens

Many endophytic species have been reported to produce antibiotic substances which can be applied in the agricultural sector against a number of fungal pathogens. Studies have shown that liquid extracts from cultures of fungal endophytes have been reported to inhibit the growth of several plant pathogenic fungi (Inácio et al., 2006; Kim et al., 2007; Zabalgogeazcoa, 2008; Ownley et al., 2010; Selim et al., 2012; Hardoim et al., 2015; Schulz et al., 2015; Terhonen et al., 2019). It has been observed that application of endophytic fungi induces the release of certain antifungal compounds that improve the defense mechanisms of the host plants to defend against pathogens. Some endophytes may be mycoparasites and act as useful alternatives for biocontrol of plant pathogenic fungi. It has been estimated that more than 20,000 bioactive metabolites are of microbial origin. These fungal secondary metabolites encompass a wide range of chemical structures and biological activities (Bérdy, 2005; Suryanarayanan and Hawksworth, 2005). As per Yu et al. (2010), about 35% of biologically active fungal endophytes are isolated from medicinal plants, while 29% are isolated from crops, and 18% from plants in special environments and other plants, respectively. Peláez (2005) has reported approximately 1500 fungal metabolites with antibiotic activities, some of which have been approved as drugs. Some metabolites such as micafungin from *Coleophoma empetri* (Frattarelli et al., 2004); mycophenolate from *Penicillium brevicompactum* (Curran and Keating, 2005); rosuvastatin from *Penicillium citrinum* and *P. brevicompactum* (Scott et al., 2004); cefditoren pivoxil from *Cephalosporium* sp. (Darkes and

Plosker, 2002); fumagillin from *Aspergillus fumigatus* (Chun et al., 2005); and illudin-S from *Omphalotus illudens* (McMorris et al., 1996) have been reported to have important agricultural and pharmaceutical applications. Some endophytes [e.g., *Acremonium strictummay* (Sánchez Márquez et al., 2007) and *Helminthosporium solani* (Rivera Varas et al., 2007)] act as mycoparasites and aid in control of plant pathogens.

Fungal endophytes also protect plants from nematode attacks. The production of certain secondary metabolites also imposes inhibitory effects against nematodes and can be used as bionematicides in agriculture. It has been observed that alkaloids were released in endophyte colonized plants and translocate to roots and cause inhibitory effects against nematodes. Some endophytic fungi, namely *Neotyphodium* spp. and *Fusarium oxysporum* have been reported to inhibit the entry of some species of migratory and sedentary endoparasitic nematodes (Timper et al., 2005; Panaccione et al., 2006; Malinowski and Belesky, 2000). Along with this inhibition, certain chemical changes may also be triggers, which is another type of plant protection strategy induced endophytic fungi against plant parasitic nematodes. Nematophagous fungi like *Beauveria bassiana*, *Torrubiella confragosa*, and *Metarhizium anisopliae* have been reported to provide protection to plants against parasitic nematodes (Bordallo et al., 2002). However, variation in the nematicidal activity and defense mechanisms of fungal endophytes against these organisms was also observed (Chomcheon et al., 2005; Giménez et al., 2007; Ali et al., 2018).

Experimental studies on fungal endophytes showed these organisms are effective against bacterial and fungal pathogens (Lehtonen et al., 2006; Wang et al., 2007; Márquez et al., 2007; Romo et al., 2007; Bamisile et al., 2018). Interaction of fungal endophytes with colonized host and primary pests have some basic effects:

- Improve health and fitness of colonized plants.
- Induce systemic resistance in colonized hosts that can also be transferred vertically from parent to their offspring, hence providing the same resistance for the next generation.
- Utilization of the nutrition of colonized plants in the production of secondary metabolites. This production induces chemical defenses among colonized plants and protects them against plant pathogens.
- Colonization of plants with fungal endophytes improves their tolerance to biotic stress such as root herbivory by plant parasitic nematodes.
- Endophytic fungi have been reported to reduce insects feeding on endophytic colonized plants. Their colonization indirectly enhances seed dispersal by ants.
- Fungal endophytes promote uptake of nutrients such as phosphorus and nitrogen in their colonized plants.
- Colonization with fungal endophytes improves crop yield, plant growth, cell division, and development.
- Fungal endophytes produce a wide range of novel antibiotics and bioactive volatile compounds such as ammonia, lipids, alkyl pyrones, hydrogen cyanide, alcohols, ketones, and esters, useful for various biology fields.

Numerous secondary metabolites produced by the symbiotic association between endophytes and host plants have been discovered, identified, and screened for their inhibitory activities against plant pathogenic bacteria, fungi, viruses, and nematodes. Bioactive compounds produced by fungal endophytes are presented in Table 14.3.

**Table 14.3** Bioactive compounds produced by fungal endophytes on their colonized host plants and the primary hosts.

| Fungal endophyte | Bioactive compounds | References |
|---|---|---|
| *Balansia obtecta* | Ergobalancine | Tintjer and Rudgers (2006) |
| *Neotyphodium coenophyalum* | Ergovaline | Tintjer and Rudgers (2006) |
| *Neotyphodium* sp. | Ergonovine | Miles et al. (1996) |
| *Aspergillus fumigatus* | Chanoclavine aldehyde, fumiclavine, A, fumiclavine B, fumiclavine C | Panaccione (2005) |
| *Helmintosporium* sp., *Phoma* sp., *Phomopsis* sp., *Xylaria* sp., *Hypoxylon* sp., *Chalara* sp., *Rhinocladiella* sp. | Bioactive cytochalasines | Isaka et al. (2001), Wagenaar et al. (2000) |
| *Penicillium* sp. | Berkeleydione | Davis et al. (2005) |
| *Morinia pestalozzioides* | Moriniafungin | Pelaez et al. (2000) |
| *Aspergillus niger* | Rubrofusarin B, fonsecinone A, asperpyrone B, aurasperone A | Song et al. (2004) |
| *Periconia* sp. | *Periconicin A* | Kim et al. (2004) |
| *Pestalotiopsis jester* | Jesterone, hydroxyjesterone | Li and Strobel (2001) |
| *Pestalotiopsis microspora* | Ambuic acid | Li et al. (2001) |
| *Epichlöe festucae* | IAA, IEtOH, methylindole-3-carboxylate, indole-3-carboxaldehyde, diacetamide, cyclonerodiol | Yue et al. (2000) |
| *Colletotrichum gloeosporides* | Colletotric acid | Zou et al. (2000) |
| *Muscodor albus* | 1-Butano, 3-methylacetate | Strobel and Daisy (2003) |
| *Chaetomium chiversii C5-36-62* | Radicicol | Turbyville et al. (2006) |
| *Pestalotiopsis adusta* | Pestalachlorides | Li et al. (2008) |
| *Phomopsis* sp. | Terpenoid antimicrobial | Nithya and Muthumary (2010) |
| *Ampelomyces* sp. | 3-Omethyl alaternin and altersolanol A | Miller et al. (1998) |
| *Xylaria* sp. *YX-28* | 7-Amino-4-methylcoumarin | Xu et al. (2008) |
| *Colletotrichum gloeosporioides* | Piperine | Chithra et al. (2014) |
| *Fusarium redolens* | Peimisine and imperialine-3β-D-glucoside | Pan et al. (2015) |
| *Acremonium coenophialum* | Chitinases | Roberts et al. (1992) |
| *Phomopsis phaseoli, Melanconium betulinum* | 3-hydroxypropionic acid | Chomcheon et al. (2005) |
| *Cladosporium delicatulum* | Plumbagin (5-hydroxyl-2-methylnaptalene-1,4-dione) | Venkateswarulu et al. (2018) |

## 14.6 Developing a successful fungal endophyte inoculant for agricultural crops

Endophytes are a class of plant associated microorganisms that have shown potential benefits in agriculture such as promoting fitness and stress tolerance as well as protection against pathogens in agricultural crops. Fungal endophytes are known to enhance abiotic and biotic plant stress tolerance (Easton and Fletcher, 2007; Murphy et al., 2018). However, the use of a fungal endophyte inoculum is still an unlikely approach due to lack of belief in the efficacy of such treatments among end users. In the present era of chemical pesticides, use of these organisms as biocontrol agents in agricultural crops still requires more research and promotion among farmers and other users. Some research is available regarding the pesticidal potential of endophyte inoculants for agricultural crops. Murphy et al. (2018) proposed a successful research pipeline for the production of a reliable, consistent, and environmentally targeted fungal endophyte seed-delivered inoculant for barley cultivars. They suggested that a careful screening and selection procedure for fungal endophyte strains followed by extensive testing in a controlled environment and multiyear field trials has resulted in the validation of an endophyte consortium suitable for barley crops grown on relatively dry sites. This work presented a single solution to a defined agricultural need (e.g., an endophyte crop treatment combined with a change in agricultural practice can give extra benefits).

The general statement that endophyte plant associations makes plants more resistant to diseases and improve growth is not always true. Therefore, it must be considered that the crop or plant host that is being used to inoculate with an endophyte species should not be its natural host plant. Significant variation in the pattern of produced metabolites must be ensured before use in field trials (Giménez et al., 2007). Inoculation with endophytic fungi changes the growth, physiology, and metabolism of plant hosts as well as beneficial fungi and bacteria. This sometimes leads to breaks in the symbiotic balance between endophyte and host plant. Moreover, it must also be determined that the endophyte to be inoculated is useful to promote crop yield and disease protection (Gao et al., 2010).

## 14.7 Diversity of fungal endophytes as a tool in the agriculture industry

Fungal endophytes are now recognized as biocontrol agents to be used against pathogens causing diseases in field crops. As discussed earlier, fungal endophytes protect plants against pathogens and pests either by improvement of plant fitness and performance or by producing a vast variety of novel secondary metabolites including volatile organic compounds (Lugtenberg et al., 2016). Developments in endophytic research have led to availability of these microbes or their products commercially, providing an alternative to chemical pesticides and also generating revenue and employment (Cocq et al., 2017).

Experimental studies revealed that the ability of fungal endophytes to confer disease, salt, and heat tolerance to grasses (Rodriguez et al., 2008) can be transferred to agricultural crops (Rodriguez et al., 2008; Redman et al., 2001). Based on this concept, endophytes have been commercialized to improve agricultural crops in relation to combat with stresses and pests. A number of products containing endophytic fungi are available commercially including BioEnsureR-Corn and BioEnsureR-Rice. These products can increase crop yield up to 80% under heavy drought and other stress conditions. The products are available in liquid form and are sprayed on to seeds before sowing. They establish a symbiotic relationship with seedlings before germination and promote plant performance during their growth after germination. The Epichloë endophytes are one another examples of commercial products based on these organisms, which are popular particularly in New Zealand, Australia, South America, and the United States (Young et al., 2013; Johnson et al., 2013). As these endophytes are able to enhance the plants' survival through protection from abiotic and biotic stresses, they are useful in agriculture. Some endophytes are being adapted by farmers as inoculants in combating various agricultural constraints including biotic and abiotic stresses. Interestingly, commercialization of these endophytes is contributing approximately US $130 million per annum to the economy of New Zealand (Johnson et al., 2013).

## 14.8 Advantages of endophytism over chemical methods of crop protection

The fungal endophytes represent a group of organisms with potential applications in the agriculture sector. The mutualistic relationship between fungal endophytes and their host plants helps plants to cope with biotic and abiotic stresses, whereas fungi obtain nutrients and protection inside the plant tissues. This interaction model explains the plant–endophyte association as endophytism. The endophytes provide protection to their host against various pathogens, are biological in origin, are environmentally friendly, and are potential sources of novel natural agents for exploitation in agricultural applications.

The role of fungal endophytes in plant protection is a fundamental but frequently overlooked aspect of plant biology. These organisms have been recognized as a repository of novel compounds of immense value in agriculture industry. They are excellent biocontrol agents and can be used to reduce the pathogenicity or survival of plant pathogens. Use of these fungi may overcome some of the challenges faced with other methods of plant disease control. There are numerous advantages of endophytism over chemical methods of crop protection.

1. Colonization with fungal endophytes induces genetic resistance against numerous plant pathogens. This reduces the need for chemical fungicides to control many foliar phytopathogens, which have very high sporulation rates and are well suited for widespread dissemination as airborne propagules.

2. Use of systemic fungicides, though helpful in controlling plant pathogens, increases the chances of developing fungicide resistance in these pathogens. The use of endophytic fungi reduces the probability of resistance development and helps in management these pathogens in a better way.
3. Use of synthetic chemicals also affects beneficial microbes in addition to inhibiting pathogens. The formation of effective root nodules with N-fixing bacteria is reported to be inhibited or slowed during pesticide treatment. Therefore, endophytic symbiosis is a useful alternative to chemical pesticides.
4. Use of endophytes as biocontrol agents also lowers the dependency of farmers on synthetic chemicals, reducing the possible exposure of users and the environment to harmful chemicals.
5. In comparison to other methods, biological control with the use of endophytic fungi is an ecofriendly and cost-effective approach. The cultivation, multiplication, handling, and use of these fungi is easy and environmental friendly.

Despite all the advantages, extensive implementation of these biocontrol agents in agriculture remains challenging. Poor marketing and lack of unawareness and belief among farmers in the efficacy of such treatments are some of the challenges preventing use of these biocontrol agents. Moreover, most studies of endophytic fungi are largely confined to laboratory experiments and exact mechanisms of these symbiotic interactions remain unclear.

## 14.9 Conclusion

Endophytic fungi are a new and unexplored area of biocontrol research in the agriculture sector. As endophytes are asymptomatic microorganisms that inhabit the interior of healthy plants and provide protection to the colonized hosts, they offer great untapped potential that can be exploited to maintain healthy crops. These organisms have been investigated to date for their interactions and associations with host plants, production of metabolites, as well as their pesticidal and antimicrobial activity against plant pests and pathogens. The advantages of endophytic fungi present over synthetic pesticides enticed researchers to investigate the potential use of these microbes more rapidly. Some success has been achieved and some fungal endophytes and their metabolites are commercially available.

It has been observed in genomic studies of endophytes that some of the gene clusters encoding selected secondary metabolites are not expressed under standard laboratory cultivation conditions. This raises the need to investigate the physiological and environmental conditions required for endophytic fungi to produce secondary metabolites more efficiently. Although scientific approaches focused on the diversity of endophytes have recently gained momentum, information on endophyte behavior and their mutualistic interactions with crops remains scanty. Additional studies are needed to explore the full potential of endophytic fungi in plant protection and to understand the roles of these fungi in suppression of plant diseases.

## Acknowledgments

The authors gratefully thank their respective organizations for providing every possible support during compilation of the present scientific work. We also express our thanks for encouragement and every possible support provided by everyone during this scientific study.


## References


Abd-Allah, E.F., Hashem, A., Alqarawi, A.A., Bahkali, A.H., Alwhibi, M.S., 2015. Enhancing growth performance and systemic acquired resistance of medicinal plant *Sesbania sesban* (L.) Merr using arbuscular mycorrhizal fungi under salt stress. Saudi J. Biol. Sci. 22 (3), 274–283.

Agostinelli, M., Cleary, M., Martín, J.A., Albrectsen, B.R., Witzell, J., 2018. Pedunculate Oaks (*Quercus robur* L.) differing in vitality as reservoirs for fungal biodiversity. Front. Microbiol. 9, 1758.

Agrios, G.N., 2005. Plant Pathology. Elsevier Academic Press Limited, UK.

Akello, J., Dubois, T., Gold, C.S., Coyne, D., Nakavuma, J., Paparu, P., 2007. ) *Beauveria bassiana* (Balsamo) vuillemin as an endophyte in tissue culture banana (*Musa* spp.). J. Invertebr. Pathol. 96 (1), 34–42.

Ali, A.H., Radwan, U., El-Zayat, S., El-Sayed, M.A., 2018. Desert plant-fungal endophytic association: the beneficial aspects to their hosts. Biol. Forum Int. J. 10 (1), 138–145.

Alvin, A., Miller, K.I., Neilan, B.A., 2014. Exploring the potential of endophytes from medicinal plants as sources of antimycobacterial compounds. Microbiol. Res. 169 (7–8), 483–495.

Alwadi, H.M., Baka, Z.A.M., 2001. Microorganisms associated with *Withania somnifera* leaves. Microbiol. Res. 156 (4), 303–309.

Aly, A.H., Debbab, A., Proksch, P., 2011. Fungal endophytes: unique plant inhabitants with great promises. Appl. Microbiol. Biotechnol. 90 (6), 1829–1845.

Arnold, A.E., 2007. Understanding the diversity of foliar endophytic fungi: progress, challenges, and frontiers. Fungal Biol. Rev. 21 (2–3), 51–66.

Arnold, A.E., 2008. In Endophytic Fungi: Hidden components of tropical community ecology. In: Schnitzer, S., Carson, W. (Eds.), Tropical Forest Community Ecology. Blackwell Publishing, Malden, USA.

Arnold, A.E., Mejía, L.C., Kyllo, D., Rojas, E.I., Maynard, Z., Robbins, N., et al., 2003. Fungal endophytes limit pathogen damage in a tropical tree. Proc. Natl. Acad. Sci. 100 (26), 15649–15654.

Arroyo García, R., Martínez Zapater, J.M., García Criado, B., Zabalgogeazcoa, I., 2002. Genetic structure of natural populations of the grass endophyte *Epichloë festucae* in semiarid grasslands. Mol. Ecol. 11 (3), 355–364.

Atugala, D.M., Deshappriya, N., 2015. Effect of endophytic fungi on plant growth and blast disease incidence of two traditional rice varieties. J. Natl. Sci. Found. Sri Lanka 43 (2), 173–187.

Bacon, C.W., White, J., 2000. Microbial Endophytes. CRC Press, New York, USA.

Bae, H., Kim, S., Sicher Jr, R.C., Kim, M.S., Strem, M.D., Bailey, B.A., 2009. The beneficial endophyte *Trichoderma hamatum* isolate DIS 219b promotes growth and delays the onset of the drought response in *Theobroma cacao*. J. Exp. Bot. 60 (11), 3279–3295.

Bailey, B.A., Bae, H., Strem, M.D., Crozier, J., Thomas, S.E., Samuels, G.J., et al., 2008. Antibiosis, mycoparasitism, and colonization success for endophytic *Trichoderma* isolates with biological control potential in *Theobroma cacao*. Biol. Contr. 46 (1), 24–35.

Bamisile, B.S., Dash, C.K., Akutse, K.S., Keppanan, R., Wang, L., 2018. Fungal endophytes: beyond herbivore management. Front. Microbiol. 9, 544. Available from: https://doi.org/10.3389/fmicb.2018.00544.

Barbara, S., Christine, B., Siegfried, D., Anne-Katrin, R., Karsten, K., 2002. Endophytic fungi: a source of novel biologically active secondary metabolites. Mycol. Res. 106 (9), 996–1004.

Behie, S.W., Jones, S.J., Bidochka, M.J., 2015. Plant tissue localization of the endophytic insect pathogenic fungi *Metarhizium* and *Beauveria*. Fungal Ecol. 13, 112–119.

Bennett, R., Hutmacher, R., Davis, R., Bennett, R., 2008. Seed transmission of *Fusarium oxysporum* f. sp. vasinfectum race 4 in California. J. Cotton Sci. 12 (2), 160–164.

Bérdy, J., 2005. Bioactive microbial metabolites: a personal view. J. Antibiot. 58 (1), 1–26.

Bhardwaj, A., Agrawal, P., 2014. A review fungal endophytes: as a store house of Bioactive compound. World J. Pharm. Pharm. Sci. 3 (9), 228–237.

Bills, G.F., Polishook, J.D., 1994. Abundance and diversity of microfungi in leaf litter of a lowland rain forest in Costa Rica. Mycologia 86 (2), 187–198.

Boddey, R.M., Urquiaga, S., Alves, B.J.R., Reis, V., 2003. Endophytic nitrogen fixation in sugarcane: present knowledge and future application. Plant Soil 252 (1), 139–149.

Bolton, M.D., 2009. Primary metabolism and plant defense fuel for the fire. Mol. Plant Microb. Interact. 22 (5), 487-49.

Bordallo, J.J., Lopez-Llorca, L.V., Jansson, H.B., Salinas, J., Persmark, L., Asensio, L., 2002. Colonization of plant roots by egg-parasitic and nematodetrapping fungi. New Phytol. 154, 491–499.

Bourgaud, F., Gravot, A., Milesi, S., Gontier, E., 2001. Production of plant secondary metabolites: a historical perspective. Plant Sci. 161 (5), 839–851.

Brem, D., Leuchtmann, A., 2001. Epichloë grass endophytes increase herbivore resistance in the woodland grass *Brachypodium sylvaticum*. Oecologia 126 (4), 522–530.

Brundrett, M.C., 2006. Understanding the roles of multifunctional mycorrhizal and endophytic fungi. In: Schulz, B.J.E., Boyle, C.J.C., Sieber, T.N. (Eds.), Microbial Root Endophytes. Springer-Verlag, Berlin, Germany, pp. 281–293.

Bultman, T.L., Murphy, J.C., 2000. Do fungal endophytes mediate wound-induced resistance?. In: Bacon, C.W., White, J.F. (Eds.), Microbial Endophytes. Marcel Dekker, New York, pp. 421–452.

Cao, L.X., You, J.L., Zhou, S.N., 2002. Endophytic fungi from *Musa acuminata* leaves and roots in South China. World J. Microbiol. Biotechnol 18 (2), 169–171.

Cao, R., Liu, X., Gao, K., Mendgen, K., Kang, Z., Gao, J., et al., 2009. Mycoparasitism of endophytic fungi isolated from reed on soilborne phytopathogenic fungi and production of cell wall-degrading enzymes in vitro. Curr. Microbiol. 59 (6), 584–592.

Cardoso, J.A., Odokonyero, K., Rao, I.M., Jimenéz, J.C., Acuña, T.B., 2017. Potential role of fungal endophytes in biological nitrification inhibition in *Brachiaria* grass species. J. Plant Biochem. Physiol. 5, 191. Available from: https://doi.org/10.4172/2329-9029.1000191.

Chareprasert, S., Piapukiew, J., Thienhirun, S., Whalley, A.J.S., Sihanonth, P., 2006. Endophytic fungi of teak leaves *Tectona grandis* L. and rain tree leaves *Samanea saman* Merr. World J. Microbiol. Biotechnol. 22 (5), 481–486.

Chen, L., Zhang, Q.Y., Jia, M., Ming, Q.L., Yue, W., Rahman, K., et al., 2016. Endophytic fungi with antitumor activities: their occurrence and anticancer compounds. Crit. Rev. Microbiol. 42 (3), 454–473.

Chin, Y.W., Balunas, M.J., Chai, H.B., Kinghorn, A.D., 2006. Drug discovery from natural sources. AAPS J. 8 (2), 239–253.

Chithra, S., Jasim, B., Sachidanandan, P., Jyothis, M., Radhakrishnan, E.K., 2014. Piperine production by endophytic fungus *Colletotrichum gloeosporioides* isolated from *Piper nigrum*. Phytomedicine 21 (4), 534–540.

Chomcheon, P., Wiyakrutta, S., Sriubolmas, N., Ngamrojanavanich, N., Isarangkul, D., Kittakoop, P., 2005. 3-Nitropropionic acid (3-NPA), a potent antimycobacterial agent from endophytic fungi: is 3-NPA in some plants produced by endophytes?. J. Nat. Prod. 68 (7), 1103–51105.

Chowdhary, K., Kaushik, N., 2015. Fungal endophyte diversity and bioactivity in the Indian medicinal plant *Ocimum sanctum* Linn. PLoS One 10 (11), e0141444.

Chun, E., Han, C.K., Yoon, J.H., Sim, T.B., Kim, Y.K., Lee, K.Y., 2005. Novel inhibitors targeted to methionine aminopeptidase 2 (MetAP2) strongly inhibit the growth of cancers in xenografted nude model. Int. J. Cancer 114 (1), 124–130.

Chutulo, E.C., Chalannavar, K.C., 2018. Endophytic mycoflora and their bioactive compounds from *Azadirachta indica*: a comprehensive review. J. Fungi 4 (2), 42. Available from: https://doi.org/10.3390/jof4020042.

Clarke, B.B., White, J.F., Hurley, H., Torres, M.S., Sun, S., Huff, D.R., 2006. Endophyte-mediated suppression of dollar spot disease in fine fescues. Plant Dis. 90 (8), 994–998.

Clay, K., Schardl, C., 2002. Evolutionary origins and ecological consequences of endophyte symbiosis with grasses. Am. Nat. 160 (4), S99–S127.

Cocq, K., Gurr, S.J., Hirsch, P.R., Mauchline, T.H., 2017. Exploitation of endophytes for sustainable agricultural intensification. Mol. Plant Pathol. 18 (3), 469–473.

Cohen, S.D., 2006. Host selectivity and genetic variation of *Discula umbrinella* isolates from two oak species: analyses of intergenic spacer region sequences of ribosomal DNA. Microb. Ecol. 52 (3), 463–469.

Curran, M.P., Keating, G.M., 2005. Mycophenolate sodium delayed release: prevention of renal transplant rejection. Drugs 65 (6), 799–805.

Dai, C.C., Yu, B.Y., Li, X., 2008. Screening of endophytic fungi that promote the growth of *Euphorbia pekinensis*. Afr. J. Biotechnol. 7 (19), 3505–3509.

Darkes, M.J.M., Plosker, G.L., 2002. Cefditoren pivoxil. Drugs 62 (2), 319–336.

Das, A., Varma, A., 2009. Symbiosis: the art of living. In: Varma, A., Kharkwal, A.C. (Eds.), Symbiotic Fungi: Principles and Practice. Springer-Verlag, Berlin, pp. 1–28.

Davis, R.A., Andjic, V., Kotiw, M., Shivas, R.G., 2005. Phomoxins B and C: polyketides from an endophytic fungus of the genus *Eupenicillium*. Phytochemistry 66 (23), 2771–2775.

De Bary, A., 1866. Morphologie und Physiologie der Pilze, Flechten und Myxomyceten. Verlag von Wilhelm Engelmann, Leipzig, UK.

Delaye, L., García-Guzmán, G., Heil, M., 2013. Endophytes versus biotrophic and necrotrophic pathogens are fungal lifestyles evolutionarily stable traits? Fungal Divers. 60 (1), 125–135.

Demain, A.L., 2000. Microbial natural products: a past with a future. In: Wrigley, S.K., Hayes, M.A., Thomas, R., Chrystal, E.J.T., Nicholson, N. (Eds.), Biodiversity: New Leads for Pharmaceutical and Agrochemical Industries. The Royal Society of Chemistry, Cambridge, UK, pp. 3–16.

Desale, M., Bodhankar, M.G., 2013. Antimicrobial activity of endophytic fungi isolated from *Vitex negundo* Linn. Int. J. Curr. Microbiol. Appl. Sci. 2 (12), 389–395.

Ding, G., Zheng, Z., Liu, S., Zhang, H., Guo, L., Che, Y., 2009. Photinides A-F, cytotoxic benzofuranone-derived $\gamma$- lactones from the plant endophytic fungus *Pestalotiopsis photiniae*. J. Nat. Prod. 72 (5), 942–945.

Dingle, J., McGee, P.A., 2003. Some endophytic fungi reduce the density of pustules of *Puccinia recondita* f. sp. *tritici* in wheat. Mycol. Res. 107 (3), 310–316.

Dissanayake, R.K., Ratnaweera, P.B., Williams, D.E., Wijayarathne, C.D., Wijesundera, R.L. C., Andersen, R.J., et al., 2016a. Antimicrobial activities of mycoleptodiscin B isolated from endophytic fungus *Mycoleptodiscus* sp. of *Calamus thwaitesii* Becc. J. Appl. Pharm. Sci. 6 (1), 1–6.

Dissanayake, R.K., Ratnaweera, P.B., Williams, D.E., Wijayarathne, C.D., Wijesundera, R.L. C., Andersen, R.J., et al., 2016b. Antimicrobial activities of endophytic fungi of the Sri Lankan aquatic plant *Nymphaea nouchali* and chaeoglobosin A and C, produced by the endophytic fungus *Chaetomium globosum*. Mycology 7 (1), 1–8.

Donadio, S., Monicardini, P., Alduina, R., Mazzaa, P., Chiocchini, C., Cavaletti, L., et al., 2002. Microbial technologies for the discovery of novel bioactive metabolites. J. Biotechnol. 99 (3), 187–198.

Dongyi, H., Kelemu, S., 2004. *Acremonium implicatum*, a seed-transmitted endophytic fungus in *Brachiaria* grasses. Plant Dis. 88 (11), 1252–1254.

Dudeja, S.S., Giri, R., 2014. Beneficial properties, colonization, establishment and molecular diversity of endophytic bacteria (review). Afr. J. Microbiol. Res. 8 (15), 1562–1572.

Dudeja, S.S., Giri, R., Saini, R., Suneja-Madan, P., Kothe, E., 2012. Interaction of endophytic microbes with legumes. J. Basic Microbiol. 52 (3), 248–260.

Dutta, D., Puzari, K.C., Gogoi, R., Dutta, P., 2014. Endophytes: exploitation as a tool in plant protection. Braz. Arch. Biol. Technol. 57 (5), 621–629.

Easton, H.S. & Fletcher, L.R. (2007) The importance of endophytes in agricultural system: changing plant and productivity. New Zealand Grassland Association: Endophyte Symposium, pp. 11–18.

Egamberdieva, D., Lugtenberg, B., 2014. Use of plant growth-promoting rhizobacteria to alleviate salinity stress in plants. In: Miransari, M. (Ed.), Use of Microbes for the Alleviation of Soil Stresses. Springer, New York, USA, pp. 73–96.

Elena, G.J., Beatriz, P.J., Alejandro, P., Lecuona, R., 2011. *Metarhizium anisopliae* (Metschnikoff) Sorokin promotes growth and has endophytic activity in tomato plants. Adv. Biol. Res. 5 (1), 22–27.

El-Nagerabi, S.A.F., Elshafie, A.E., Alkhanjari, S.S., 2014. Endophytic fungi associated with endogenous *Boswellia sacra*. Biodiversitas 15 (1), 24–30.

Evans, H.C., Holmes, K.A., Thomas, S.E., 2003. Endophytes and mycoparasites associated with an indigenous forest tree, *Theobroma gileri*, in Ecuador and a preliminary assessment of their potential as biocontrol agents of cocoa diseases. Mycol. Progr. 2 (2), 149–160.

Ezra, D., Hess, W.M., Strobel, G.A., 2004. New endophytic isolates of *Muscodor albus*, a volatile-antibiotic-producing fungus. Microbiology 150 (Pt 12), 4023–4031.

Faeth, S.H., Fagan, W.F., 2002. Fungal endophytes: common host plant symbionts but uncommon mutualists. Integr. Comp. Biol. 42 (2), 360–368.

Firáková, S., Šturdíková, M., Múčková, M., 2007. Bioactive secondary metabolites produced by microorganisms associated with plants. Biologia 62 (3), 251–257.

Frattarelli, D.A.C., Reed, M.D., Giacoia, G.P., Aranda, J.V., 2004. Antifungals in systemic neonatal candidiasis. Drugs 64 (9), 949–968.

Freeman, S., Rodrigues, R.J., 1993. Genetic conversion of a fungal plant pathogen to a nonpathogenic, endophytic mutualist. Science 260 (5104), 75–78.

Gangadevi, V., Muthumary, J., 2008. Taxol, an anticancer drug produced by an endophytic fungus *Bartalinia robillardoides* Tassi, isolated from medicinal plant, *Aegle marmelos* Correa ex Roxb. World J. Microbiol. Biotechnol. 24 (717), 717–724.

Ganley, R.J., Sniezko, R.A., Newcombe, G., 2008. Endophyte-mediated resistance against white pine blister rust in *Pinus monticola*. Forest Ecol. Manag. 255 (7), 2751–2760.

Gao, K., Mendgen, K., 2006. Seed-transmitted beneficial endophytic *Stagonospora* sp. can penetrate the walls of the root epidermis, but does not proliferate in the cortex of *Phragmites australis*. Can. J. Bot. 84 (6), 981–988.

Gao, F., Dai, C., Liu, X., 2010. Mechanisms of fungal endophytes in plant protection against pathogens. Afr. J. Microbiol. Res. 4 (13), 1346–1351.

Gashgari, R., Gherbawy, Y., Ameen, F., Alsharari, S., 2016. Molecular characterization and analysis of antimicrobial activity of endophytic fungi from medicinal plants in Saudi Arabia. Jundishapur J. Microbiol. 9 (1), e26157.

Gautam, A.K., 2014. Diversity of fungal endophytes in some medicinal plants of Himachal Pradesh. India. Arch. Phytopathol. Plant Protect. 47 (5), 537–544.

Gautam, A.K., Kant, M., Thakur, Y., 2013. Isolation of endophytic fungi from *Cannabis sativa* and study their antifungal potential. Arch. Phytopathol. Plant Protect. 46 (6), 627–635.

Geris dos Santos, R.M., Rodrigues-Fo, E., Rocha, W.C., Teixeira, M.F.S., 2003. Endophytic fungi from *Melia azedarach*. World J. Microbiol. Biotechnol 19 (8), 767–770.

Gherbawy, Y.A., Gashgari, R.M., 2013. Molecular characterization of fungal endophytes from *Calotropis procera* plants in Taif region (Saudi Arabia) and their antifungal activities. Plant Biosyst. 148 (6), 1085–1092.

Giménez, C., Cabrera, R., Reina, M., González-Coloma, A., 2007. Fungal endophytes and their role in plant protection. Curr. Org. Chem. 11 (8), 707–720.

Gokul Raj, K., Sundaresan, N., Ganeshan, E.J., Rajapriya, P., Muthumary, J., Sridhar, J., et al., 2014. Phylogenetic reconstruction of endophytic fungal isolates using internal transcribed spacer 2 (ITS2) region. Bioinformation 10 (6), 320–328.

Gonzalez-Teuber, M., Jimenez-Aleman, G.H., Boland, W., 2014. Foliar endophytic fungi as potential protectors from pathogens in myrmecophytic *Acacia* plants. Commun. Integr. Biol. 7 (5), 1–4.

González-Teuber, M., Vilo, C., Bascuñán-Godoy, L., 2017. Molecular characterization of endophytic fungi associated with the roots of *Chenopodium quinoa* inhabiting the Atacama Desert, Chile. Genom. Data 11, 109–112.

Gouda, S., Das, G., Sen, S.K., Shin, H.-S., Patra, J.K., 2016. Endophytes: a treasure house of bioactive compounds of medicinal importance. Front. Microbiol. 7, 1538. Available from: https://doi.org/10.3389/fmicb.2016.01538.

Goyal, S., Ramawat, K.G., Mérillon, J.M., 2017. Different shades of fungal metabolites: An Overview. In: Mérillon, J.M., Ramawat, K.G. (Eds.), Fungal Metabolites. Springer International Publishing, Switzerland.

Grosch, R., Scherwinski, K., Lottmann, J., Berg, G., 2006. Fungal antagonists of the plant pathogen *Rhizoctonia solani*: selection, control efficacy and influence on the indigenous microbial community. Mycol. Res. 110 (Pt12), 1464–1474.

Gunatilaka, A.A.L., 2006. Natural products from plant-associated microorganisms: Distribution, structural diversity, bioactivity, and implications of their occurrence. J. Nat. Prod. 69 (3), 509–526.

Guo, B., Wang, Y., Sun, X., Tang, K., 2008. Bioactive natural products from endophytes: a review. Appl. Biochem. Microbiol. 44 (2), 136–142.

Halo, B.A., Maharachchikumbura, S., Al-Yahyai, R., Al-Sadi, A.M., 2019. *Cladosporium omanense*, a new endophytic species from *Zygophyllum coccineum* in Oman. Phytotaxa 388 (1), 145.

Hamayun, M., Khan, S.A., Iqbal, I., Na, C.-I., Khan, A.L., Hwang, Y.H., et al., 2009. *Chrysosporium pseudomerdarium* produces gibberellins and promotes plant growth. J. Microbiol. 47 (4), 425–430.

Hardoim, P.R., Van Overbeek, L.S., Berg, G., Pirttilä, A.M., Compant, S., Campisano, A., et al., 2015. The hidden world within plants: ecological and evolutionary considerations for defining functioning of microbial endophytes. Microbiol. Mol. Biol. Rev. 79 (3), 293–320.

Harmon, G.E., Howell, C.R., Viterbo, A., Chet, I., Lorito, M., 2004. *Trichoderma* species–opportunistic, avirulent plant symbionts. Nat. Rev. Microbiol. 2 (1), 43–56.

Hartley, S.E., Gange, A.C., 2009. Impacts of plant symbiotic fungi on insect herbivores: mutualism in a multitrophic context. Annu. Rev. Entomol. 54 (20), 323–342.

Higgins, K.L., Arnold, A.E., Miadlikowska, J., Sarvate, S.D., Lutzoni, F., 2007. Phylogenetic relationships, host affinity, and geographic structure of boreal and arctic endophytes from three major plant lineages. Mol. Phylogenet. Evol. 42 (2), 543–555.

Hipol, R.M., 2012. Molecular identification and phylogenetic affinity of two growth promoting fungal endophytes of sweet potato (*Ipomea batatas* (L.) Lam.) from Baguio City, Philippines. Electron. J. Biol. 8 (3), 57–61.

Hipol, R.M., Tamang, S.M.A., Gargabite, B.F., Broñola-Hipol, R.L.C., 2015. Diversity of fungal endophytes isolated from *Marchantia polymorpha* populations from Baguio City, Philippines. Bull. Environ. Pharmacol. Life Sci. 4 (3), 87–91.

Huang, W.Y., Cai, Y.Z., Surveswaran, S., Hyde, K.D., Corke, H., Sun, M., 2009. Molecular phylogenetic identification of endophytic fungi isolated from three *Artemisia* species. Fungal Divers. 36, 69–88.

Huang, Y., Zimmerman, N.B., Arnold, A.E., 2018. Observations on the early establishment of foliar endophytic fungi in leaf discs and living leaves of a model woody angiosperm, *Populus trichocarpa* (Salicaceae). J. Fungi 4 (2), E58. Available from: https://doi.org/10.3390/jof4020058.

Hyde, K.D., Soytong, K., 2008. The fungal endophyte dilemma. Fungal Divers. 33, 163–173.

Ilyas, M., Kanti, A., Jamal, Y., Hertina, Agusta, A., 2009. Biodiversity of endophytic fungi associated with *Uncaria gambier* Roxb. (Rubiaceae) from west Sumatra. Biodiversitas 10 (1), 23–28.

Impullitti, A., Malvick, D., 2013. Fungal endophyte diversity in soybean. J. Appl. Microbiol. 114 (5), 1500–1506.

Inácio, M.L., Silva, G.H., Teles, H.L., Trevisan, H.C., Cavalheiro, A.J., Bolzani, V.S., et al., 2006. Antifungal metabolites from *Colletotrichum gloeosporioides*, an endophytic fungus in *Cryptocarya mandioccana* Nees (Lauraceae). Biochem. Syst. Ecol. 34 (11), 822–824.

Isaka, M., Jaturapat, A., Rukseree, K., Danwisetkanjana, K., Tanticharoen, M., Thebtaranonth, Y., 2001. Phomoxanthones A and B, novel xanthone dimers from the endophytic fungus *Phomopsis* species. J. Nat. Prod. 64 (8), 1015–1018.

Jaber, L.R., Enkerli, J., 2016. Effect of seed treatment duration on growth and colonization of *Vicia faba* by endophytic *Beauveria bassiana* and *Metarhizium brunneum*. Biol. Contr 103, 187–195.

Jaber, L.R., Enkerli, J., 2017. Fungal entomopathogens as endophytes: can they promote plant growth? Biocontr. Sci. Technol. 27 (1), 28–41.

Jia, M., Chen, L., Xin, H.-L., Zheng, C.-J., Rahman, K., Han, T., et al., 2016. A friendly relationship between endophytic fungi and medicinal plants: a systematic review. Front. Microbiol. 7, 906.

Johnson, L.J., De Bonth, A.M., Briggs, L.R., Caradus, J.R., Finch, S.C., Fleetwood, D.J., et al., 2013. The exploitation of *Epichloae* endophytes for agricultural benefit. Fungal Divers. 60 (1), 171–188.

Johnson, J.M., Alex, T., Oelmüller, R., 2014. *Piriformospora indica*: theversatile and multifunctional root endophytic fungus for enhanced yield and tolerance to biotic and abiotic stress in crop plants. J. Trop. Agric. 52 (2), 103–122.

Junker, C., Draeger, S., Schulz, B., 2012. A fine line—endophytes or pathogens in *Arabidopsis thaliana*. Fungal Ecol. 5 (6), 657–662.

Khan, R., 2007. Isolation, identification and cultivation of endophytic fungi from medicinal plants for the production and characterization of bioactive fungal metabolites. Ph.D. Thesis. University of Karachi, Karachi, Pakistan.

Khan, A.L., Kang, S.M., Dhakal, K.H., Hussain, J., Adnan, M., Kim, J.G., et al., 2013. Flavonoids and amino acid regulation in *Capsicum annuum* L. by endophytic fungi under different heat stress regimes. Sci. Horticult. 155, 1–7.

Khiralla, A., Spina, R., Yagi, S., Mohamed, I., Laurain-Mattar, D., 2017. Endophytic fungi: occurrence, classification, function and natural products. In: Hughes, E. (Ed.), Endophytic Fungi Diversity, Characterization and Biocontrol. Nova Publishers, New York, USA, pp. 1–38.

Kim, S., Shin, D., Lee, T.O.K., 2004. Periconicins, two new fusicoccane diterpenes produced by an endophytic fungus *Periconia* sp. with antibacterial activity. J. Nat. Prod. 67 (3), 448–450.

Kim, H.Y., Choi, G.J., Lee, H.B., Lee, S.W., Kim, H.K., Jang, K.S., et al., 2007. Some fungal endophytes from vegetable crops and their anti-oomycete activities against tomato late blight. Lett. Appl. Microbiol. 44 (3), 332–337.

Király, L., Barna, B., Király, Z., 2007. Plant resistance to pathogen infection: forms and mechanisms of innate and acquired resistance. J. Phytopathol. 155 (7–8), 385–396.

Kock, J.L.F., Strauss, T., Pohl, C.H., Smith, D.P., Botes, P.J., Pretorius, E.E., et al., 2001. Bioprospecting for novel oxylipins in fungi: the presence of 3-hydroxy oxylipins in *Pilobolus*. Anton. Leeuw. 80 (1), 93–99.

Krings, M., Taylor, T.N., Hass, H., Kerp, H., Dotzler, N., Hermsen, E.J., 2007. Fungal endophytes in a 400-millionyr-old land plant: infection pathways, spatial distribution, and host responses. New Phytol. 174 (3), 648–657.

Kuldau, G., Bacon, C., 2008. Clavicipitaceous endophytes: their ability to enhance resistance of grasses to multiple stresses. Biol. Contr 46 (1), 57–71.

Lee, S., Flores-Encarnación, M., Contreras-Zentella, M., Garcia-Flores, L., Escamilla, J.E., Kennedy, C., 2004. Indole-3-acetic acid biosynthesis is deficient in Gluconacetobacter diazotrophicus strains with mutations in *Cytochrome cbiogenesis* genes. J. Bacteriol. 186 (16), 5384–5391.

Lehtonen, P.T., Helander, M., Siddiqui, S.A., Lehto, K., Saikkonen, K., 2006. Endophytic fungus decreases plant virus infections in meadow ryegrass (*Lolium pratense*). Biol. Lett., 22 2 (4), 620–623.

Leuchtmann, A., Schmidt, D., Bush, L., 2000. Different levels of protective alkaloids in grasses with stroma-forming and seed-transmitted *Epichloë/Neotyphodium* endophytes. J. Chem. Ecol 26 (4), 1025–1036.

Lewis, G.C., 2004. Effects of biotic and abiotic stress on the growth of three genotypes of *Lolium perenne* with and without infection by the fungal endophyte *Neotyphodium lolii*. Ann. Appl. Biol. 144 (1), 53–63.

Li, J.Y., Strobel, G.A., 2001. Jesterone and hydroxy-jesterone antioomycete cyclohexenenone epoxides from the endophytic fungus *Pestalotiopsis jesteri*. Phytochemistry 57 (2), 261–265.

Li, X., Zhang, L., 2015. Endophytic infection alleviates $Pb^{2+}$ stress effects on photosystem II functioning of *Oryza sativa* leaves. J. Hazard. Mater. 295, 79–85.

Li, J.Y., Strobel, G., Harper, J., Lobkovsky, E., Clardy, J., 2000. Cryptocin, a potent tetramic acid antimycotic from the endophytic fungus *Cryptosporiopsis cf. quercina*. Org. Lett. 2 (6), 767–770.

Li, J.Y., Harper, J.K., Grant, D.M., Tombe, B.O., Bashyal, B., Hess, W.M., et al., 2001. Ambuic acid, a highly functionalized cyclohexenone with antifungal activity from *Pestalotiopsis* spp. and *Monochaetia* sp. Phytochemistry 56 (5), 463–468.

Li, H.M., Sullivan, R., Moy, M., Kobayashi, D.Y., Belanger, F.C., 2004. Expression of a novel chitinase by the fungal endophyte in Poaampla. Mycologia 96 (3), 526–536.

Li, E., Jiang, L., Guo, L., Zhang, H., Che, Y., 2008. Pestalachlorides A–C, antifungal metabolites from the plant endophytic fungus *Pestalotiopsis adusta*. Bioorg. Med. Chem. 16 (17), 7894–7899.

Liao, X., O'brien, T.R., Fang, W., Leger, R.J.S., 2014. The plant beneficial effects of Metarhizium species correlate with their association with roots. Appl. Microbiol. Biotechnol. 98 (16), 7089–7096.

Limsuwan, S., Trip, E.N., Kouwenc, T.R.H.M., Piersmac, S., Hiranrat, A., Mahabusarakam, W., et al., 2009. Rhodomyrtone: a new candidate as natural antibacterial drug from *Rhodomyrtus tomentosa*. Phytomedicine 16 (6–7), 645–651.

Liu, C.H., Zou, W.X., Lu, H., Tan, R.X., 2001. Antifungal activity of *Artemisia annua endophyte* cultures against phytopathogenic fungi. J. Biotechnol. 88 (3), 277–282.

Lockwood, J.L., 1992. Exploitation competition. In: Carroll, G.C., Wicklow, D.T. (Eds.), The Fungal Community—Its Organization and Role in the Ecosystem. CRC Press, Dekker, New York, pp. 243–263.

Loiret, F.G., Ortega, E., Kleiner, D., Ortega-Rodes, P., Rodes, R., Dong, Z., 2004. A putative new endophytic nitrogen–fixing bacterium *Pantoea* sp. from sugarcane. J. Appl. Microbiol. 97 (3), 504–511.

Lopez, D.C., Sword, G.A., 2015. The endophytic fungal entomopathogens *Beauveria bassiana* and *Purpureocillium lilacinum* enhance the growth of cultivated cotton (*Gossypium hirsutum*) and negatively affect survival of the cotton bollworm (*Helicoverpa zea*). Biol. Contr 89, 53–60.

Lu, H., Zou, W.X., Meng, J.C., Hu, J., Tan, R.X., 2000. New bioactive metabolites produced by *Colletotrichum* sp., an endophytic fungus in *Artemisia annua*. Plant Sci. 151 (1), 67–73.

Ludwig-Müller, J., 2015. Plants and endophytes: equal partners in secondary metabolite production? Biotechnol. Lett. 37 (7), 1325–1334.

Lugtenberg, B.J.J., Caradus, J.R., Johnson, L.J., 2016. Fungal endophytes for sustainable crop production. FEMS Microbiol. Ecol. 92 (12), fiw194. Available from: https://doi.org/10.1093/femsec/fiw194.

Malinowski, D.P., Belesky, D.P., 2000. Adaptations of endophyte-infected cool-season grasses to environmental stresses: mechanisms of drought and mineral stress tolerance. Crop Sci 40 (4), 923–940.

Malinowski, D.P., Zuo, H., Belesky, D.P., Alloush, G.A., 2004. Evidence for copper binding by extracellular root exudates of tall fescue but not perennial ryegrass infected with *Neotyphodium* spp. endophytes. Plant Soil 267 (1–2), 1–12.

Mandyam, K., Jumpponen, A., 2005. Seeking the elusive function of the root-colonising dark septate endophyte. Stud. Mycol. 53, 173–189.

Mane, R.S., Vedamurthy, A.B., 2018. The fungal endophytes: sources and future prospects. J. Med. Plants Stud. 6 (2), 121–126.

Márquez, L.M., Redman, R.S., Rodriguez, R.J., Roossinck, M.J., 2007. A virus in a fungus in a plant: three way symbiosis required for thermal tolerance. Science 315 (5811), 513–515.

McMorris, T.C., Kelner, M.J., Wang, W., Yu, J., Estes, L.A., Taetle, R., 1996. Hydroxymethyl) acylfulvene: an illudin derivative with superior antitumour properties. J. Nat. Prod. 59 (9), 896–899.

Meenatchi, A., Ramesh, V., Bagyalakshmi, Kuralarasi, R., Shanmugaiah, V., Rajendran, A., 2016. Diversity of endophytic fungi from the ornamental plant—*Adenium obesum*. Stud. Fungi 1 (1), 34–42.

Mehrotra, R.S., Aggarwal, A., 2003. Plant Pathology, second ed Mc Graw Hill Education (India) Private Limited, New Delhi, India.

Mejía, L.C., Rojas, E.I., Maynard, Z., Bael, S.V., Arnold, A.E., Hebbar, P., et al., 2008. Endophytic fungi as biocontrol agents of *Theobroma cacao* pathogens. Biol. Contr. 46 (1), 4–14.

Miles, C.O., Lane, G.A., di Menna, M.E., Garthwaite, I., Piper, E.L., Ball, O.J.-P., et al., 1996. High levels of ergonovine and lysergic acid amide in toxic *Achnatherum inebrians* accompany infection by an *Acremonium*-like endophytic fungus. J. Agric. Food Chem. 44 (5), 1285–1290.

Miller, C.M., Miller, R.V., Garton-Kenny, D., Redgravc, B., Sears, J., Condron, M.M., et al., 1998. Ecomycin unique antimycotics from *Pseudomonas viridiflava*. J. Appl. Microbiol. 84, 937–944.

Mirjalili, M.H., Farzaneh, M., Bonfill, M., Rezadoost, H., Ghassempour, A., 2012. Isolation and characterization of *Stemphylium sedicola* SBU-16 as a new endophytic taxol-producing fungus from *Taxus baccata* grown in Iran. FEMS Microbiol. Lett. 328 (2), 122–129.

Mitchell, A.M., Strobel, G.A., Moore, E., Robison, R., Sears, J., 2010. Volatile antimicrobials from *Muscodor crispans*, a novel endophytic fungus. Microbiology, 156 (Pt- 1, 270–277.

Monali, G. Desale, Bodhankar, M.G., 2013. Antimicrobial activity of endophytic fungi isolated from *Vitex negundo* Linn. Int. J. Curr. Microbiol. Appl. Sci. 2 (12), 389–395.

Montesinos, E., 2003. Plant-associated microorganisms: a view from the scope of microbiology. Int. Microbiol. 6 (4), 221–223.

Moon, C.D., Tapper, B.A., Scott, B., 1999. Identification of Epichloëendophytes in planta by a microsatellite-based PCR fingerprinting assay with automated analysis. Appl. Environ. Microbiol. 65 (3), 1268–1279.

Murphy, B.R., Doohan, F.M., Hodkinson, T.R., 2015. Fungal root endophytes of a wild barley species increase yield in a nutrient-stressed barley cultivar. Symbiosis 65 (1), 1–7.

Murphy, B.R., Doohan, F.M., Hodkinson, T.R., 2018. From concept to commerce: developing a successful fungal endophyte inoculant for agricultural crops. J. Fungi 4 (1), E24.

Neubert, K., Mendgen, K., Brinkmann, H., Wirsel, S.G.R., 2006. Only few fungal species dominate highly diverse mycofloras associated with the common reed. Appl. Environ. Microbiol. 72 (2), 1118–1128.

Nisa, H., Kamili, A.N., Nawchoo, I.A., Shaf, S., Shameem, N., Bandh, S.A., 2015. Fungal endophytes as prolific source of phytochemicals and other bioactive natural products: a review. Microb. Pathol. 82, 50–59.

Nithya, K., Muthumary, J., 2010. Secondary metabolite from *Phomopsis* sp. isolated from *Plumeria acutifolia*. Recent Res. Sci. Technol. 2 (4), 99–103.

Owen, N.L., Hundley, N., 2004. Endophytes-the chemical synthesizers inside plants. Sci. Progr. 87 (Pt2), 79–99.

Ownley, B.H., Gwinn, K.D., Vega, F.E., 2010. Endophytic fungal entomopathogens with activity against plant pathogens: ecology and evolution. BioControl 55 (1), 113–128.

Pan, B.F., Su, X., Hu, B., Yang, N., Chen, Q., Wu, W., 2015. *Fusarium redolens* 6WBY3, an endophytic fungus isolated from *Fritillaria unibracteata var. wabuensis*, produces peimisine and imperialine-3β-d-glucoside. Fitoterapia 103, 213–221.

Panaccione, D.G., 2005. Origins and significance of ergot alkaloid diversity in fungi. FEMS Microbiol. Lett. 251 (1), 9–17.

Panaccione, D.G., Kotcon, J.B., Schardl, C.L., Johnson, R.D., Morton, J.B., 2006. Ergot alkaloids are not essential for endophytic fungus-associated population suppression of the lesion nematode, *Pratylenchus scribneri*, on perennial ryegrass. Nematology 8 (4), 583–590.

Pandya, U., Saraf, M., 2009. Application of fungi as a biocontrol agent and their bio fertilizer potential in agriculture. Int. J. Adv. Res. Dev. 1 (1), 90–99.

Park, J.H., Choi, G.J., Lee, H.B., Kim, K.M., Jung, H.S., Lee, S.W., et al., 2005. Griseofulvin from *Xylaria* sp. Strain F0010, an endophytic fungus of *Abies holophylla* and its antifungal activity against plant pathogenic fungi. J. Microbiol. Biotechnol. 15 (1), 112–117.

Park, J.H., Park, J.H., Choi, G.J., Lee, S.W., Jang, K.S., Choi, Y.H., et al., 2013. Screening for antifungal endophytic fungi against six plant pathogenic fungi. Mycobiology 31 (3), 179–182.

Patle, P.N., Navnage, N.P., Ramteke, P.R., 2018. Endophytes in plant system: roles in growth promotion, mechanism and their potentiality in achieving agriculture sustainability. Int. J. Chem. Stud. 6 (1), 270–274.

Paul, N.C., Lee, H.B., Lee, J.H., Shin, K.S., Ryu, T.H., Kwon, H.R., et al., 2014. Endophytic fungi from *Lycium chinense* Mill and characterization of two new Korean records of *Colletotrichum*. Int. J. Mol. Sci. 15 (9), 15272–15286.

Peláez, F., 2005. Biological activities of fungal metabolites. In: An, Z. (Ed.), Handbook of Industrial Mycology. Marcel Dekker, New York, USA.

Pelaez, F., Cabello, A., Platas, G., Diez, M.T., Del Val, A.G., Basilio, A., et al., 2000. The discovery of enfumafungin, a novel antifungal compound produced by an endophytic *Hormonema* species biological activity and taxonomy of the producing organisms. Syst. Appl. Microbiol. 23 (3), 333–343.

Petrini, O., 1991. Fungal endophytes of tree leaves. In: Andrews, J.H., Hirano, S.S. (Eds.), Microbial Ecology of Leaves. Springer, New York, USA, pp. 179–197.

Photita, W., Lumyong, S., Lumyong, P., Hyde, K.D., 2001. Endophytic fungi of wild banana (*Musa aciminata*) at Doi Suthep Pui National Park, Thailand. Mycol. Res. 105 (12), 1508–1514.

Pimentel, I.C., Glienke-Blanco, C., Gabardo, J., Stuart, R.M., Azevedo, J.L., 2006. Identification and colonization of endophytic fungi from soybean (*Glycine max* (L.) Merril) under different environmental conditions. Braz. Arch. Biol. Technol. 49 (5), 705–711.

Pinto, L.S.R.C., Azevedo, J.L., Pereira, J.O., Vieira, M.L.C., Labate, C.A., 2000. Symptomless infection of banana and maize by endophytic fungi impairs photosynthetic efficiency. New Phytol. 147 (3), 609–615.

Pirttila, A.M., Laukkanen, H., Hohtola, A., 2002. Chitinase production in pine callus (*Pinus sylvestris* L.): a defense reaction against endophytes?. Planta 214 (6), 848–852.

Porras-Alfaro, A., Bayman, P., 2011. Hidden fungi, emergent properties: endophytes and microbiomes. Ann. Rev. Phytopathol 49, 291–315.

Potshangbam, M., Devi, S.I., Sahoo, D., Strobel, G.A., 2017. Functional characterization of endophytic fungal community associated with *Oryza sativa* L. and *Zea mays* L. Front. Microbiol. 8, 325.

Powthong, P., Jantrapanukorn, B., Thongmee, A., Suntornthiticharoen, P., 2013. Screening of antimicrobial activities of the endophytic fungi isolated from *Sesbania grandiflora* (L.) Pers. J. Agric. Sci. Technol. 15, 1513–1522.

Pradeep, S.M., Mahmood, R., Jagadeesh, K.S., 2010. Screening and characterization of Lasparaginase producing microorganisms from Tulsi (*Ocimum sanctum* L.), Karnataka. J. Agric. Sci. 23 (4), 660–661.

Premalatha, K., Kalra, A., 2013. Molecular phylogenetic identification of endophytic fungi isolated from resinous and healthy wood of *Aquilaria malaccensis*, a red listed and highly exploited medicinal tree. Fungal Ecol. 6 (3), 205–211.

Promputtha, I., Lumyong, S., Dhanasekaran, V., Mckenzie, E.H.C., Hyde, K.D., Jeewon, R., 2007. A phylogenetic evaluation of whether endophytes become saprotrophs at host senescence. Microb. Ecol. 53 (4), 579–590.

Purahong, W., Hyde, K.D., 2011. Effects of fungal endophytes on grass and non-grass litter decomposition rates. Fungal Divers. 47, 1–7.

Quesada-Moraga, E., López-Díaz, C., Landa, B.B., 2014. The hidden habit of the entomopathogenic fungus *Beauveria bassiana*: first demonstration of vertical plant transmission. PLoS One 9 (2), e89278. Available from: https://doi.org/10.1371/journal.pone.0089278.

Rajamani, T., Suryanarayanan, T.S., Murali, T.S., Thirunavukkarasu, N., 2018. Distribution and diversity of foliar endophytic fungi in the mangroves of Andaman Islands, India. Fungal Ecol. 36, 109–116.

Rajamanikyam, M., Vadlapudi, V., Amanchy, R., Upadhyayula, S.M., 2017. Endophytic fungi as novel resources of natural therapeutics. Braz. Arch. Biol. Technol. 60, e17160542.

Ratnaweera, P.B., de Silva, D.E., Williams, E.D., Anderson, A.J., 2015. Antimicrobial activities of endophytic fungi obtained from the arid zone invasive plant *Opuntia dillenii* and the isolation of equisetin, from endophytic *Fusarium* sp. BMC Complem. Alternat. Med 15, 220.

Redecker, D., Kodner, R., Graham, L.E., 2000. Glomalean fungi from the Ordovician. Science 289 (5486), 920–1921.

Redman, R.S., Dunigan, D.D., Rodríguez, R.J., 2001. Fungal symbiosis from mutualism to parasitism: who controls the outcome, host or invader? New Phytol. 151 (3), 705–716.

Redman, R.S., Sheehan, K.B., Stout, R.G., Rodriguez, R.J., Henson, J.M., 2002. Thermo tolerance conferred to plant host and fungal endophyte during mutualistic symbiosis. Science 298 (5598), 1581.

Remy, W., Taylor, T.N., Hass, H., Kerp, H., 1994. Four hundred-million-year-old vesicular arbuscular mycorrhizae. Proc. Natl. Acad. Sci. USA 91 (25), 11841–11843.

Ren, C.G., Dai, C.C., 2012. Jasmonic acid is involved in the signaling pathway for fungal endophyte-induced volatile oil accumulation of *Atractylodes lancea* plantlets. BMC Plant Biol. 12, 128.

Richmond, D.S., Grewal, P.S., Cardina, J., 2004. Influence of Japanese beetle *Popillia japonica* larvae and fungal endophytes on competition between turfgrasses and dandelion. Crop Sci 44 (2), 600–606.

Rivera Varas, V.V., Freeman, T.A., Gusmestad, N.C., Secor, G.A., 2007. Mycoparasitism of *Helminthosporium solani* by *Acremonium strictum*. Phytopathology 97 (10), 1331–1337.

Roberts, C.A., Marek, S.M., Niblack, T.L., Karr, A.L., 1992. Parasitic *Meloidogyne* and mutualistic *Acremonium* increase chitinase in tall fescue. J. Chem. Ecol. 18 (7), 1107–1116.

Rodriguez, R.J., Redman, R.S., Henson, J.M., 2004. The role of fungal symbioses in the adaptation of plants to high stress environments. Mitigat. Adapt. Strategies Global Change 9 (3), 261–272.

Rodriguez, R.J., Henson, J., Van Volkenburgh, E., Hoy, M., Wright, L., Beckwith, F., et al., 2008. Stress tolerance in plants via habitat-adapted symbiosis. ISME J. 2 (4), 404–416.

Rodriguez, R., White Jr., J., Arnold, A., Redman, R., 2009. Fungal endophytes: diversity and functional roles. New Phytol 182 (2), 314–330.

Romo, M., Leuchtmann, A., García, B., Zabalgogeazcoa, I., 2007. A totivirus infecting the mutualistic fungal endophyte *Epichloë festucae*. Virus Res. 124 (1–2), 38–43.

Saikkonen, K., Ion, D., Gyllenberg, M., 2002. The persistence of vertically transmitted fungi in grass metapopulations. Proc. Biol. Sci. 2669 (1498), 1397–1403.

Saithong, P., Panthavee, W., Stonsaovapak, S., Congfa, L., 2010. Isolation and primary identification of endophytic fungi from *Cephalotaxus mannii* trees. Maejo Int. J. Sci. Technol. 4 (3), 446–453.

Sánchez Márquez, S., Bills, G.F., Zabalgogeazcoa, I., 2007. The endophytic mycobiota of the grass *Dactylis glomerata*. Fungal Divers. 27, 171–195.

Sanders, I.R., 2004. Plant and arbuscular mycorrhizal fungal diversity—are we looking at the relevant levels of diversity and are we using the right techniques? New Phytol. 164, 415–418.

Sandhiya, G.S., Sugitha, T.C., Balachandar, D., Kumar, K., 2005. Endophytic colonization and in planta nitrogen fixation by a diazotrophic *Serratia* sp. in rice. Indian J. Exp. Biol. 43, 802–807.

Sansanwal, R., Ahlawat, U., Priyanka Wati, L., 2017. Role of endophytes in agriculture. Chem. Sci. Rev. Lett 6 (24), 2397–2407.

Schardl, C., Craven, K., 2003. Interspecific hybridization in plant-associated fungi and oomycetes: a review. Mol. Ecol 12 (11), 2861–2873.

Schardl, C.L., Leuchtmann, A., Spiering, M.J., 2004. Symbioses of grasses with seed borne fungal endophytes. Annu. Rev. Plant Biol. 55, 315–340.

Schardl, C.L., Young, C.A., Pan, J., Florea, S., Takach, J.E., Panaccione, D.G., et al., 2013. Currencies of mutualisms: sources of alkaloid genes in vertically transmitted epichloae. Toxins 5 (6), 1064–1088.

Schulz, B., Boyle, C., 2005. The endophytic continuum. Mycol. Res. 109 (6), 661–686.

Schulz, B., Haas, S., Junker, C., Andree, N., Schobert, M., 2015. Fungal endophytes are involved in multiple balanced antagonisms. Curr. Sci. 109 (1), 39–45.

Scott, L.J., Curran, M.P., Figgitt, D.P., 2004. Rosuvastatin: a review of its use in the management of dyslipidemia. Am. J. Cardiovasc. Drugs 4 (2), 117–138.

Selim, K.A., El-Beih, A.A., AbdEl-Rahman, T.M., El-Diwany, A.I., 2012. Biology of endophytic fungi. Curr. Res. Environ. Appl. Mycol 2 (1), 31–82.

Shankar, N.B., Krishnamurthy, Y.L., 2010. Endophytes: the real untapped high energy biofuel resource. Curr. Sci. 98 (7), 883.

Sharma, N., Gautam, A.K., 2018. Pathogenicity events in plant pathogenic bacteria: a brief note. J. New Biol. Rep. 7 (3), 141–147.

Sharma, N., Gautam, A.K., 2019. Early pathogenicity events in plant pathogenic fungi: a comprehensive review. Biol. Forum. Int. J. 11 (1), 24–34.

Sibanda, E.P., Mabandla, M., Chisango, T., Nhidza, A.F., Mduluza, T., 2018. Endophytic fungi from Vitex payos: identification and bioactivity. Acta Mycol. 53 (2), 1111.

Sieber, T.N., 2007. Endophytic fungi in forest trees: are they mutualists? Fungal Biol. Rev. 21 (2–3), 75–89.

Silva, G.H., Teles, H.L., Trevisan, H.C., Bolzani, V.D.S., Young, M.C.M., Pfenningc, L.H., et al., 2005. New bioactive metabolites produced by *Phomopsis cassiae*, an endophytic fungus in *Cassia spectabilis*. J. Braz. Chem. Soc. 16 (6B), 1463–1466.

Sinclair, J.B., Cerkauskas, R.F., 1996. Latent infection vs. endophytic colonization by fungi. In: Redlin, S.C., Carris, L.M. (Eds.), Endophytic Fungi in Grasses and Woody Plants: Systematics, Ecology, and Evolution. APS Press, St. Paul, MN, pp. 3–29.

Smalla, K., Wieland, G., Buchner, A., Zock, A., Parzy, J., Roskot, N., et al., 2001. Bulk and rhizosphere soil bacterial communities studied by denaturing gradient gel electrophoresis: plant dependent enrichment and seasonal shifts. Appl. Environ. Microbiol. 67 (10), 4742–4751.

Sonaimuthu, V., Krishnamoorthy, S., Johnpaul, M., 2010. Taxol producing endophytic fungus *Fusarium culmorum* SVJM072 from medicinal plant of *Tinospora cordifolia*—a first report. J. Biotechnol. 150, S425.

Song, Y.C., Li, H., Ye, Y.H., Shan, C.Y., Yang, Y.M., Tan, R.X., 2004. Endophytic naphthopyrone metabolites are co-inhibitors of xanthine oxidase, SW1116 cell and some microbial growths. FEMS Microbiol. Lett. 241 (1), 67–72.

Stadler, M., Keller, N.P., 2008. Paradigm shifts in fungal secondary metabolite research. Mycol. Res. 112 (Pt2), 127–130.

Stone, J.K., Bacon, C.W., White, J.F., 2000. An overview of endophytic microbes: endophytism defined. In: Bacon, C.W., White, J.F. (Eds.), Microbial Endophytes. Marcel Dekker, New York, USA, pp. 3–29.

Stone, J., White, J., Polishook, J., 2004. Endophytic fungi. In: Mueller, G., Foster, M., Bills, G. (Eds.), Measuring and Monitoring Biodiversity of Fungi. Inventory and Monitoring Methods. Elsevier Academic Press, Boston, MA, pp. 241–270.

Strobel, G., 2003. Endophytes as sources of bioactive products. Microb. Infect. 5 (6), 535–544.

Strobel, G., 2018. The emergence of endophytic microbes and their biological promise. J. Fungi 4 (2). Available from: https://doi.org/10.3390/jof4020057.

Strobel, G., Daisy, B., 2003. Bioprospecting for microbial endophytes and their natural products. Microbiol. Mol. Biol. Rev. 67 (4), 491–502.

Strobel, G.A., Dirkse, E., Sears, J., Markworth, C., 2001. Volatile antimicrobials from *Muscodor albus*, a novel endophytic fungus. Microbiology 147 (Pt11), 2943–2950.

Strobel, G., Daisy, B., Castillo, U., Harper, J., 2004. Natural products from endophytic microorganisms. J. Nat. Prod. 67 (2), 257–268.

Sturz, A.V., Nowak, J., 2000. Endophytic communities of rhizobacteria and the strategies required to create yield enhancing associations with crops. Appl. Soil Ecol. 15 (2), 183–190.

Sudha, V., Govindaraj, R., Baskar, K., Al-Dhabi, N.A., Duraipandiyan, V., 2016. Biological properties of endophytic fungi. Braz. Arch. Biol. Technol. 59, e16150436.

Suryanarayanan T.S., Rajagopal K. (1998). Fungal endophytes in leaves of some south Indian tree species. Proceedings of the Asia—Pacific Mycological Conference on Biodiversity and Biotechnology, Hua-Hin, Thailand, 252–256.

Suryanarayanan, T.S., Hawksworth, D.L., 2005. Fungi from little explored and extreme habitats. In: Deshmukh, S.K., Rai, M.K. (Eds.), Biodiversity of Fungi: Their Role in Human Life. Oxford & IBH Publishing Co. Pvt. Ltd, New Delhi, India, pp. 33–48.

Tan, R.X., Zou, W.X., 2001. Endophytes: a rich source of functional metabolites. Nat. Prod. Rep. 18 (4), 448–459.

Tanaka, A., Tapper, B.A., Popay, A., Parker, E.J., Scott, B., 2005. A symbiosis expressed non-ribosomal peptide synthetase from a mutualistic fungal endophyte of perennial ryegrass confers protection to the symbiotum from insect herbivory. Mol. Microbiol 57 (4), 1036–1050.

Taylor, T.N., Taylo, E.L., 2000. The Rhynie Chert ecosystem: a model for understanding fungal interactions. In: Bacon, C.W., White, J.F. (Eds.), Microbial Endophytes. Marcel Dekkker, New York, pp. 33–45.

Tayung, K., Barik, B.P., Jha, D.K., Deka, D.C., 2011. Identification and characterization of antimicrobial metabolite from an endophytic fungus, *Fusarium solani* isolated from bark of Himalayan yew. Mycosphere 2 (3), 203–213.

Terhonen, E., Blumenstein, K., Kovalchuk, A., Asiegbu, F.O., 2019. Forest tree microbiomes and associated fungal endophytes: functional roles and impact on forest health. Forests 10 (1), 42. Available from: https://doi.org/10.3390/f10010042.

Timper, P., Gates, R.N., Bouton, J.H., 2005. Response of *Pratylenchus* spp. in tall fescue infected with different strains of the fungal endophyte *Neotyphodium coenophialum*. Nematology 7 (1), 105–110.

Tintjer, T., Rudgers, J.A., 2006. Grass-herbivore interactions altered by strains of a native endophyte. New Phytol. 170 (3), 513–521.

Tomita, F., 2003. Endophytes in Southeast Asia and Japan: their taxonomic diversity and potential applications. Fungal Divers. 14, 187–204.

Trivedi, G., Shah, R., Patel, P., Saraf, M., 2017. Role of endophytes in agricultural crops under drought stress: current and future prospects. J. Appl. Microbiol. 3 (4), 174–188.

Turbyville, T.J., Wijeratne, E.M., Liu, M.X., Burns, A.M., Seliga, C.J., Luevano, L.A., et al., 2006. Search for HSP90 inhibitors with potential anticancer activity: isolation and SAR studies of radicicol and monocillin I from two plant associated fungi of the Sonoran desert. J. Nat. Prod. 69 (2), 178–189.

Vandenkoornhuyse, P., Baldauf, S.L., Leyval, C., Straczek, J., Young, J.P.W., 2002. Extensive fungal diversity in plant roots. Science 295 (5562), 2051. Available from: https://doi.org/10.1126/science.295.5562.2051.

Vaz, A.B.M., Fonseca, P.L.C., Badotti, F., Skaltsas, D., Tomé, L.M.R., Silva, A.C., et al., 2018. A multiscale study of fungal endophyte communities of the foliar endosphere of native rubber trees in Eastern Amazon. Sci. Rep. 8, 16151. Available from: https://doi.org/10.1038/s41598-018-34619-w.

Vega, F.E., Posada, F., Aime, M.C., Pava-Ripoll, M., Infante, F., Rehner, S.A., 2008. Entomopathogenic fungal endophytes. Biol. Contr 46 (1), 72–82.

Venkateswarulu, N., Shameer, S., Bramhachari, P.V., Thaslim Basha, S.K., Nagaraju, C., Vijaya, T., 2018. Isolation and characterization of plumbagin (5-hydroxyl-2-methylnaptalene-1,4-dione) producing endophytic fungi *Cladosporium delicatulum* from endemic medicinal plants isolation and characterization of plumbagin producing endophytic fungi from endemic medicinal plants. Biotechnol. Rep. 20, e00282.

Wagenaar, M.M., Corwin, J., Strobel, G., Clardy, J.J., 2000. Three new cytochalasins produced by an endophytic fungus in the genus *Rhinocladiella*. J. Nat. Prod. 63 (12), 1692–1695.

Waller, F., Achatz, B., Baltruschat, H., Fodor, J., Becker, K., Fischer, M., et al., 2005. The endophytic fungus *Piriformospora indica* reprograms barley to salt-stress tolerance, disease resistance, and higher yield. PNAS J. 102 (38), 13386–13391.

Wang, F.W., Jiao, R.H., Cheng, A.B., Tan, S.H., Song, Y.C., 2007. Antimicrobial potentials of endophytic fungi residing in *Quercus variabilis* and brefeldin A obtained from *Cladosporium* sp. World J. Microbiol. Biotechnol. 23 (1), 79–83.

Wang, J.G.H., Chuan, C., 2011. Fungi in their own right. Fungal Genet. Biol. 27 (2–3), 134–145.

Wani, M.C., Taylor, H.L., Wall, M.E., Coggon, P., McPhail, A.T., 1971. Plant antitumor agents. VI. The isolation and structure of taxol, a novel antileukemic and antitumor agent from *Taxus brevifolia*. J. Am. Chem. Soc 93 (9), 2325–2327.

Webber, J., 1981. A natural control of Dutch elm disease. Nature 292, 449.

Wei, Y.K., Gao, Y.B., Zhang, X., Su, D., Wang, Y.H., Xu, H., et al., 2007. Distribution and diversity of *Epichloe/Neotyphodium* fungal endophytes from different populations of *Anchnatherum sibiricum* (Poaceae) in the inner Mangolia Steppe, China. Fungal Divers. 24, 329–345.

White, J.F., Tadych, M., Torres, M.S., Bergen, M.S., Irizarry, I., Chen, Q., et al., 2016. Endophytic microbes, evolution and diversification. In: Kliman, R.M. (Ed.), Encyclopedia of Evolutionary Biology., Vol. 1. Academic Press, Oxford, pp. 505–510.

Wilberforce, E., Boddy, L., Griffiths, R., Griffith, G., 2003. Agricultural management affects communities of culturable root-endophytic fungi in temperate grasslands. Soil Biol. Biochem. 35 (8), 1143–1154.

Wilkinson, H.H., Siegel, M.R., Blankenship, J.D., Mallory, A.C., Bush, L.P., Schardl, C.L., 2000. Contribution of fungal loline alkaloids to protection from aphids in a grass-endophyte mutualism. Mol. Plant Microb. Interact. 13 (10), 1027–1033.

Wilson, D., 1995. Endophyte: the evolution of a term, and clarification of its use and definition. Oikos 73 (2), 274–276.

Worapong, J., Strobel, G., Ford, E.J., Li, J.Y., Baird, G., Hessl, W.M., 2001. *Muscodor albusanam.* gen. et sp. nov., and endophyte from *Cinnamomum zeylanicum*. Mycotaxon 79, 67–79.

Wyrebek, M., Huber, C., Sasan, R.K., Bidochka, M.J., 2011. Three sympatrically occurring species of *Metarhizium* show plant rhizosphere specificity. Microbiology 157 (Pt 10), 2904–2911.

Xu, L., Zhang, Y., Wang, J., Pang, J., Huang, C., Wu, X., et al., 2008. Benzofuran derivatives from the mangrove fungus *xylaria* sp. J. Nat. Prod. 71 (7), 1251–1253.

Yan, J., Broughton, S., Yang, S., Gange, A., 2015. Do endophytic fungi grow through their hosts systemically? Fungal Ecol. 13, 53–59.

Yang, B., Wang, X.-M., Ma, H.-Y., Yang, T., Jia, Y., Zhou, J., et al., 2015. Fungal endophyte *Phomopsis liquidambari* affects nitrogen transformation processes and related microorganisms in the rice rhizosphere. Front. Microbiol. 6, 982. Available from: https://doi.org/10.3389/fmicb.2015.00982.

Yao, Y.Q., Lan, F., Qiao, Y.M., Wei, J.G., Huang, R.S., Li, L.B., 2017. Endophytic fungi harbored in the root of *Sophora tonkinensis* Gapnep: diversity and biocontrol potential against phytopathogens. Microbiol. Open 6 (3), e437. Available from: https://doi.org/10.1002/mbo3.437.

Yong, Y.H., Dai, C.C., Gao, F.K., Yang, Q.Y., Zhao, M., 2009. Effects of endophytic fungi on growth and two kinds of terpenoids for *Euphorbia pekinensis*. Chin. Trad. Herbal Drugs 40, 18–22.

Young, C.A., Hume, D.E., McCulley, R.L., 2013. Forages and pastures symposium: fungal endophytes of tall fescue and perennial ryegrass: pasture friend or foe? J. Anim. Sci. 91 (5), 2379–2394.
Yu, H., Zhang, L., Li, L., Zheng, C., Guo, L., Li, W., et al., 2010. Recent developments and future prospects of antimicrobial metabolites produced by endophytes. Microbiol. Res. 165 (6), 437–449.
Yuan, Z.L., Dai, C.C., Li, X., Tian, L.S., Wang, X.X., 2007. Extensive host range of an endophytic fungus affects the growth and physiological functions in rice (*Oryza sativa* L.). Symbiosis 43 (1), 21–28.
Yuan, Y., Feng, H., Wang, L., Li, Z., Shi, Y., Zhao, L., et al., 2017. Potential of endophytic fungi isolated from cotton roots for biological control against *Verticillium* wilt disease. PLoS One 12 (1), e0170557.
Yue, Q., Miller, C.J., White, J.F., Richardson, M.D., 2000. Isolation and characterization of fungal inhibitors from *Epichloë festucae*. J. Agric. Food Chem. 48 (10), 4687–4692.
Zabalgogeazcoa, I., 2008. Fungal endophytes and their interaction with plant pathogens. Span. J. Agric. Res. 6 (special issue), 138–146.
Zakaria, L., Jamil, M.I.M., Anuar, I.S.M., 2016. Molecular characterisation of endophytic fungi from roots of wild banana (*Musa acuminata*). Trop. Life Sci. Res. 27 (1), 153–162.
Zhang, H.W., Song, Y.C., Tan, R.X., 2006. Biology and chemistry of endophytes. Nat. Prod. Rep. 23 (5), 753–771.
Zheng, R.Y., Jiang, H., 1995. *Rhizomucor endophyticus* sp. nov., an endophytic zygomycetes from higher plants. Mycotaxon 56, 455–466.
Zhou, D., Hyde, K.D., 2001. Host-specificity, host-exclusivity, and host-recurrence in saprobic fungi. Mycol. Res. 105 (12), 1449–1457.
Zou, W.X., Meng, J.C., Lu, H., Chen, G.X., Shi, G.X., Zhang, T.Y., et al., 2000. Metabolites of *Colletotrichum gloeosporioides*, an endophytic fungus in *Artemisia mongolica*. J. Nat. Prod. 63 (11), 1529–1530.

## Further reading

Brakhage, A.A., 2013. Regulation of fungal secondary metabolism. Nat. Rev. Microbiol. 11 (1), 21–32.
Jaber, L.R., Araj, S.E., 2017. Interactions among endophytic fungal entomopathogens (Ascomycota: Hypocreales), the green peach aphid *Myzus persicae* Sulzer (Homoptera: Aphididae), and the aphid endoparasitoid *Aphidius colemani* Viereck (Hymenoptera: Braconidae). Biol. Contr. 116, 53–61.
Netzker, T., Fischer, J., Weber, J., Mattern, D.J., König, C.C., Valiante, V., et al., 2015. Microbial communication leading to the activation of silent fungal secondary metabolite gene clusters. Front. Microbiol. 6, 299. Available from: https://doi.org/10.3389/fmicb.2015.00299.

# Index

*Note*: Page numbers followed by "*f*" and "*t*" refer to figures and tables, respectively.

Lightning Source UK Ltd.
Milton Keynes UK
UKHW022250060320
359930UK00007B/168